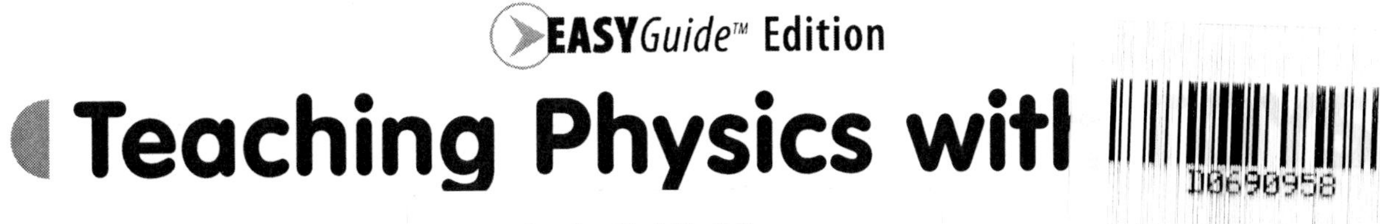
EASYGuide™ Edition
Teaching Physics witl
D0690958

Art Credits

Photo of runners provided by the Impington Swimming Club (UK).
So that his children will believe him, the winner is indeed Ivan Scott.

Photo of 1 kg standard mass provided by the Bureau International des Poids et Mesures (Paris, France).

Photo of Amsterdam buildings provided by the University of the Virgin Islands.

EASYGuide™ Edition

Teaching Physics with TOYS

Hands-On Investigations for Grades 3–9

Authors

Beverley A.P. Taylor, Department of Physics, Miami University
Dwight J. Portman, Department of Physics, Miami University
Susan Gertz, Center for Chemistry Education, Miami University
Lynn Hogue, Center for Chemistry Education, Miami University

Series Editor

Mickey Sarquis, Director
Center for Chemistry Education, Miami University

Terrific Science Press
Miami University Middletown
Middletown, Ohio USA

Terrific Science Press
Miami University Middletown
4200 East University Blvd.
Middletown, Ohio 45042
cce@muohio.edu
www.terrificscience.org

Printed in the United States of America

10 9 8 7 6 5 4 3 2 1

ISBN 1-883822-40-8

This material is based upon work supported by the National Science Foundation under grant numbers TEI-8751244, TPE-9055448, and ESI-9355523 and by the Ohio Board of Regents under grant numbers 8-37, 9-38, 00-34, and 01-40. This project was supported, in part, by the National Science Foundation and the Ohio Board of Regents. Any opinions, findings, and conclusions or recommendations expressed in this material are those of the authors and do not necessarily reflect the views of the National Science Foundation or the Ohio Board of Regents.

Contents

Introduction

Activities

Appendix

Acknowledgments

The authors and editor wish to thank the following individuals who have contributed to the development of *Teaching Physics with TOYS, EASYGuide Edition*.

Terrific Science Press Design and Production Team

Document Production Managers: Susan Gertz, Amy Stander
Production Coordinator: Dot Lyon
Technical Writing and Editing: Dot Lyon, Amy Stander, Amy Hudepohl, Tom Schaffner, Don Robertson
Production: Dot Lyon, Tom Schaffner, Don Robertson, Anita Winkler, Elizabeth Kramer
Illustrations: Carole Katz, Tom Schaffner, Don Robertson, Tom Nackid
Photography: Susan Gertz
Cover Design and Layout: Susan Gertz

Reviewers and Classroom Testers

Sandy Van Natta, White Oak Middle School (retired), White Oak, OH
Cris Cornelssen, Cameron Park Elementary School, Fairfield, OH
Lee Ann Ellsworth, Northwestern Middle School, Springfield, OH
Linda Jester, John XXIII Catholic Elementary School (retired), Middletown, OH

Participants in the Teaching Science with TOYS Program

By providing feedback during Teaching Science with TOYS program sessions, testing TOYS activities in their classrooms, and sharing their ideas, many dedicated professionals have contributed to the first edition of *Teaching Physics with TOYS* and to this *Teaching Physics with TOYS, EASYGuide Edition.* We thank them.

Foreword

Toys...reunite the fun/hands-on and mental/minds-on aspects of science teaching and learning while developing process skills, attitudes, and content. Thus, thoughtful explorations with toys are recommended for bringing out the playful, investigative side of children of all ages.—Thomas O'Brien, 1993

As a child or as an adult, most of us find it difficult to walk past a colorful display of toys without pausing, smiling, and taking a closer look. The urge to roll the truck down the hill, bounce the Silly Putty™, or wind up the walking dinosaur is nearly irresistible. We typically associate toys with fun, discovery, and creativity. In contrast, if presented with a display of physics and chemistry laboratory equipment, "fun, discovery, and creativity" would not be the words that came to most peoples' minds.

For nearly 20 years, a group of faculty at Miami University (OH) has worked together to give teachers (and through teachers, students) the opportunity to find out that "fun, discovery, and creativity" are words that very much describe the exploration of physics and chemistry principles. Our idea is to teach basic physics and chemistry principles using toys, thus capitalizing on the natural attractiveness of toys and also showing that physical science is an integral part of our everyday experiences. Thomas O'Brien discusses these ideas in "Teaching Fundamental Aspects of Science Toys" (*School Science and Mathematics*, volume 93, number 4).

- Toys build on and extend students' out-of-school experiences, intersecting with children's mental frameworks and enabling them to interpret new information in light of their pre-existing conceptions.
- Toys are readily available and have a low cost in comparison with conventional science equipment.
- Toys are inherently motivational and interactive, enabling teachers to engage students' attention.
- Toys help make otherwise abstract principles concrete and relevant to students' lives.
- Toys simultaneously involve both students and teachers in the fun and mental aspects of science.

With funding from state and federal institutions, private foundations, and industry, we developed and have taught the Teaching Science with TOYS program for teachers across the United States and in countries around the world. The first edition of *Teaching Physics with TOYS* enabled teachers who could not attend our programs to share in the fun and learning. Now, the completely revised *Teaching Physics with TOYS—EASYGuide Edition* provides new activities in collaboration with K'NEX® Education along with many new features to guide and support your teaching.

Mickey Sarquis

EASYGuide™ Edition Features

With the *Teaching Physics with TOYS—EASYGuide Edition*, bringing hands-on, inquiry learning into your classroom is easier than ever before.

GUIDE students' inquiry with step-by-step teaching notes.

Teacher notes provide a detailed guide to teaching the activity, including information such as

- what to expect students to do or learn during the step,
- reasonable plans for student-designed experiments, and
- expected student responses to the questions associated with the step.

As needed, teacher notes include activity introductions, class discussions, and optional steps to further help you guide students through the learning process.

Balance This!

Students build and explore balance toys and discover how varying the amount and position of we

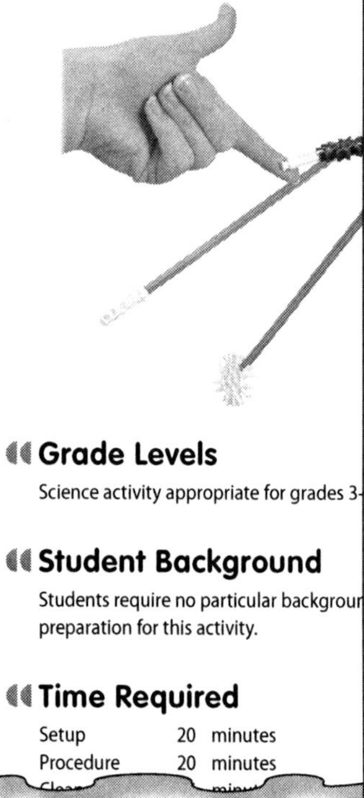

Grade Levels

Science activity appropriate for grades 3-

Student Background

Students require no particular backgrou preparation for this activity.

Time Required

Setup 20 minutes
Procedure 20 minutes

Procedure

This section provides teacher notes corresponding to each step of the student procedure. The procedure without teacher notes is included in the reproducible Student Notebook pages at the end of this activity and at www.terrificscience.org/physicsez/.

	Student Procedure	Teacher Notes
	Activity Introduction	• Give students opportunities to play with an assortment of commercial and homemade balance toys prior to the activity. • As a class, have students discuss the balance toys, sharing their ideas about how the toys balance. Record their ideas on a chart or chalkboard. Point out each toy's support point.
	Make a Balance Toy	
1	Try to balance a craft stick or K'NEX rod (we'll just say "stick" for both) so that it stands up on your finger. What happens?	If using K'NEX, avoid distributing the longer rods (such as red and gray) in this step. Students will need to use the longer rods later. Students should discover that balancing a stick upright on the finger is very difficult.
2	Try to balance a balance toy on your finger. Does it stay on your finger if you move your hand around? Try some other balance toys. How do these balance toys behave? How is this different than what the stick did in step 1?	Balance toys should easily balance on the finger, even when the toys are tipped.
3	Compare the balance toys and the stick to find out why one may be harder to balance than the other. TIP: On the items you were able to balance, look below your finger. What do the balance toys all have in common that is different than the stick?	Students will see that balance toys placed on the finger have some parts that hang or extend below the finger. Although students may not realize it yet, these parts lower the center of gravity so that the toys balance.

Balance This! Student Notebook

Make a Balance Toy

(1) Try to balance a craft stick or K'NEX® rod (we'll just say "stick" for both) so that it stands up on your finger.

What happens?

(2) Try to balance a balance toy on your finger. Does it stay on your finger if you move your hand around? Try some other balance toys.

How do these balance toys behave? How is this different than what the stick did in step 1?

(3) Compare the balance toys and the stick to find out why one may be harder to balance than the other.

TIP: On the items you were able to balance, look below your finger.

What do the balance toys all have in common that is different than the stick?

PREPARE quickly with reproducible student pages (in book and at *www.terrificscience.org/physicsez/*).

- Numbered steps lead students through the procedure.
- Scientific inquiry and scientific ways of knowing are embedded throughout the activity.
- Student questions provide opportunities for predicting, observing, and reflecting on results.
- Space is provided for most student answers. Longer responses or drawings may require extra paper. For graphing, reproducible graph paper is provided at *www.terrificscience.org/physicsez/*.

CUSTOMIZE with editable student pages (at *www.terrificscience.org/physicsez/*).

- Adapt activities for different student abilities.
- Copy fewer pages. (Editable versions of activities have fewer pages because lines for student answers are not included.)

KNOW what to expect.

- Review sample data for procedures.
- Look at examples of possible student drawings and designs.
- Read complete explanations of the science concepts.

Explanation

This section is intended for teachers. Modify the explanation for students as needed.

Center of Gravity

The center of gravity of an object or connected group of objects is the point at which the force of gravity acts on the object as if all the weight were concentrated at that point. The center of gravity may be located where no actual material exists.

An object balances when its center of gravity is below or directly above the point of support. An object with its center of gravity below its point of support is more stable. Balancing objects that have their cent
gravity below the point of support act like sin
pendulums. When the object is tipped, the ce
of gravity swings to one side like the ball of a
pendulum. Like a pendulum, the forces acting
object bring it back to equilibrium, and the ce
gravity is once again directly below the suppo

An object can balance with its center of gravit
the point of support, but the object is unstab
the area of support is large. If the object is tip
even a little, gravity pulls it away from the bal
position. For example, tightrope walkers and
on a balance beam are unstable because thei
of gravity is above their point of support (the
balance beam). They need to be very skilled a
their center of gravity directly over the rope o
The bananas on the girl's head (shown in step
more stable than the woman on the log beca
support area is larger compared to the size of
object.

Making Balance Toys

In step 1, students try to balance the stick on
with no other weight added. Under these con
the stick's center of gravity is well above the p
pport (th nd the stick falls. The

In step 7, changing the positio f the added weight changes the behavior of the ba cing stick. If part of the additional weight is move upward and outward, the center of gravity of e whole system is redistributed and moves upwa and outward. In response, the whole system tilts o otates to move the center of gravity back below the s k (again, like the swing of a pendulum). If the stick b anced upright on the support point before moving t weight, it may balance at an angle after moving th weight. If the center of gravity is moved upward t point above the point of support, the whole system m y rotate off its support point.

Sample Answers

Where students follow the same procedure using the same materials, these answers are close to answers you can expect. Where students design their own experiment or model, students' results will vary.

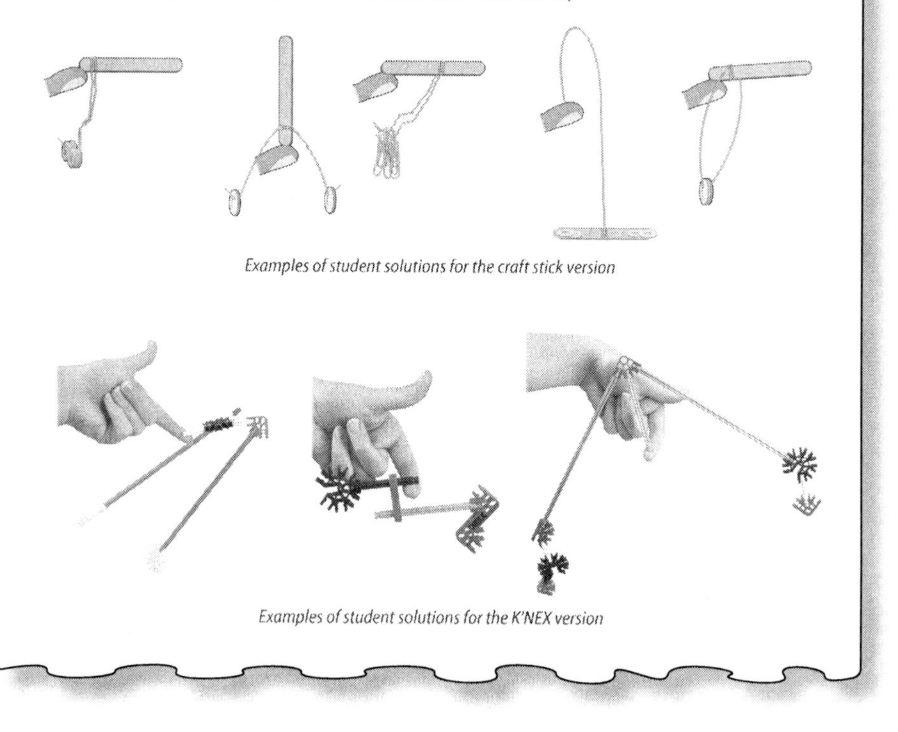

Examples of student solutions for the craft stick version

Examples of student solutions for the K'NEX version

Model Building with K'NEX®

Many of the activities in this book incorporate K'NEX building materials. The same set of K'NEX pieces are used to build assorted levers and pulley systems, a balance, a crank fan, tops, cars, and more, making K'NEX a fun and economical alternative to single-use equipment. If you already use K'NEX in the classroom, you probably have many of the pieces needed to build these models. If you need K'NEX, *www.terrificscience.org/physicsez/* lists K'NEX Education kits that have the parts needed for the models in this book. To give you more flexibility, we've also offered alternatives to K'NEX in most activities.

In some of the activities using K'NEX (such as "Six-Cent Top," "Magnet Cars," and "Balance This!"), students build their own K'NEX creations. For activities such as these, no particular set of K'NEX pieces or building instructions are needed.

Other activities in this book (such as "Seesaw Forces," "Gear Up, Gear Down," and "Levers at Work") require specific K'NEX models. The website *www.terrificscience.org/physicsez/* provides color-coded K'NEX assembly diagrams and parts lists for these models. Note that some K'NEX Education kits may have assembly diagrams for models *similar* to those in this book, but they are *not identical*. **You must use the assembly diagrams provided at *www.terrificscience.org/physicsez/*.** Since K'NEX pieces are color coded, the assembly diagrams should be printed in color. We recommend that you print one set of assembly instructions for each group of students. As an alternative to color printing, you can make color overhead transparencies and project them as students assemble the models.

Each activity that uses specific K'NEX models includes the approximate time required to build them. Most of the models took our middle-school-aged student testers about 15 minutes to complete. While younger or less experienced K'NEX users may take a little longer, building times will decrease as students gain experience. You may want to schedule a day before the activity to have students build the models. Another time-saver is to put all of the pieces each group will need into a small box or plastic zipper-type bag before class. This method will also shorten cleanup time and help keep the K'NEX pieces organized. Alternatively, you can have students from a higher grade level come into the classroom to help build the models. Or, you may choose to build them yourself in advance.

Teaching Students to Be Scientists

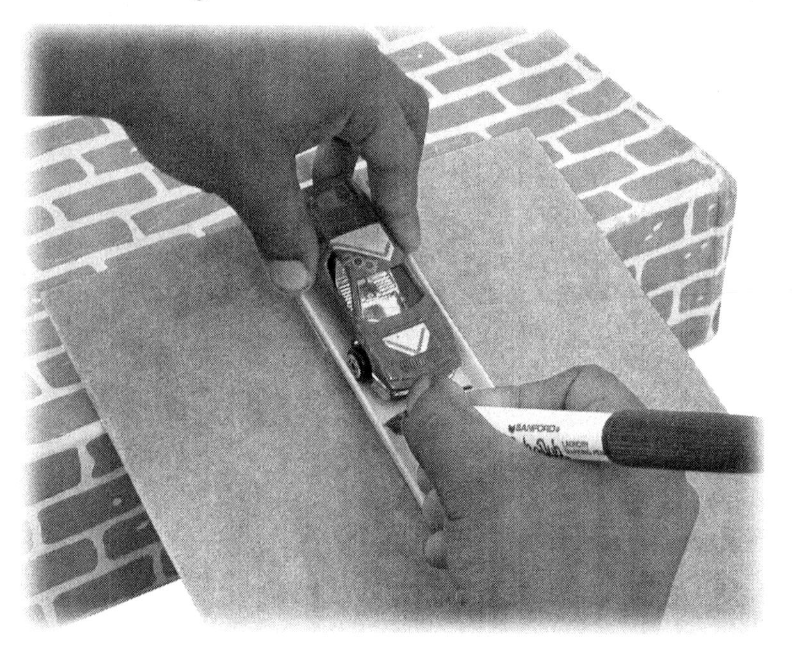

> ***"Scientific inquiry** refers to the diverse ways in which scientists study the natural world and propose explanations based on the evidence derived from their work. Inquiry also refers to the activities of students in which they develop knowledge and understanding of scientific ideas, as well as an understanding of how scientists study the natural world."*
>
> National Science Education Standards

We've all heard the phrase "inquiring minds want to know," and we hope that our students do have inquiring minds. But what do we mean by teaching with inquiry? While the word "inquiry" may be associated with particular styles of teaching (and may be further classified as structured, guided, or open), we define teaching with inquiry simply as the following: Give students ample opportunities to apply the reasoning and procedural skills of scientists while learning the principles and concepts of science along the way.

So, how do we teach students to think and work as scientists do? The most familiar model is usually called the scientific method — an ordered list of steps, typically beginning with "observation" and ending with "conclusion." The scientific method, as traditionally represented, reflects a linear process: Follow each step in sequence until the task is completed. Here's the problem: A linear process implies that one cannot do the steps in a different order or go back to repeat a step later in the process. Although this simple linear process is easier to teach to students, it just doesn't reflect the many nonlinear ways in which scientists go about their work.

As an alternative to the scientific method, science educator William Harwood proposes the *activity model of the scientific inquiry process.* (See the figure on the next page and a more detailed description on the following pages.) The model describes 10 activities that must be done (often more than once) to develop and carry out inquiry—to do the work of scientists. While this model encompasses the activities that are part of the scientific method, its circular shape emphasizes the nonlinear nature of science inquiry. The model more closely reflects the way that many scientists work: engaging in an activity as often as necessary and moving among activities as needed to solve a problem or answer a question. "Ask questions" is shown at the center of the circle because asking questions is the central feature of any scientific inquiry.

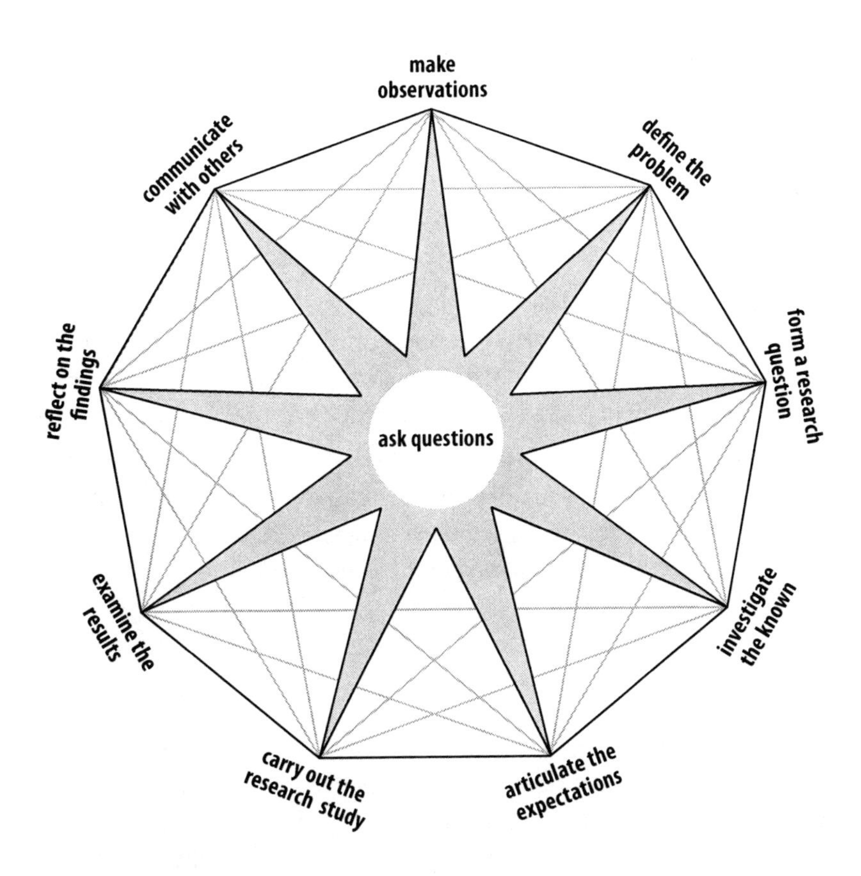

This model is intended to help you in framing inquiry teaching and learning in your classroom. The message of this model is that "doing science" is not limited to a step-by-step execution of the scientific method.

The investigations in *Teaching Physics with TOYS* provide many ideas for you and your students to put this model into action while learning physics concepts. (See Appendix for a matrix of National Science Education Standards met by the activities in this book.) Keep in mind that every science activity need not involve every facet of inquiry. In fact, early in the school year you may want to have students specifically focus on one or two facets as described in the examples listed below.

- Give students a data table and charge them with finding a pattern in the data.
- Do a series of demonstrations in which students just focus on making observations. This experience trains them to be careful observers and to distinguish between observations and inferences they draw from their observations.
- Give students practice starting from a general question and developing several specific questions that could be answered by performing an experiment.
- If your current goal is to have students practice making observations, gathering and recording data, and clearly communicating their results, let them use an experimental design you provide rather than asking them to design their own experiments.

Having practiced the various facets of inquiry, your students will be ready to carry out the entire process on their own. The Experiment Planning Guide included at *www.terrificscience.org/physicsez/* is designed to help them do so. The following pages provide other resources to help you guide your students in their learning.

Activity Model of the Scientific Inquiry Process

In the course of an inquiry, scientists move among these activities in unique paths and repeat activities as often as they find necessary. The following description of inquiry activities (adapted from "A New Model for Inquiry" by William Harwood) is provided as a reproducible student handout at *www.terrificscience.org/physicsez/*.

Ask questions	Asking general and divergent questions is the central feature of any scientific inquiry. Divergent questions focus on imagining new possibilities and usually include words or phrases such as imagine, suppose, if...then, how might, could you, and what are.
Define the problem	Limit the arena that you will explore. For example, your general question might be, "What causes global climate change?" And then you might define the problem by focusing on "How do ocean currents affect the global climate?"
Form a research question	Develop a question that can drive a research study.
Investigate the known	Consult books and articles that have been published regarding your area of interest. Scientists often consult experts in the field (scientists or other people with specific expertise). The need to investigate the known comes up frequently throughout the course of a scientific inquiry.
Articulate the expectations	Develop an expectation for your study. • This sometimes may be a formal hypothesis, often worded as an if...then statement (expected cause and effect) based on an observation, experience, or scientific reason. • More likely, however, the expectation will be a prediction (a statement about what may happen in the investigation based on prior knowledge) or a simple goal such as, "If I vary the initial temperature of a glass of water and measure how long it takes to come to room temperature, I will be able to find a pattern that describes the relationship between the two quantities."
Carry out the research study	Technically, this is the most involved activity. • Choose the means to investigate your question, gather or create materials, and collect data while carrying out your investigation. • Be sure to follow a procedure that will allow a person not familiar with the project to duplicate the tasks and get the same results.

Carry out the research study (continued)	• Carefully identify independent, dependent, and controlled variables. The independent variable is the one variable the investigator chooses to change. The dependent variable changes as a result of, or in response to, the change in the independent variable. Controlled variables are the variables that must not change. The following question can help you identify the dependent and independent variables in your experiment: How does the independent variable affect the dependent variable?

Question	Independent Variable	Dependent Variables	Controlled Variables
Does the height of a ramp affect how far a toy car travels?	number of blocks under one end of the ramp	distance the car travels	• use the same car each time • blocks are all the same size • track material and floor surface are the same
Do all balls bounce equally?	types of balls	height ball bounces	• use same surface each time • only measure the first bounce • drop from the same height

Examine the results	Data can be obtained in various forms, depending on the type of study. You must be confident that your data are reliable—ask yourself if the results make sense. If you are uncertain, then you should repeat the study or engage in other activities to determine whether you can trust the study results.
Reflect on the findings	Spend considerable time thinking about what your results mean. Ask yourself how your results connect with what is known and how you explain them to colleagues and other interested people. Can you identify a pattern (repeating cycle) or trend (general drift, tendency, or direction of a set of data) in the data you collected?
Communicate with others	Scientists rarely work in isolation, so you shouldn't either. Throughout the course of an inquiry, scientists communicate with peers and colleagues elsewhere. Many inquiries involve collaborative efforts of several scientists. Good communication among these scientists is an essential feature of inquiry. When the study is completed, the last activity will be formal communication through oral or written presentations. Discussion leads to the identification of more testable questions, clarifies understanding, and addresses misconceptions.
Make observations	Like asking questions, scientists point to making observations as something they do at many different times in a scientific inquiry. Observations may be your starting point for some inquiries, but you also make observations when you carry out the study and investigate the known.

Components of a Well-Designed Graph

Representing data in the form of a graph helps scientists examine results, reflect on findings, and communicate with others. Presenting data as a graph rather than as a table of numbers helps to

- communicate the information quickly, and
- identify trends and relationships which may not be readily apparent in a table.

The following rules for good graphing will help students create meaningful, easy-to-read graphs. The rules are provided as a reproducible student handout at *www.terrificscience.org/physicsez/*.

Choose an appropriate format for the graph

The style of graph for experimental data depends on the type of data collected and the ideas the graph is meant to emphasize.

- *Bar graphs* represent data values as vertical bars. They are often used when the independent variable is qualitative (non-numerical)—such as types of balls; however, bar graphs can also be used when the independent variable is quantitative (numerical)—such as the height of a ramp in centimeters. The bars create a bold visual representation of a trend in the data, such as the distance a car travels getting gradually longer as the ramp height increases.

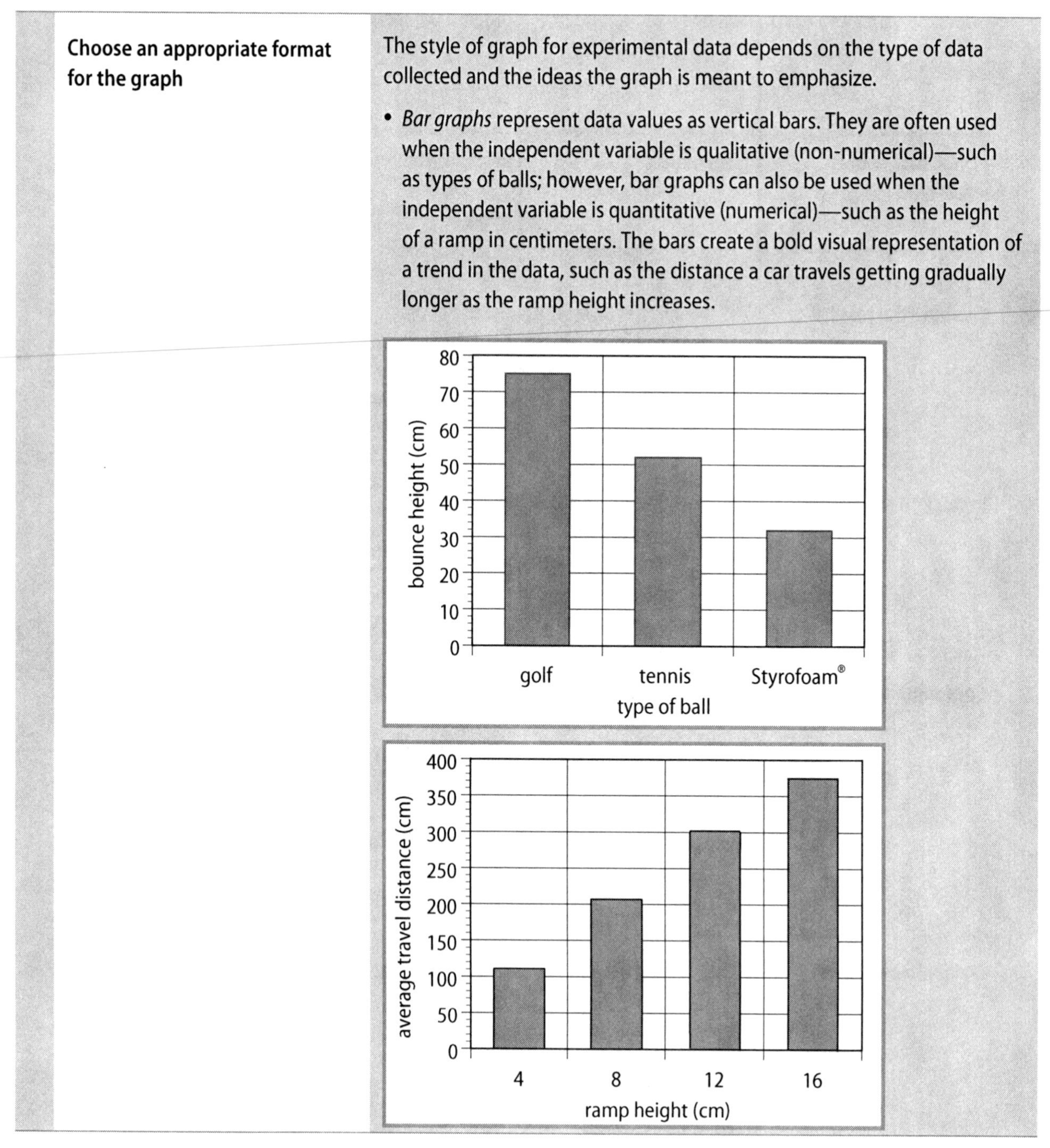

Choose an appropriate format for the graph (continued)	• *Line graphs* record individual data values as points on the graph. They provide an excellent way to represent data where both independent and dependent variables are quantitative (numerical) and where a definite relationship between the variables is anticipated. Often a best-fit line or curve is drawn to illustrate this relationship. Not all data points will lie on the line because they represent experimental measurements, which will have some variation no matter how carefully the data was taken. 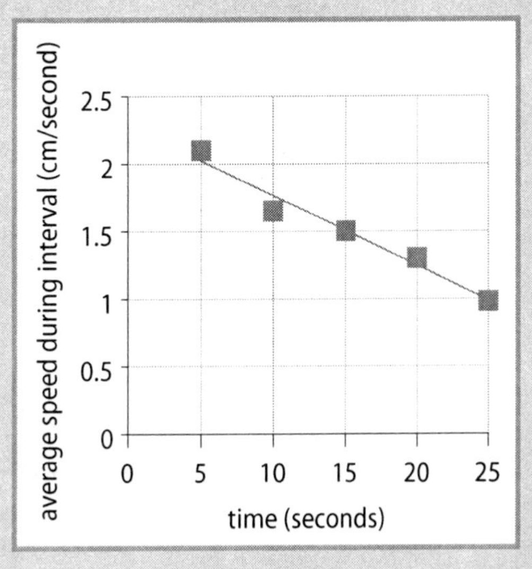 • *Scatter graphs* are similar to line graphs in that they record individual data values as points on the graph. The term scatter graph is often used to describe graphs where a large data set has been plotted and the data points have a scattered appearance. This scatter can be a result of multiple variables that cannot all be controlled. The term is also used to describe graphs made to investigate a possible relationship between two variables. For example, you might plot a scatter graph to see if a relationship exists between the outdoor temperature in the morning and the number of students absent from class. The overall shape of the scattered points indicates what relationship may exist between the variables. However, keep in mind that a relationship indicated by a scatter graph does not provide any evidence that one variable is actually causing changes in the other. No scatter graphs are used in this book.
Write a descriptive title	One purpose of the graph is to communicate information in a concise manner. A graph conveys little information if it is not titled properly.
Label each axis and indicate the units used	The independent variable is always placed on the x-axis (horizontal axis). The dependent variable is always placed on the y-axis (vertical axis).

Choose an appropriate scale for each variable	Examine your data to find the largest and smallest values. This will help you determine the range of each axis and the size of each increment. The scales should begin with a value lower than your lowest data point. (Graphs do not always need to start at zero.) A graph is easier to read (and plot) when each square represents a value of 1, 2, 5, or a multiple of 10 times those numbers (such as 10, 20, 50). A graph constructed on an inappropriate scale can be misleading, in addition to being difficult to read. The importance of choosing an appropriate scale is shown in the example below. You may want to have students create graphs of the same data using different scales to illustrate this point. 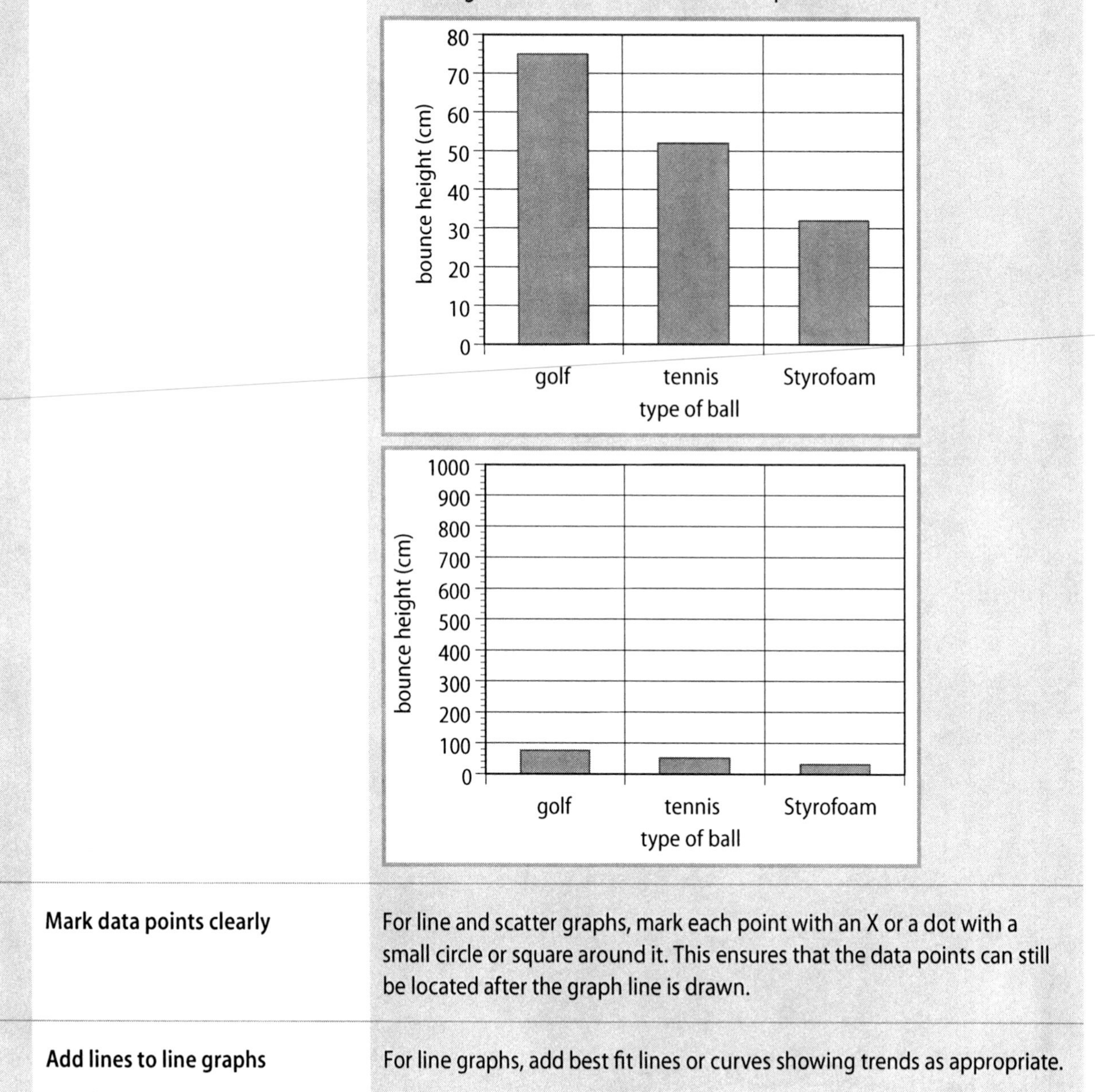
Mark data points clearly	For line and scatter graphs, mark each point with an X or a dot with a small circle or square around it. This ensures that the data points can still be located after the graph line is drawn.
Add lines to line graphs	For line graphs, add best fit lines or curves showing trends as appropriate.

Writing to Inform

The Activity Model of the Scientific Inquiry Process discussed previously emphasizes the important role of communication in science inquiry. Because of this, writing is a logical and natural part of the science classroom. In science, much of the writing is "writing to inform;" in other words, you share what you know about a topic or subject with another person. Students have many opportunities to write while engaging in scientific inquiry: writing questions, hypotheses, and predictions; recording observations and data; documenting experimental procedures; and explaining conclusions are all part of the process. The following suggestions for helping students become better writers are summarized from the book *How to Write to Learn Science* (Robert Tierney and John Dorroh, 2004).

Writing the procedure	Writing a procedure that others can follow to repeat an experiment is an essential part of scientific inquiry. Give students opportunities to practice writing instructions for something they do every day, such as opening their school lockers. Then, have students actually try using someone else's instructions exactly as written. They will soon learn the level of attention to detail that is required for writing instructions.
Learning to observe	Students' ability to observe carefully and write down what they see can be sharpened through practice. For example, give pairs of students something to observe such as the toy called magnetic marbles. Instruct one of each pair to play with the toy and make observations while the other acts as a recorder. Have students divide their observations into separate lists labeled "fact" and "inference." A fact would be "the marbles sometimes repel each other and sometimes attract each other." An inference would be "the marbles have magnets in them." Another example is found in the Operation® game. The fact would be "if the tweezers touch the metal edges of the holes, then the light goes on." An inference would be "the tweezers are completing an electric circuit." In this context, older students may enjoy reading about the occupation of a Fair Witness in chapters 11 and 12 of *Stranger in a Strange Land,* by Robert Heinlein. In this story, a Fair Witness is a person rigorously trained to observe, remember, and report without prejudice, distortion, lapses in memory, or personal involvement.
Writing conclusions	Understanding the difference between results and conclusions can be difficult for students. The results are what happened in terms of both numerical data and observations. Conclusions are generalizations that experimenters make based on the results. The book *How to Write to Learn Science* suggests having students divide the conclusion about an experiment into these three parts: "the best, most truthful response to the original question; a reason for that response; and questions that remain."

Communication to an audience	This part of writing to inform can take many forms. For example, students can assume the role of a reporter writing about the investigation for the general public. Writing such an article improves students' writing and thinking. When writing a newspaper or magazine article, students should consider the following guidelines: • Tell the reader what he or she needs, wants, or would like to know. • Supply answers to the reader's 5W+H question(s): who, what, why, where, when, and how. • Aim to be balanced and fair. • Allow readers to make up their own minds. • Offer unbiased facts or a balanced range of opinions from different sources. • Provide clear, interesting, and sufficient information.

Writing to Learn

While writing to inform is an important part of science, writing helps students do much more than inform others: writing can also help them to transform knowledge and make it their own. Why does writing promote thinking? Through experience, we all develop personal theories of how the world works, theories that we constantly consult, modify, act upon, and modify again. One way we think about these theories is to develop an "inner speech" that explains our experiences to ourselves. Through writing—particularly informal, risk-taking writing that will not be evaluated as right or wrong—we reflect upon that inner speech and discover something about ourselves.

> *Anyone who has ever written a letter, a page in a diary, or an entry in a journal realizes that the physical act of putting pen to paper uncovers ideas the writer was not aware of previously. Writing exercises both sides of the brain. More importantly, if the writer has ownership of the subject, the writer understands more deeply and retains what he or she has discovered for a longer period of time.*—Robert Tierney and John Dorroh, 2004

Writing to learn experiences can be incorporated into the science classroom in a variety of ways. Some ideas follow:

KWL	One useful form of writing for the science classroom is a KWL (Know, Wonder, Learn) graphic organizer. The exact format varies, with a fourth category often added. Here is an example:

K: what I know	W: what I want to know	H: how to find out	L: what I have learned

Freewriting	Through freewriting assignments, students become involved with a topic by relating it to their own experiences. This approach encourages students to free associate while writing as fast as they can. In the science classroom, students should be encouraged to be personally expressive while incorporating course material. For example "Choose an activity you do frequently, such as riding your bike or skateboard. Give as many examples as you can think of that show Newton's laws at work while doing that activity."
Microthemes	A microtheme is a brief essay limited to one side of a 5-inch x 8-inch index card. The four formats described here are offered in "Microtheme Strategies for Developing Cognitive Skills" (John C. Bean et al,1982), each challenging and cultivating writing and cognitive skills in a different way. • *Summary-Writing Microtheme:* Students read a body of material, discuss its structure (main idea, supportive points, connections among its parts), condense it, and write a summary. This exercise strengthens reading comprehension and writing ability. In this writing, students must avoid imposing personal opinion on data or distorting an author's perspective. • *Thesis-Support Microtheme:* Students must take a stand and defend it. This exercise strengthens the ability to discover, state, and defend an issue, using clear evidence and logical reasoning. • *Data-Provided Microtheme:* Students are provided with data in the form of tables or factual statements. The student must draw conclusions based on evidence. Selecting, arranging, connecting, and generalizing about data develops inductive reasoning. Students thus progress from merely listing facts to making assertions. • *Quandary-Posing Microtheme:* Students are presented with a real-life problem or puzzling situation. Students must explain the underlying scientific principles in clear terms and pose a solution. By moving students from rote learning to application, concept comprehension and abstract reasoning are stengthened.
Simile reviews	Similes vividly compare things and point out what is similar about them, generally using the word "like" or "as." Here's an example: "The fog was like a blanket in front of my eyes." Students are asked to write a paragraph completing the following pattern: [Science topic] is/are like _______________ because _______________ . For example, "A battery is like a spring because they both store energy and can transfer that energy to other objects. In the battery, chemical potential energy is stored in the arrangement of the parts of the molecules, while in a spring, elastic potential energy is stored in the arrangement of the molecules relative to one another."

Safety First

Experiments, demonstrations, and hands-on activities add relevance, fun, and excitement to science education at any level. However, even the simplest experiment can become dangerous when the proper safety precautions are ignored or when the experiment is done incorrectly or performed by students without proper supervision. While the experiments in this book include cautions, warnings, and safety reminders from sources believed to be reliable and while the text has been extensively reviewed, it is your responsibility to develop and follow procedures for the safe execution of any experiment you choose to do. You are also responsible for the safe handling, use, and disposal of materials used in accordance with local and state regulations and requirements.

- Read each experiment carefully and observe all safety precautions and disposal procedures. Determine and follow all local and state regulations and requirements.

- Always practice experiments yourself before using them with your class. This is the only way to become thoroughly familiar with an experiment, and familiarity will help prevent potentially hazardous (or merely embarrassing) mishaps. In addition, you may find variations that will make the experiment more meaningful to your students.

- You, your assistants, and any students participating in the preparation or performance of an experiment must wear appropriate personal protective equipment.

- Special safety instructions are not given for everyday classroom materials being used in a typical manner. Use common sense when working with hot, sharp, or breakable objects. Keep tables or desks covered to avoid stains. Keep spills cleaned up to avoid falls.

- Remember that you are a role model for your students—your attention to safety will help them develop good safety habits while assuring that everyone has fun with these experiments.

References

Ambron, J. "Writing to Improve Learning in Biology," *Journal of College Science Teaching.* 1987, *16,* 263–266.

Bean, J.C.; Drenk, D.; and Lee, F. D. "Microtheme Strategies for Developing Cognitive Skills," *Teaching Writing in All Disciplines;* Griffin, C.W., Ed.; Jossey-Bass: San Francisco, 1982; pp 27–38.

Carle, M.A.; Sarquis, M.; Nolan, L.M. *Physical Science: The Challenge of Discovery;* D.C. Heath: Lexington, MA, 1991.

Elbow, P. *Writing Without Teachers,* 2nd ed.; Oxford University Press: New York, 1998.

Harwood, W.S. "A New Model for Inquiry," *Journal of College Science Teaching.* 2004, *33*(7), 29–33.

Heinlein, Robert A. *Stranger in a Strange Land;* Ace: New York, 1987.

Holliday, W.G.; Yore, L.; and Alvermann, D.E. "The Reading-Science Learning-Writing Connection: Breakthroughs, Barriers, and Promises," *Journal of Research in Science Teaching.* 1994, *31,* 877–894.

Madigan, C. "Writing as a Means, Not an End," *Journal of College Science Teaching.* 1987, *16,* 245–249.

National Research Council. *National Science Education Standards: Observe, Interact, Change, Learn;* National Academy Press: Washington, DC, 1996.

O'Brien, T. "Teaching Fundamental Aspects of Science with Toys," *School Science and Mathematics.* 1993, *93*(4), 203–207.

Tierney, R.; Dorroh, J. *How to Write to Learn Science;* National Science Teachers Association: Arlington, VA, 2004.

Six-Cent Top

What makes a top keep spinning? Students experiment with rotational inertia and the variables affecting it.

Grade Levels

Science activity appropriate for grades 3–8

Student Background

Students should understand the concept of mass. Practice in data recording would be helpful. A prior introduction to the concept of variables would also be helpful, although the concept can be introduced here.

Time Required

Setup	30	minutes
Part A	15	minutes
Part B	30	minutes
Part C	15–30	minutes
Cleanup	5	minutes

Assessment time is not included.

Key Science Topics

- mass
- rotational inertia
- rotational motion
- variables

National Science Education Standards Overview

See *www.terrificscience.org/physicsez/* for details of how these standards relate to the activity.

Science as Inquiry

Abilities Necessary to Do Scientific Inquiry

K–4 *Plan and conduct a simple investigation.*
K–4 *Employ simple equipment and tools to gather data and extend the senses.*
K–4 *Use data to construct a reasonable explanation.*

5–8 *Design and conduct a scientific investigation.*
5–8 *Use appropriate tools and techniques to gather, analyze, and interpret data.*
5–8 *Develop descriptions, explanations, predictions, and models using evidence.*
5–8 *Use mathematics in all aspects of scientific inquiry.*

Physical Science

K–4 *Position and motion of objects*

5–8 *Motions and forces*

Science and Technology

Abilities of Technological Design

K–4 *Evaluate a product or design.*
K–4 *Communicate a problem, design, and solution.*

5–8 *Implement a proposed design.*
5–8 *Evaluate completed technological designs or products.*
5–8 *Communicate the process of technological design.*

Materials

For Getting Ready

- oaktag or paperboard (enough for an 8-cm disk per group)

Cereal boxes can be used instead.

- scissors
- (optional) compass
- (optional) protractor

For the Procedure

Activity Introduction, per class

- assorted commercial tops

Parts A and B, per group

- pencil (about 10–12 cm long) with a blunt point

Hexagonal pencils keep the disk in place better than round pencils.

- piece of paper
- 8-cm disk prepared in Getting Ready
- ruler
- 2 small rubber bands
- stopwatch
- 6 small paper clips
- 6 pennies
- tape or glue stick

Part C, per group

- material to make a top, such as
  - assorted K'NEX® rods and connectors (including tan clip connector)
  - assorted pencils, oaktag, cardboard, poster board, unwanted CDs, tape, pennies, paper clips, washers, and small rubber bands
- ruler
- stopwatch
- (optional) scissors

For the Assessment

Per group

- 2 sticks (any type)
- paperboard (enough to make 2 disks)
- 4 small rubber bands
- 12 washers with holes large enough to fit the sticks through

Safety and Disposal

No special safety or disposal procedures are required.

Getting Ready

Use the template below or a compass to make each group an 8-cm-diameter round disk. Poke a small starter hole, approximately 2 mm in diameter, in the center of each circle. Make pencil marks every 60 degrees by using the template or a protractor. Make these disks in advance or make a few patterns and have students make their own disks. Older students can use compasses and protractors to make the circles and mark every 60 degrees.

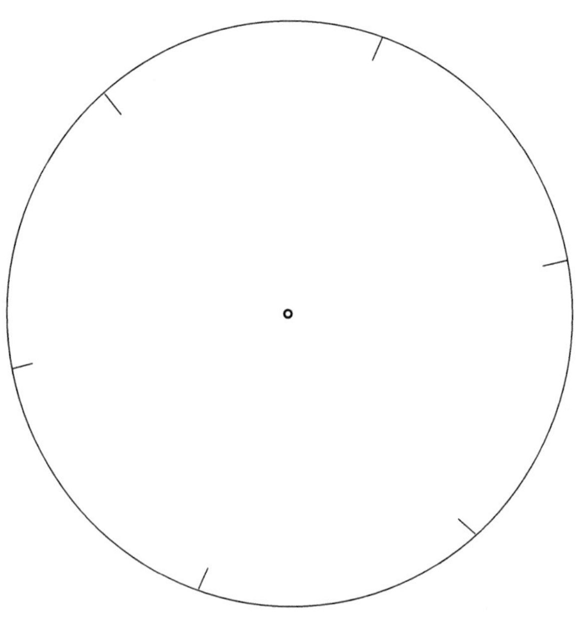

Template for 8-cm disk

Procedure

This section provides teacher notes corresponding to each step of the student procedure. The procedure without teacher notes is included in the reproducible Student Notebook pages at the end of this activity and at www.terrificscience.org/physicsez/.

	Student Procedure	Teacher Notes
	Activity Introduction	• Bring out a variety of spinning tops for demonstration and student experimentation. • Ask students to describe the various tops according to their size and shape. Find what is common about all of the tops—most likely that they all have a wide middle and a narrow or pointed bottom.
	Part A: A Simple Top	
1	Put the point of the pencil down on the table or floor and spin it like a top. TIP: Spin the pencil on a piece of paper to prevent marking the table or floor. What happens? What could you do to make the pencil spin longer?	When students try to spin the pencil on its point, the pencil will quickly fall over on its side. Students should come up with the idea of adding a wider middle to the pencil or adding more mass around the middle of the pencil.
2	Make a top by poking the pencil through the starter hole in a paperboard disk without tearing the hole any bigger than necessary. The disk should sit about 2.5 cm above the tip of the pencil. Prevent the disk from sliding by placing one rubber band around the pencil below the disk and another around the pencil above the disk.	Younger students may need help with the rubber bands.
3	Spin your top by twirling the pencil with your fingers. Practice for a few minutes until you get the top to spin on most tries. Compare your top to the pencil in step 1. Which can you spin better? Measure the spin time of the top (the length of time that the top spins before its disk touches the table or floor). Record the results in the data table at the end of Part B. Repeat nine times and calculate an average. TIP: Don't record the spin times when the top doesn't spin well. Just try again.	Adding the disk to the pencil allows the unit to spin a little bit more like a top. Rather than falling immediately, the pencil with the disk spins for about 1 second. (See Sample Answers.) If students are having trouble getting the top to spin at all, have them lower the disk on the pencil a bit.

	Student Procedure	Teacher Notes
	Part B: Does Mass Matter?	
1	Think about what would happen if you added mass to the disk and then spun the top. ✎ Write your prediction.	Depending on their previous experience with tops, students may predict that adding mass to the disk will cause the top to spin longer.
2	Add six paper clips at the pencil marks on the disk. Practice spinning the top until you are consistent. TIP: Evenly distribute the paper clips around the disk for proper balance. ✎ Measure the spin time of the top. Record the results in the data table at the end of Part B. Repeat nine times and calculate an average. ✎ Which needs a bigger push from your fingers to start spinning, the top with paper clips or the top without paper clips? ✎ Which has a longer spin time, the top with paper clips or the top without paper clips?	A top with paper clips typically requires a bigger force from your fingers to start spinning, but then it spins longer. (See Sample Answers.)
3	Remove the six paper clips and hold them in one hand. Hold six pennies in the other hand. ✎ Which feels heavier, the paper clips or the pennies? ✎ How well do you think the top would spin if the pennies were attached to the top instead of paper clips?	The pennies have more mass than the paper clips. Based on the results when mass was added with paper clips, students may realize that adding mass by replacing the paper clips with the pennies will allow the top to spin even longer.
4	Tape or glue six pennies at the pencil marks on the disk. Practice spinning the top until you are consistent. ✎ Measure the spin time of the top. Record the results in the data table at the end of Part B. Repeat nine times and calculate an average. ✎ Which needs a bigger push from your fingers to start spinning, the top with the pennies or the top with the paper clips? ✎ Which has a longer spin time, the top with the pennies or the top with the paper clips? ✎ How does mass affect the spinning of the top?	A top with pennies typically requires a larger force from your fingers to start spinning, but then it spins longer than a top with paper clips. Students should conclude that mass added to the edge of the disk makes the top spin longer. (See Sample Answers.) Introduce the idea that once in motion, an object that rotates about an axis (such as a top) tends to keep rotating. When mass is added near the edge of the disk, it becomes harder to start the top turning. However, once this top has started turning, it will continue to spin longer. Depending on the age of the students, you may want to introduce the term rotational inertia. (See Explanation.)

Student Procedure	Teacher Notes
Part C: Design Your Own Top	
① Make a top with the materials provided. TIP: If you use K'NEX, the special tan clip connector will help lock the disk and keep it from twirling around on the stick. On another piece of paper, draw a picture of your top. Label the stick and the disk.	See Sample Answers for examples of designs. If students use K'NEX, they often select a rod to serve the same role as the pencil and use several connectors and rods to make the disk. Students need to attach the tan clip connector below the disk to prevent the disk from twirling around on the stick. Snap the tan clip connector to the stick with the pin facing upwards. The pin fits through one of the holes in a connector, keeping the stick and the disk locked together so they spin as one unit. 
② Use your top to do the following challenges. Create data tables for each challenge on another piece of paper. Include at least three trials per design. In Parts A and B, the top's disk was about 2.5 cm above the point of the pencil. Try spinning your top using three different disk heights. Measure and record the height of the disk and the spin time for each trial. Which disk height has the longest spin time? Try three disk diameters. Measure and record the diameter of the disk and the spin time for each trial. What effect does disk diameter have on spin time? Do you see a pattern? Explain your results.	If class time is limited, assign one challenge to each group. Have groups report their results to the class. If the disk is placed too high on the pencil or rod, the spinning top is generally less stable. (See Sample Answers.) Generally, tops with larger disk diameters spin better, and disks with very small diameters are usually less stable than larger ones. However, when the diameter of the disk gets very large, it may touch the ground when the top tips even a little, causing the top to stop spinning sooner. (See Sample Answers.)
Assessment	
❶ Make two identical, simple tops using pencils, paperboard, and rubber bands. Stack six washers over the pencil of one top as shown in the left photo. Tape or glue six washers around the disk perimeter of the other top as shown in the right photo.	Students can make simple tops like in Part A.

Student Procedure	Teacher Notes
2 You will be testing how the position of mass on the top affects its spin. Based on your work with tops, predict which top will average the longest spin time and which will average the shortest spin time. ✎ Write your predictions in the data table. Explain your thinking. ✎ Time and record in the data table three good spin times for each top. ✎ Calculate and record in the data table the average spin time for each top.	Students should apply their knowledge and what they learned in Part B to predicting how location of mass affects the spinning of a top.
3 On another piece of paper, explain how your predictions compare with your spin time results.	The top with washers at the perimeter of the disk should spin longer than the other top. Some of the students may have correctly predicted the results.

Sample Answers

Where students follow the same procedure using the same materials, these answers are close to answers you can expect. Where students design their own experiment or model, students' results will vary.

	Spin Time for Top with Paper Disk Only	Spin Time for Top with Paper Clips Added	Spin Time for Top with Pennies Added
Trial 1	0.8 seconds	2.3 seconds	4.6 seconds
Trial 2	1.9 seconds	2.5 seconds	3.8 seconds
Trial 3	0.6 seconds	2.9 seconds	4.6 seconds
Trial 4	0.9 seconds	2.5 seconds	3.1 seconds
Trial 5	1.2 seconds	2.7 seconds	5.2 seconds
Trial 6	1.0 seconds	2.8 seconds	4.1 seconds
Trial 7	0.7 seconds	3.1 seconds	3.6 seconds
Trial 8	0.7 seconds	2.4 seconds	4.7 seconds
Trial 9	1.0 seconds	2.2 seconds	4.2 seconds
Trial 10	0.8 seconds	2.9 seconds	4.7 seconds
Average	1.0 seconds	2.6 seconds	4.3 seconds

Example data for Parts A and B

Examples of student tops for Part C

Effect of Disk Height on Spin Time				
Disk Height	**Trial 1**	**Trial 2**	**Trial 3**	**Average**
2.5 cm	1.1 seconds	1.1 seconds	1.0 seconds	1.1 seconds
5.0 cm	0.6 seconds	0.4 seconds	0.3 seconds	0.4 seconds
7.5 cm	0.2 seconds	0.1 seconds	0.2 seconds	0.2 seconds

Example data for Part C

Effect of Disk Diameter on Spin Time				
Disk Diameter	**Trial 1**	**Trial 2**	**Trial 3**	**Average**
5 cm	0.2 seconds	0.2 seconds	0.1 seconds	0.2 seconds
8 cm	0.6 seconds	0.4 seconds	0.3 seconds	0.4 seconds
12 cm	2.1 seconds	2.7 seconds	2.3 seconds	2.4 seconds

Example data for Part C

Effect of Mass Position on Spin Time					
Position of Mass	**Prediction and Reason**	**Spin Times**			
		Trial 1	**Trial 2**	**Trial 3**	**Average**
washers stacked over stick	varies	0.8 seconds	1.1 seconds	0.7 seconds	0.9 seconds
washers attached around disk perimeter	varies	2.1 seconds	2.4 seconds	2.1 seconds	2.2 seconds

Example data for Assessment

Explanation

This section is intended for teachers. Modify the explanation for students as needed.

Tops are fun to play with because they keep spinning for some time after being set in motion. In this activity, students observe that the duration of the spin is affected by the amount of mass near the edge of the disk. Why? Once in motion, an object (such as a top) that rotates about an axis (in this case, the stick) tends to keep rotating about its axis. The resistance of an object to change its rotational state of motion is called rotational inertia. When the top is spun, it tends to keep spinning because of this inertia. When more mass is added near the edge of the disk, the rotational inertia is increased. The increased rotational inertia makes it harder to start the top turning, but once it has started turning it will continue to spin longer. Friction eventually causes the top to slow down.

The rotational inertia depends not only on the mass of the spinning top but also on the location of the mass. The farther the mass is located from the axis of rotation, the greater the rotational inertia. A top with six washers attached near the disk perimeter will have more rotational inertia than a top with the same mass but with the washers stacked over the stick. Many real machines have the mass located far from the axis of rotation in order to increase the rotational inertia and rotational kinetic energy. For example, the Ford Museum in Dearborn, Michigan, features a large display of old steam engines. Each engine has a flywheel constructed as a wheel with spokes and most of the mass located near the perimeter.

Another application for the importance of rotational inertia is in space flight. Every space shuttle and space ship is equipped with an inertial guidance system. This system is a set of specially designed and arranged gyroscopes (spinning tops). The gyroscopes are set spinning before the flight takes off. Once in space, the spinning gyroscopes are used as a reference by the ship's computers and commander to determine the ship's orientation in space. Also, some satellites, when launched into space, are set spinning like tops. This spinning motion prevents the satellites from tumbling out of control.

Cross-Curricular Integration

Language arts:

- Have students write a list poem about tops. Explain that a list poem is made up of a list of items or events. You may want to read examples from Walt Whitman. Although the list poem can be any length and may be rhymed or unrhymed, each item in the list should be written the same basic way to create a pattern of repetition. To begin, have students review their completed student notebook pages and list characteristics and observations of tops. Then, they can select the best items from this list to use as a basis for the poem. Encourage students to edit and rewrite by juggling the word order, playing with the pattern, and paying careful attention to the sounds of words. (See example below.)

Technology:

- Research how the space program uses inertial guidance systems in space. Create a poster and present your findings to the class.

Reference

Hewitt, P. *Conceptual Physics*, 9th ed.; Addison Wesley: San Francisco, 2002; pp 125–146.

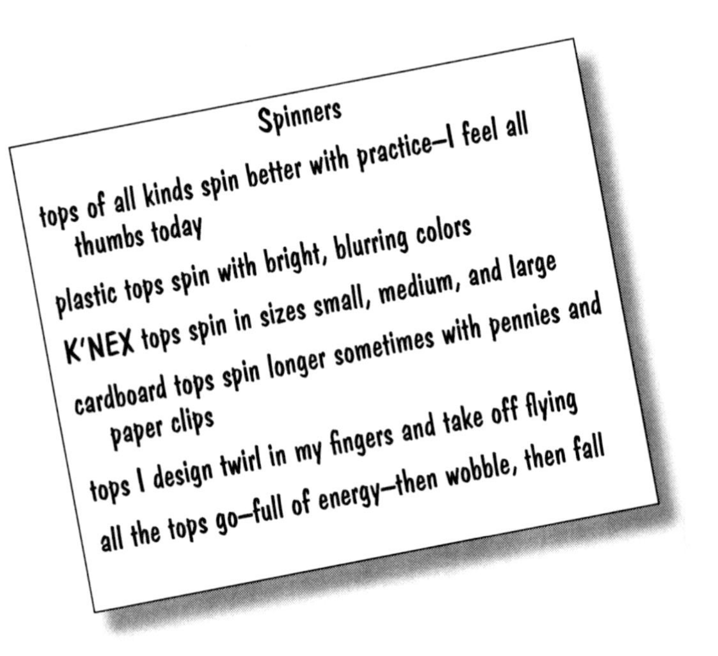

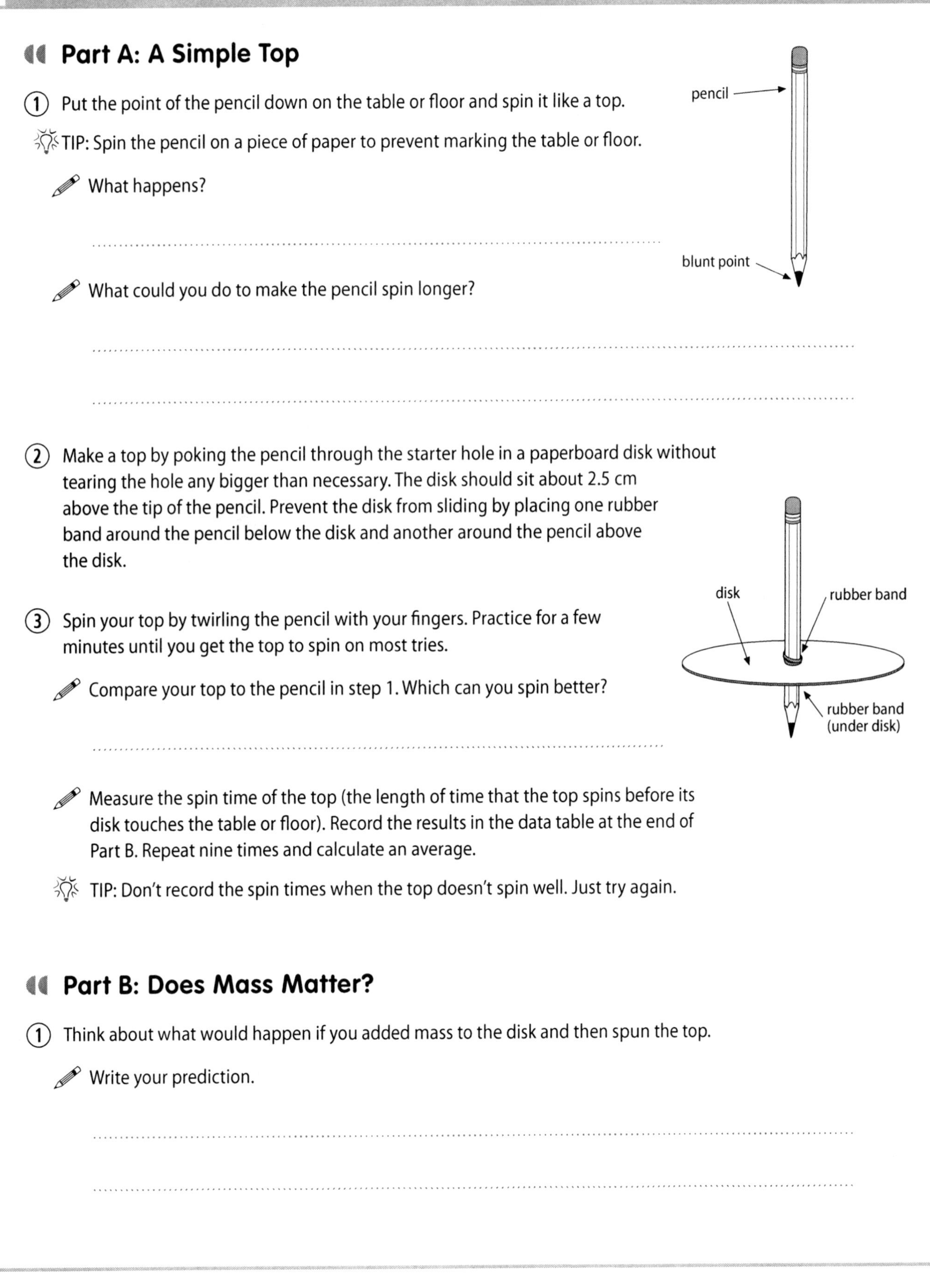

Part A: A Simple Top

1. Put the point of the pencil down on the table or floor and spin it like a top.

TIP: Spin the pencil on a piece of paper to prevent marking the table or floor.

What happens?

....

What could you do to make the pencil spin longer?

....

....

2. Make a top by poking the pencil through the starter hole in a paperboard disk without tearing the hole any bigger than necessary. The disk should sit about 2.5 cm above the tip of the pencil. Prevent the disk from sliding by placing one rubber band around the pencil below the disk and another around the pencil above the disk.

3. Spin your top by twirling the pencil with your fingers. Practice for a few minutes until you get the top to spin on most tries.

Compare your top to the pencil in step 1. Which can you spin better?

....

Measure the spin time of the top (the length of time that the top spins before its disk touches the table or floor). Record the results in the data table at the end of Part B. Repeat nine times and calculate an average.

TIP: Don't record the spin times when the top doesn't spin well. Just try again.

Part B: Does Mass Matter?

1. Think about what would happen if you added mass to the disk and then spun the top.

Write your prediction.

....

....

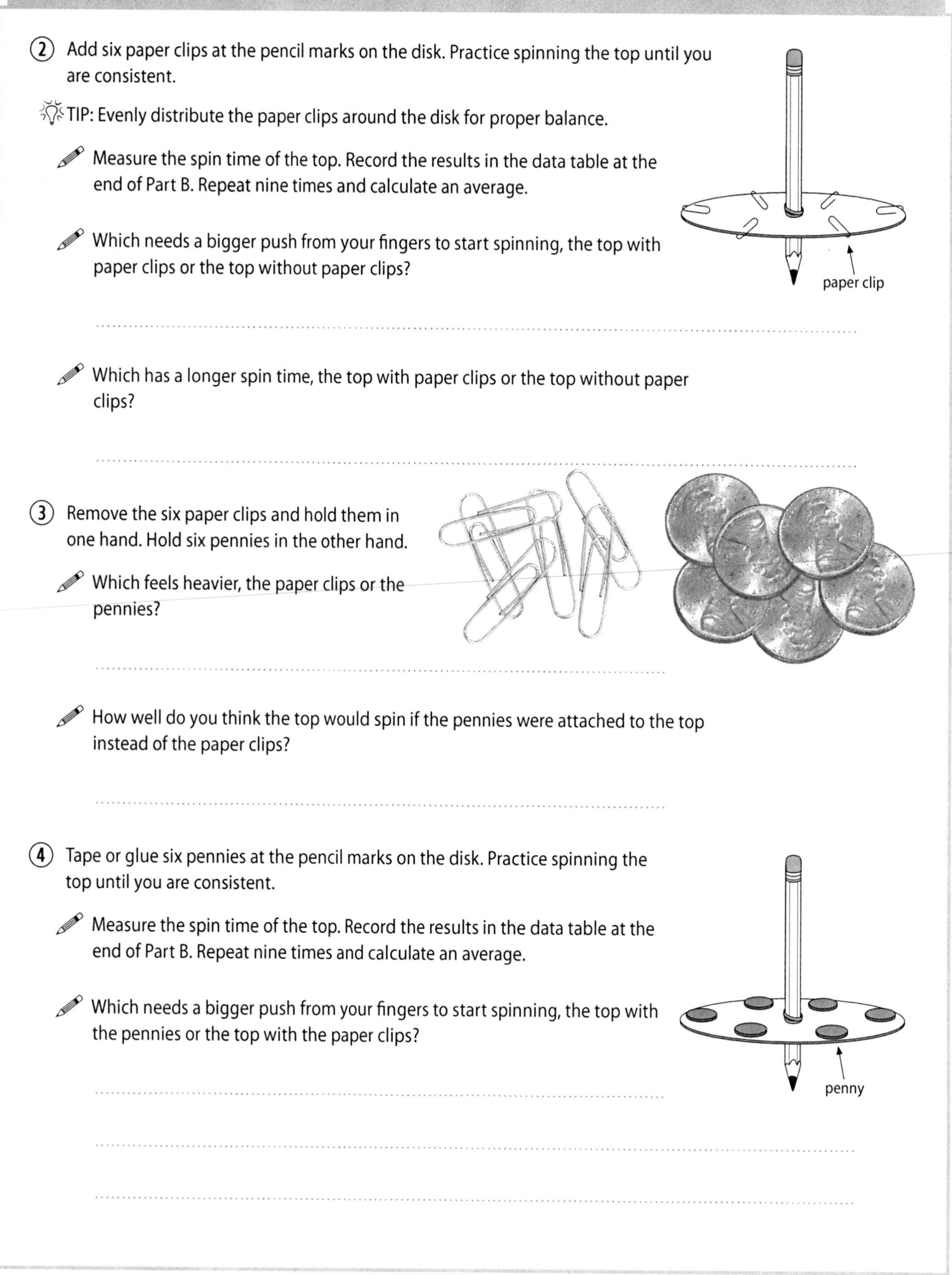

② Add six paper clips at the pencil marks on the disk. Practice spinning the top until you are consistent.

TIP: Evenly distribute the paper clips around the disk for proper balance.

Measure the spin time of the top. Record the results in the data table at the end of Part B. Repeat nine times and calculate an average.

Which needs a bigger push from your fingers to start spinning, the top with paper clips or the top without paper clips?

....

Which has a longer spin time, the top with paper clips or the top without paper clips?

....

③ Remove the six paper clips and hold them in one hand. Hold six pennies in the other hand.

Which feels heavier, the paper clips or the pennies?

....

How well do you think the top would spin if the pennies were attached to the top instead of the paper clips?

....

④ Tape or glue six pennies at the pencil marks on the disk. Practice spinning the top until you are consistent.

Measure the spin time of the top. Record the results in the data table at the end of Part B. Repeat nine times and calculate an average.

Which needs a bigger push from your fingers to start spinning, the top with the pennies or the top with the paper clips?

....

....

....

Which has a longer spin time, the top with the pennies or the top with the paper clips?

How does mass affect the spinning of the top?

Data Table for Parts A and B			
	Spin Time for Top with Paper Disk Only	Spin Time for Top with Paper Clips Added	Spin Time for Top with Pennies Added
Trial 1			
Trial 2			
Trial 3			
Trial 4			
Trial 5			
Trial 6			
Trial 7			
Trial 8			
Trial 9			
Trial 10			
Average			

Part C: Design Your Own Top

① Make a top with the materials provided.

TIP: If you use K'NEX®, the special tan clip connector will help lock the disk and keep it from twirling around on the stick.

On another piece of paper, draw a picture of your top. Label the stick and the disk.

② Use your top to do the following challenges. Create data tables for each challenge on another piece of paper. Include at least three trials per design.

In Parts A and B, the top's disk was about 2.5 cm above the point of the pencil. Try spinning your top using three different disk heights. Measure and record the height of the disk and the spin time for each trial. Which disk height has the longest spin time?

....

....

....

Try three disk diameters. Measure and record the diameter of the disk and the spin time for each trial. What effect does disk diameter have on spin time? Do you see a pattern? Explain your results.

....

....

....

....

....

....

Assessment

1. Make two identical, simple tops using pencils, paperboard, and rubber bands. Stack six washers over the pencil of one top as shown in the left photo. Tape or glue six washers around the disk perimeter of the other top as shown in the right photo.

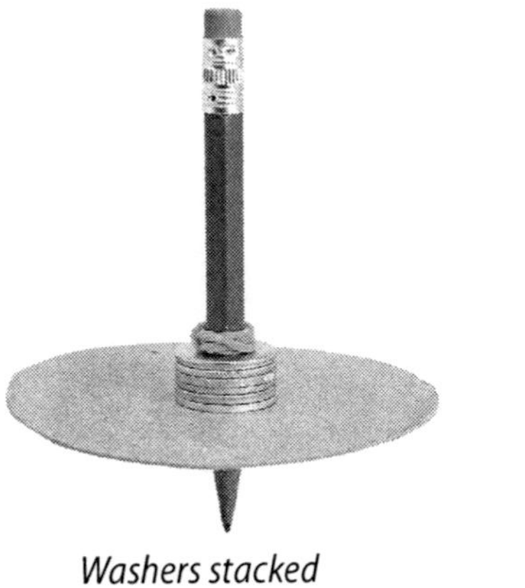

Washers stacked over pencil

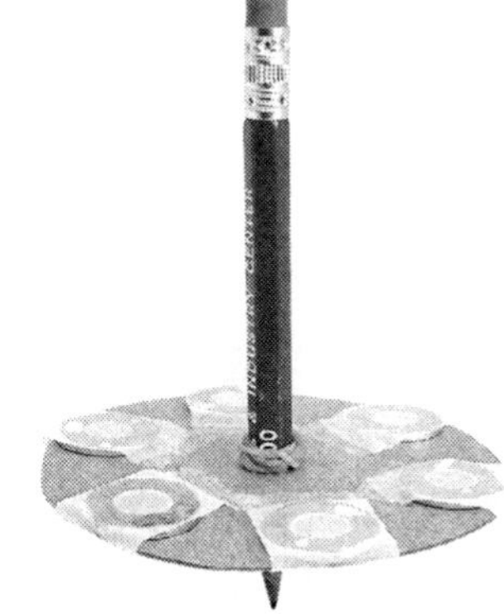

Washers attached around disk perimeter

2. You will be testing how the position of mass on the top affects its spin. Based on your work with tops, predict which top will average the longest spin time and which will average the shortest spin time.

✎ Write your predictions in the data table. Explain your thinking.

Effect of Mass Position on Spin Time

Position of Mass	Prediction and Reason	Spin Times			
		Trial 1	Trial 2	Trial 3	Average
washers stacked over pencil					
washers attached around disk perimeter					

✎ Time and record in the data table three good spin times for each top.

✎ Calculate and record in the data table the average spin time for each top.

3. On another piece of paper, explain how your predictions compare with your spin time results.

Crash Test

Students observe the law of inertia in action and apply it to automobile safety.

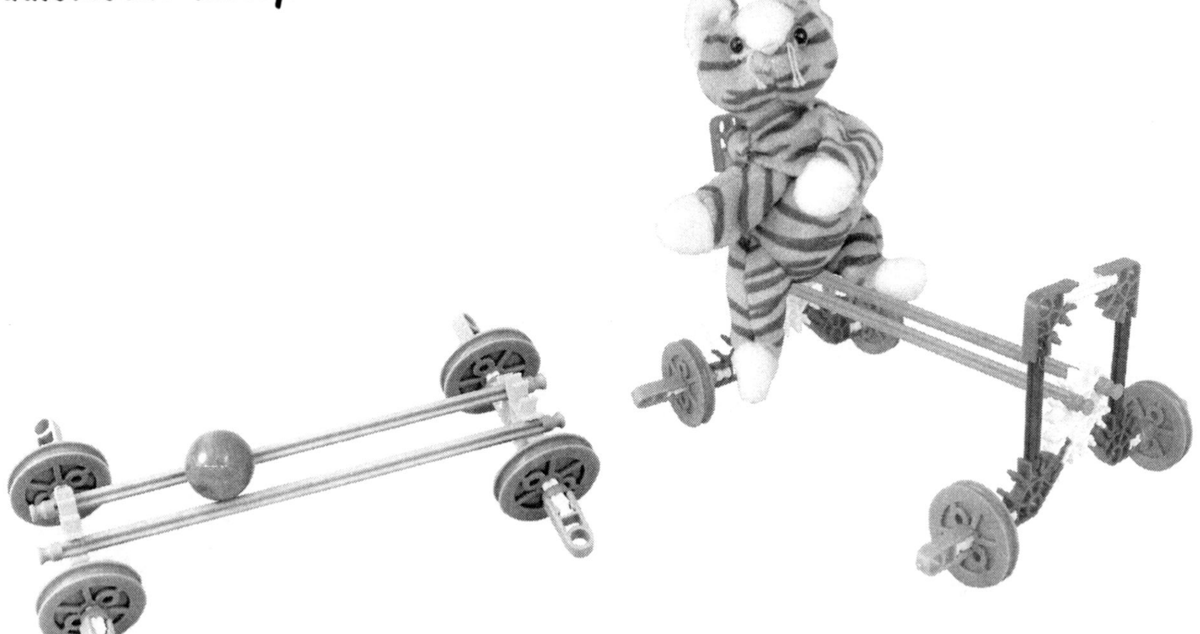

Grade Levels

Science activity appropriate for grades 3–7

Student Background

This activity is designed to introduce the concept of inertia, so students need no previous knowledge of the topic. Some previous knowledge of forces and motion is recommended.

Time Required

Setup		negligible
Procedure	20	minutes
Cleanup		negligible

Assessment time is not included.

Key Science Topics

- force
- inertia
- laws of motion

National Science Education Standards Overview

See *www.terrificscience.org/physicsez/* for details of how these standards relate to the activity.

Science as Inquiry

Abilities Necessary to Do Scientific Inquiry

K–4 *Plan and conduct a simple investigation.*
K–4 *Communicate investigations and explanations.*

5–8 *Develop descriptions, explanations, predictions, and models using evidence.*
5–8 *Think critically and logically to make the relationships between evidence and explanations.*
5–8 *Communicate scientific procedures and explanations.*

Physical Science

K–4 *Properties of objects and materials*
K–4 *Position and motion of objects*

5–8 *Motions and forces*

Materials

For the Procedure

Per class

- (optional) clip-on roller skates

If you have a pair of these around, you can offer them to students to explore along with their K'NEX® cars.

Per group

- assorted K'NEX rods, connectors, and wheels
- marble or steel ball bearing

The marble or ball bearing needs to be large enough to sit on an assembled car between two parallel K'NEX rods without falling through.

- large, solid object (such as a book or wall)
- assorted objects that won't easily fall off the car (such as small blocks of wood, boxes of paper clips, pencils sitting crosswise, and dolls)

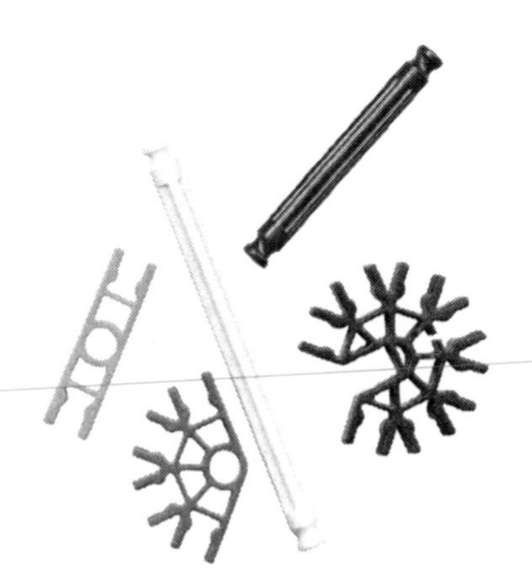

For the Assessment

Per group

- car made in Procedure or another toy car
- object to act as a passenger (such as small doll or stuffed animal)
- something to act as a seat belt (such as string, shoestring, or ribbon)

Safety and Disposal

No special safety or disposal procedures are required.

Getting Ready

Experiment to determine whether the objects you offer the students (marbles, steel ball bearings, objects, dolls, or stuffed animals) will work in the procedure.

Procedure

This section provides teacher notes corresponding to each step of the student procedure. The procedure without teacher notes is included in the reproducible Student Notebook pages at the end of this activity and at www.terrificscience.org/physicsez/.

	Student Procedure	Teacher Notes
	Activity Introduction	To arouse interest, pose some questions that you can return to for discussion later. For example: • Have you ever ridden in a car, bus, or train that slowed down or stopped suddenly? • What happened to the people and objects inside the vehicle? Students can probably point out that people and objects in a vehicle that stops suddenly appear to be thrown forward. Tell students that they are going to play with some toys to find out why this happens.
	Investigating Inertia	
1	**Use K'NEX pieces to build a car that can roll and carry a marble. The marble must be able to roll back and forth from the front to the back of the car without falling off, even when the car rolls and stops.** **TIP: You can use long K'NEX rods to create a track for the marble.**	Let the students come up with their own designs. Two K'NEX rods placed parallel along the length of the car make a good place for the marble to roll. Make sure the students use K'NEX pieces to prevent the marble from rolling off the ends of the car. (See Sample Answers for some examples of car designs.)
2	**Investigate the movement of your car without putting the marble on it. Try rolling the car around while keeping your hand on it. Try pushing the car gently and releasing it.** **Why does the car move when you are pushing it with your hand?** **Why does the car keep moving even after you push it and let it go?** **Why does the car slow down and then stop?**	Students should establish that the cars move because their hands exert a force on the cars. Some students might be able to explain that the car moves even after they let go because a moving object keeps moving unless some force causes it to stop. Students may recognize that friction causes the car to eventually stop.

Student Procedure	Teacher Notes
Optional ☞ Push car.	Have students place the marble at the front of the car and push the car from the back to see what happens to the marble. Students have to look closely because the movements happen fast. Placing a pencil next to the car so it points to the marble's starting position may help them see what happens. Some students may not see the movements unless they are told what to look for. The car rolls forward because there is a force on it, but there is no force on the marble. Therefore, although the marble appears to roll towards the back of the car, it actually rotates in approximately the same spot until the back of the car catches up with the marble and pushes it forward. (Because of friction, the marble moves forward with the car for a small distance.) So the car, which had been set in motion, keeps moving but the marble, which had not been set in motion, doesn't move until a force—from the car—acts upon the marble.
③ **Place a marble at the very back of your car and gently push the car into a large, solid object such as a book or wall.** ✎ **What happens to the marble when the car hits the object? Why do you think this happens?**	When the car hits the object, the car stops but the marble keeps moving. Students should understand that when the object exerts a force on the car and stops it, that force is not exerted on the marble so the marble keeps moving.
Class Discussion ☞	• Give students a chance to share their observations about the marble's motion. Guide a class discussion from the students' work to introduce or review the concepts of inertia and friction as they apply to the car. (See Explanation.) • Ask students what would happen if a surface was made smoother and smoother, to eventually eliminate friction totally. State that the object would travel a longer and longer distance. In the absence of friction, it would continue forever. Explain that in outer space, there is no friction and so an object, once started, keeps moving indefinitely.

Student Procedure	Teacher Notes
4 Replace the marble with other cargo (such as pencils, dolls, or something else) and repeat step 3 to see what happens. ✎ What cargo did you try? Compare how your new cargo moves with how the marble moved. ✎ How do the car, marble, and other cargo you tested show Newton's first law of motion?	Students repeat step 3 with other objects to dispel the common misconception that, because the marble rolls, the behavior is special and that the results would be different for other objects. To help students write about how the objects in the activity follow Newton's first law of motion (provided on the Student Notebook page), make sure your students understand that the words "at rest" mean sitting still. You may need to discuss the law further before students can answer the question.
Assessment	
1 Work with your group to design an experiment to show whether wearing seat belts is important. Make sure to test whether wearing shoulder and lap seat belts is better than just wearing lap seat belts. ✎ On another piece of paper, write your hypothesis and experimental design. Draw and label your planned setup.	Before students begin brainstorming, introduce or review the characteristics of good experiments. Depending on their experience, students may benefit from using the Experiment Planning Guide provided at *www.terrificscience.org/physicsez/*.
2 Perform the experiment after your teacher reviews and approves your experimental design. ✎ Explain whether wearing seat belts is important. How did you come to this conclusion? ✎ Describe the difference between wearing just lap seat belts and wearing shoulder and lap seat belts. ✎ Use Newton's first law of motion to explain what happens with and without seat belts. ✎ How do your results compare with your prediction?	A reasonable plan is for students to: • Place a doll or other passenger on the car and roll it into a wall or book. The doll will move to the front of the car just as the marble did. Students should be able to explain that this is why you seem to fall forward when a driver slams on the brakes. In case of such a slam (or worse, a collision), you could keep on moving straight out of the windshield if there were not something exerting a force on you to make you stop when the car stops—like a seat belt or an air bag. • Test the doll on the car after creating just lap seat belts and then shoulder and lap seat belts. Students should be able to explain why the results are different when the doll "buckles up." When the doll is tied to the car, the "belt" exerts a force on the doll and stops the doll from moving separately from the car—the doll and the car are forced to act as one unit. Wearing a shoulder and lap belt is preferable to only a lap belt, because the shoulder belt restrains the upper body.

Sample Answers

Where students follow the same procedure using the same materials, these answers are close to answers you can expect. Where students design their own experiment or model, students' results will vary.

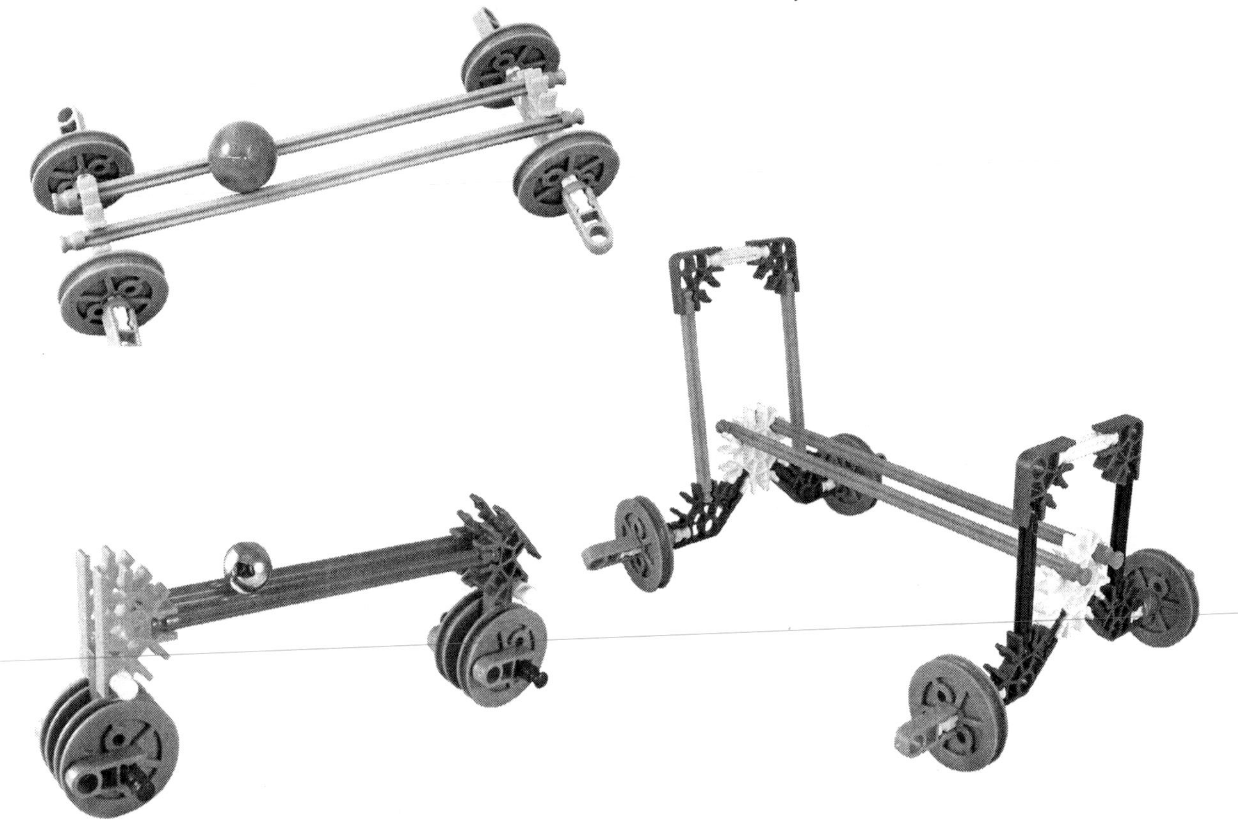

Examples of K'NEX cars

Roll each of these into a wall.

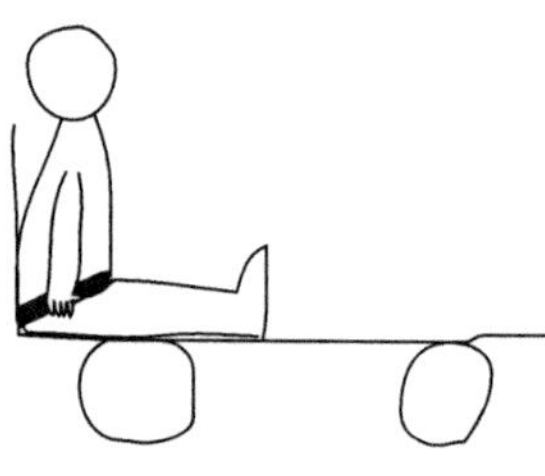

1. Doll on car with no belt
2. Doll on car with lap belt
3. Doll on car with lap and shoulder belts

Example of student drawing for Assessment

Explanation

This section is intended for teachers. Modify the explanation for students as needed.

Newton's First Law of Motion

The foundation for Newton's first law of motion was the work of Galileo Galilei (1564–1642). Galileo was the first scientist to tell us that an object at rest remains at rest, and an object in motion with no forces acting on it will continue moving forever, rejecting Aristotle's earlier notion that a force was required to keep an object moving. Galileo called this property of matter inertia. Sir Isaac Newton (1642–1727) took the property of matter that Galileo called inertia and restated the description in terms of a law he called the law of inertia. This scientific law is now popularly known as Newton's first law of motion. This law states that if the net force on an object is zero, the object continues in its original state of motion. In other words, an object at rest remains at rest, and an object moving with some velocity continues with that same velocity. Net force is defined as the sum of all of the forces acting on an object. Students should understand that having a net force of zero is equivalent to having no forces acting on the object.

Many objects illustrate the law of inertia. One example is the NASA spacecraft Pioneer 10, the first manufactured object to leave the solar system. Pioneer 10 was launched from Earth on March 2, 1972, on a mission to explore the outer planets in our solar system. As it passed Jupiter, the planet's motion through space accelerated the ship to a speed of approximately 82,000 miles per hour. Pioneer 10 is outside the known boundary of our solar system, coasting (obeying the law of inertia) toward a star known as Aldebaran. Aldebaran is about 68 light years away and it will take Pioneer over 2 million years to reach it. Visit the NASA website for more information about Pioneer 10 and her sister ship Pioneer 11.

The Toy Car and the Marble

The toy car begins at rest, sitting still on the table before being pushed. It is obeying the law of inertia (first law of motion). While being pushed, the toy car moves because a force is applied to it. As the force is applied, the car accelerates from being at rest to moving along the table at some speed. This is Newton's second law in action. The car continues to move after one's hand is no longer applying a force because the car is obeying the law of inertia. How fast the car accelerates and how far it goes after being released depend on the size of the force applied to the car and on its mass. The car eventually slows down and stops because the force of friction between the wheels and the surface creates a force in a direction opposite to the direction of motion.

When a marble is placed in the front of the car, the marble and car are both at rest. When a force is applied to the car, the car moves forward and the marble appears to roll to the back of the car. It appears that a force pushed the marble to the back of the car. However, there was no backward force on the marble. It remains nearly at rest. (There is just a little frictional force acting on the marble.) The marble stays at rest, rotating in place until the back of the car catches up and pushes the marble forward with the car.

Once the marble and car are in motion, what happens when the car hits an object? The car stops because a force acted on it. No force acted on the marble, so the marble keeps moving until it hits the front of the car.

Seat Belts and Air Bags

People in cars wear seat belts and cars are equipped with air bags because of the laws of inertia and force. When riding in a car, passengers are moving forward with the car. When the brakes are applied or if the car stops suddenly in a collision, the force is applied to the car, not to the passenger. If the passenger is not wearing a seat belt, the passenger continues to move forward. If the passenger is wearing a seat belt, then the belt exerts a force on the passenger so that he or she remains in place and slows down with the car. The passenger in turn exerts an equal and opposite force on the seat belt. Wearing a shoulder and lap belt is preferable to only a lap belt, because the shoulder belt restrains the upper body. If the air bag activates, the passenger's forward movement exerts a force on the bag as the air bag exerts an equal and opposite force on the passenger.

Cross-Curricular Integration

Home, safety, and career:

- Have students create a poster or other visual aid about seat belt safety emphasizing the "do's and don'ts." Posters (or other selections) could be displayed in stores, restaurants, and public buildings, particularly in conjunction with National Child Passenger Safety Week or other theme promotions.
- In addition to the seat belt assessment, you can discuss other safety implications. For example, the reason you trip when you catch your foot on something is that your foot is stopped but the rest of you keeps going. Also, the concept of friction helps explain why you will slide on a low-friction surface such as wet smooth concrete next to a pool—and why it is so dangerous to run there.

Language arts:

- After you discuss inertia as defined scientifically, you may wish to discuss a more philosophical definition of inertia. Inertia may be thought of as the tendency of the couch potato to stay on the couch, and the tendency of the active person to stay active. Ask the students if they think people act in accordance with this type of law of inertia. Ask what forces get people to get moving, or to stop. Ask if they can think of anything else that tends to stay the way it is, if no force acts on it. Some ideas are reputations, governments, and habits. You might wish to brainstorm a list of things that act like they are affected by inertia. The students could choose one of these to write about, explaining why they believe—or don't believe—a law of inertia could be written about the item they chose.
- Read and discuss one or more of the following stories by Seymour Simon:
    - "The Broken Window" in *Einstein Anderson Shocks His Friends* (grades 4–7)
      In this story, some students claim the rear bus window was broken by a book that was thrown backwards when the driver hit the brakes. Einstein knows they must be lying, because the book would continue to move forward instead of backward when the bus suddenly stopped.
    - "The Challenge of the Space Station" in *Einstein Anderson Shocks His Friends* (grades 4–7)
      In this story, Margaret is writing a story about the construction of a space station. Einstein points out several scientific errors in her story. One of them is that the construction workers can easily move larger girders around because they are weightless. Einstein knows that they still have a large mass and thus will require a large force to start them moving.
    - "Thinking Power" in *Einstein Anderson Sees Through the Invisible Man* (grades 4–7)
      In this story, Einstein is able to help a friend win a contest because he understands inertia.
- After reading "The Broken Window" in *Einstein Anderson Shocks His Friends,* have students use a problem/solution graphic analyzer. (See *www.terrificscience.org/physicsez/* for reproducible master.) Students should identify the problem of the story and use what they learned in the story to come up with a solution. Have students write a paragraph about their ideas.
- Students can create a TV commercial about seat belt safety.

Investigating Inertia

① Use K'NEX® pieces to build a car that can roll and carry a marble. The marble must be able to roll back and forth from the front to the back of the car without falling off, even when the car rolls and stops.

TIP: You can use long K'NEX rods to create a track for the marble.

② Investigate the movement of your car without putting the marble on it. Try rolling the car around while keeping your hand on it. Try pushing the car gently and releasing it.

Why does the car move when you are pushing it with your hand?

....

....

Why does the car keep moving even after you push it and let it go?

....

....

Why does the car slow down and then stop?

....

....

③ Place a marble at the very back of your car and gently push the car into a large, solid object such as a book or wall.

What happens to the marble when the car hits the object? Why do you think this happens?

....

....

....

....

....

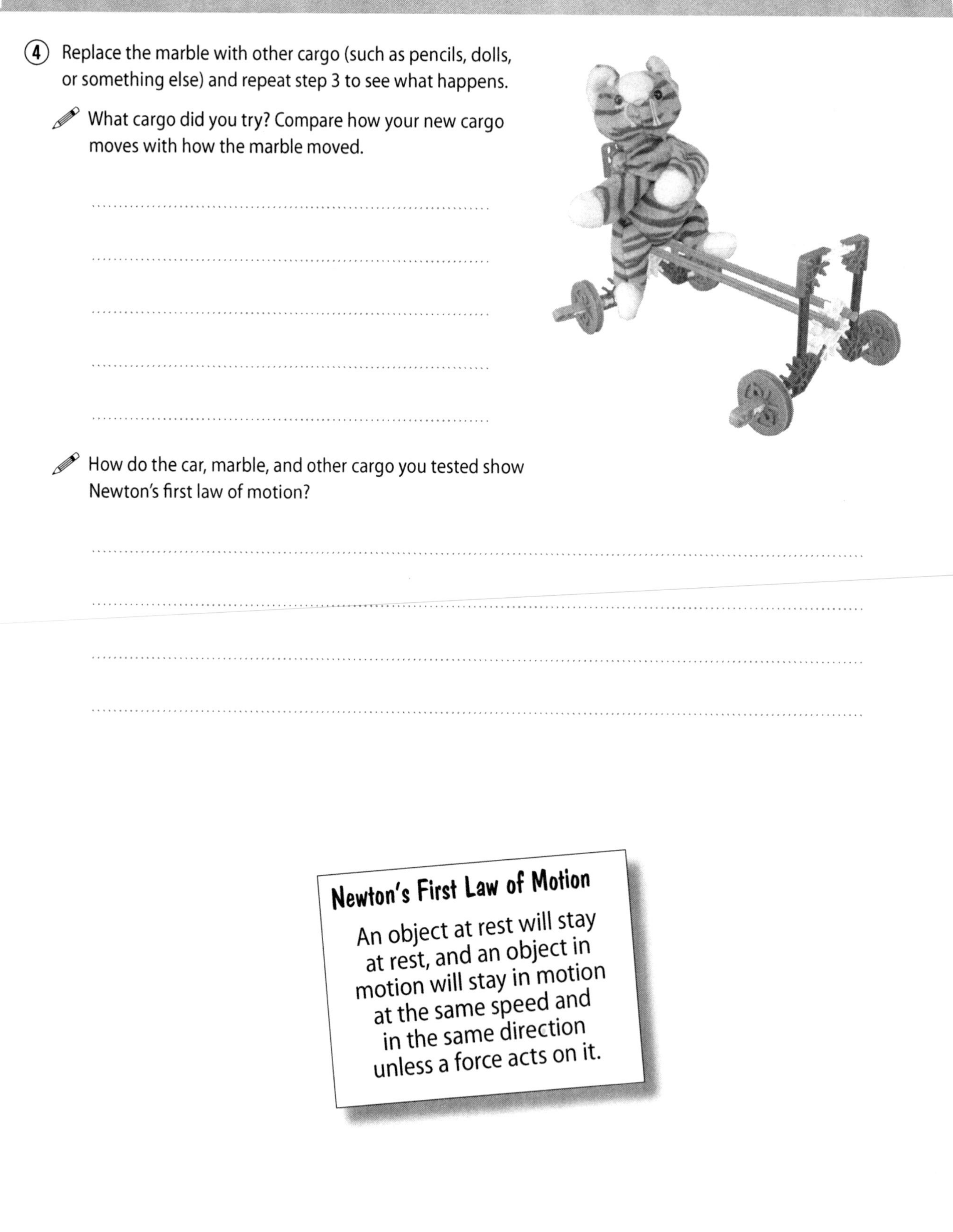

④ Replace the marble with other cargo (such as pencils, dolls, or something else) and repeat step 3 to see what happens.

What cargo did you try? Compare how your new cargo moves with how the marble moved.

How do the car, marble, and other cargo you tested show Newton's first law of motion?

Newton's First Law of Motion

An object at rest will stay at rest, and an object in motion will stay in motion at the same speed and in the same direction unless a force acts on it.

Assessment

1. Work with your group to design an experiment to show whether wearing seat belts is important. Make sure to test whether wearing shoulder and lap seat belts is better than just wearing lap seat belts.

   On another piece of paper, write your hypothesis and experimental design. Draw and label your planned setup.

2. Perform the experiment after your teacher reviews and approves your experimental design.

   Explain whether wearing seat belts is important. How did you come to this conclusion?

   ....

   ....

   ....

   Describe the difference between wearing just lap seat belts and wearing shoulder and lap seat belts.

   ....

   ....

   ....

   ....

   Use Newton's first law of motion to explain what happens with and without seat belts.

   ....

   ....

   ....

   How do your results compare with your prediction?

   ....

   ....

Ramps and Cars

Ready, set, go! Students explore how different ramp angles affect the travel distance of toy cars.

Grade Levels

Science activity appropriate for grades 3–6

Student Background

Students should have a basic understanding of gravity and falling objects.

Time Required

Setup	10 minutes
Procedure	30 minutes
Cleanup	5 minutes

Assessment time is not included.

Key Science Topics

- distance
- gravity
- inclined planes
- motion

National Science Education Standards Overview

See *www.terrificscience.org/physicsez/* for details of how these standards relate to the activity.

Science as Inquiry

Abilities Necessary to Do Scientific Inquiry

K–4 *Plan and conduct a simple investigation.*
K–4 *Employ simple equipment and tools to gather data and extend the senses.*

5–8 *Design and conduct a scientific investigation.*
5–8 *Use appropriate tools and techniques to gather, analyze, and interpret data.*
5–8 *Develop descriptions, explanations, predictions, and models using evidence.*
5–8 *Use mathematics in all aspects of scientific inquiry.*

Physical Science

K–4 *Position and motion of objects*

5–8 *Motions and forces*

Materials

For the Procedure

Per group

- 4 same-sized blocks (or books)
- at least 1 m of straight plastic race car track

  For supply source suggestions, see www.terrificscience.org/supplies/. Students may also have track at home to share.
- firm surface to support the track (such as stiff cardboard or a wooden board)
- toy car (such as Hot Wheels® or Matchbox®)
- meterstick, ruler, measuring wheel, or measuring tape
- graph paper

  Graph paper masters for copying are available at www.terrificscience.org/physicsez/.
- (optional) colored stickers

For the Assessment

All materials listed for the Procedure

Safety and Disposal

No special safety or disposal procedures are required.

Getting Ready

- Test all cars in advance to determine how much each one veers off a straight pathway while traveling. Some veering is not a problem, but measuring travel distances becomes difficult when cars veer a great deal. Replace cars that veer too much with other cars. Also, add as much track as possible beyond the ramp to reduce veering.
- Choose a method for marking and measuring the predicted and trial distances. Below is one idea:
  - Have students use stickers marked with "P" (for predicted) and "A" (for actual). Have students predict how far the car will go by placing a "P" sticker on the floor. After releasing the car, have them mark the car's actual stopping position on the floor with the "A" sticker. Then students can measure the distances.

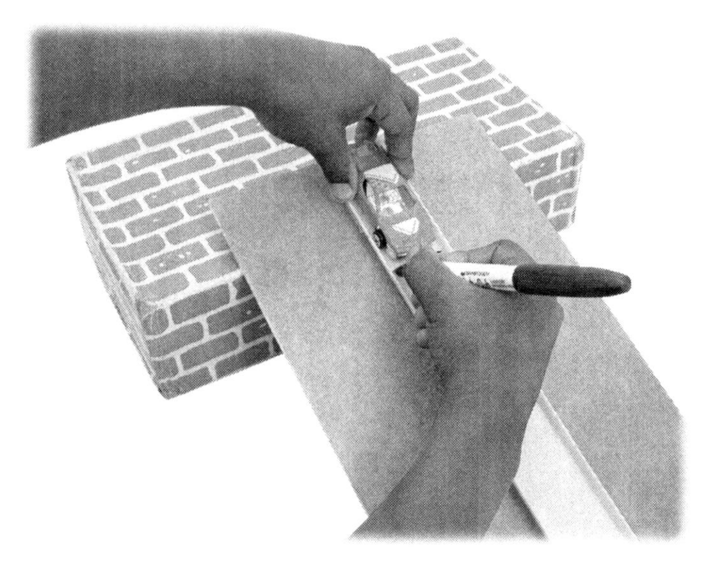

Procedure

This section provides teacher notes corresponding to each step of the student procedure. The procedure without teacher notes is included in the reproducible Student Notebook pages at the end of this activity and at www.terrificscience.org/physicsez/.

	Student Procedure	Teacher Notes
	Activity Introduction	Provide time for free exploration to allow the students to examine and play with the cars and ramps they will use in the activity. This play forms the basis for the predictions they will make in step 2.
	Raise the Ramp	
1	**Place one block under one end of the ramp and lay the track along the ramp. Hold your car at the top of the ramp (on the track) and mark a starting line just in front of it. All members of your group will use the same starting line and the same car to carry out the experiment.**	To smooth the transition from the ramp to the floor, make sure that the track extends beyond the edge of the ramp onto the floor.
2	**Have each group member mark where he or she predicts the front wheels of the car will stop after the car is released from the starting line. Measure the distance from the starting line to the place each group member predicts the car will stop.** **Record each student's predicted travel distance in the data table.**	Students should mark and measure the predicted travel distance based on the method being used.
3	**Have the first student place the car at the starting line and then release it. Be sure not to push the car; otherwise, the speed at which the car leaves the ramp may not be the same in each trial.**	Before students do this step, you may want to review the idea of controlling variables in an experiment.
4	**After the car comes to rest, mark the location of the front wheels. Measure the distance the car traveled.** **Record the actual travel distance in the data table.** **Calculate the difference between the first student's predicted distance and the actual distance. Record this number in the data table.**	Students can calculate the difference between the predicted and actual values in different ways depending on their mathematical abilities. • If students understand the concept of positive and negative numbers, have them subtract the predicted from the actual distance traveled to determine how much variation exists. • Alternatively, students can express the difference in the two values in terms of less than and greater than.

Student Procedure	Teacher Notes
⑤ Give each group member a turn to do steps 3 and 4. ✎ Calculate the average actual travel distance for your group and record this number in the data table.	See Sample Answers for example data.
⑥ Raise the ramp by adding a second block. Repeat steps 2–5. This time create a data table on another piece of paper and record the results.	Students can create a data table like the one used for the one block height.
⑦ Raise the ramp again by adding two more blocks. Repeat steps 2–5. Create a data table on another piece of paper and record the results. TIP: Notice that this time you are using four blocks, not three.	Students skip using three blocks so that predictions about three blocks can be made in the Assessment. Students can create a data table like the one used for the one block height.
⑧ On graph paper, make a graph relating the number of blocks under the ramp to the average distances the car moved. ✎ Explain what this graph shows.	Help students plan their graphs, making sure that students have space on their graphs for data about three blocks (which will be gathered in Assessment). You may want younger students to create a bar graph. More advanced classes can create a line graph in which the height of the ramp in centimeters is plotted against the distance the car traveled. (See Sample Answers for example graphs.) Students should conclude that when the ramp is higher, the car travels farther after leaving the ramp. Lead students to understand that the reason the car travels farther is that it is traveling faster when it leaves the ramp.
⑨ For each of the heights you tested in this experiment, look at your data tables for the differences between your predicted and actual travel distances. ✎ How did the accuracy of your predictions change as you went through the experiment?	Students' predictions will probably improve as they proceed through the experiment because they gain experience with the system and the concept. As travel distances increase, student predictions may be farther off in terms of numbers but closer in terms of percentages. For example, if a car goes 2 m with a 1-block ramp height and the prediction is 2.5 m, the prediction is off by 25%. However, if the car travels 10 m at a more inclined ramp and the prediction is 11 m, the prediction is off by only 10%. (Your students' mathematical abilities will determine whether or not you want to bring this up.)

Student Procedure	Teacher Notes
Assessment	
1 Look at your step 8 graph of the average distances the car traveled with one, two, and four blocks. ✎ Use this graph to predict the distance your car would travel if three blocks were placed under the ramp.	
2 Test your prediction. ✎ Create a data table below and record the results. ✎ How do your results compare with your prediction?	

Sample Answers

Where students follow the same procedure using the same materials, these answers are close to answers you can expect. Where students design their own experiment or model, students' results will vary.

Ramp Height of 1 Block			
Name of Student	**Predicted Travel Distance**	**Actual Travel Distance**	**Difference Between Predicted and Actual Travel Distances**
1. Carlos	varies	96 cm	varies
2. Dionne	varies	117 cm	varies
3. Jamal	varies	105 cm	varies
4. Amy	varies	114 cm	varies
Average		108 cm	

Example of data for one block height

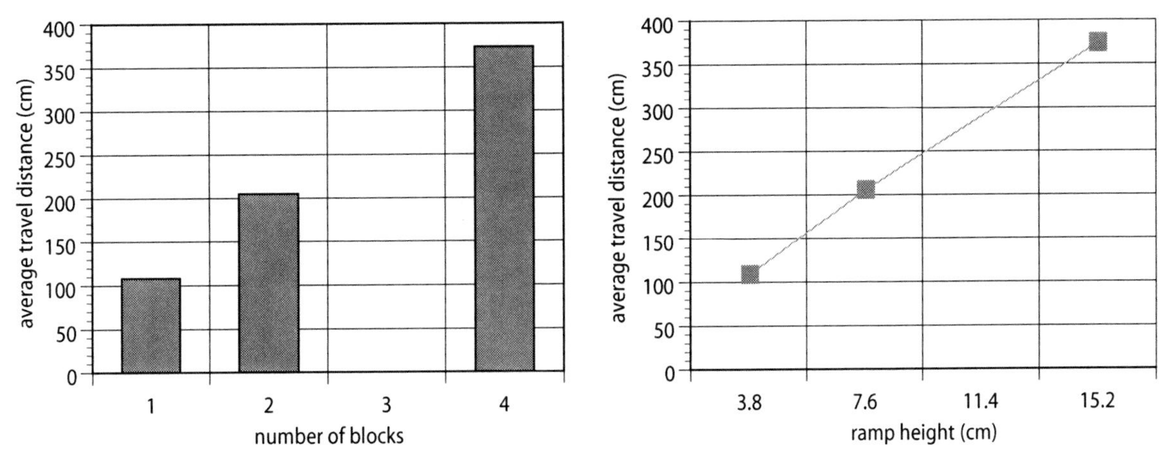

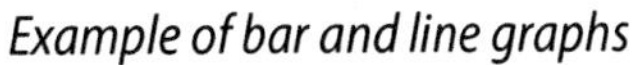
Example of bar and line graphs

Ramp Height of 3 Blocks		
Predicted Travel Distance	**Actual Travel Distance**	**Difference Between Predicted and Actual Travel Distances**
varies	290 cm	varies

Example of data for Assessment

Explanation

This section is intended for teachers. Modify the explanation for students as needed.

Force of Gravity and Vectors

In this activity, students observe that the toy car rolls farther as the incline of the track becomes steeper. The car rolls farther with a steeper incline because it is moving faster when it reaches the bottom of the ramp. The faster the car is moving when it leaves the ramp, the more distance it can cover before friction slows and stops it.

So why is the car moving faster when it reaches the bottom of a steeper incline? To understand, we need to think about how the force of gravity acts on objects. If you hold a toy car in the air and drop it, it falls to the ground. When you place the same toy car at the top of an inclined ramp and let go, the car rolls down the ramp. In both cases the car moves because the force of gravity acts on it. The amount of gravitational force acting on an object depends on the object's mass. The mass of the car is the same whether the car is dropped or rolled down a ramp, so the force of gravity acting on the car has to be the same in both cases. But, the car falls to the ground much faster than it rolls down the ramp. Why? In addition to a small effect from friction, there is another important factor.

When you hold the car in the air and drop it, the entire force of gravity acts in the direction of motion—straight down—causing the car to accelerate and reach the ground quickly. (See Example 1 in the figure below.) But in this experiment, the downward force of gravity is divided into components, called vectors, that act in more than one direction. When the car is on the ramp, part of the force of gravity acts parallel to the ramp. This force component parallel to the ramp causes the car to accelerate. This acceleration is not as great as when the car falls straight down, because the force causing the acceleration is not as great. In the figure below, the arrows labeled "B" represent the forces causing acceleration. (Note the size difference of the arrows labeled "B" in Examples 2 and 3 below.) The rest of the force of gravity acts at right angles to the ramp and holds the car against the surface. (See the arrows labeled "A" in the examples below.)

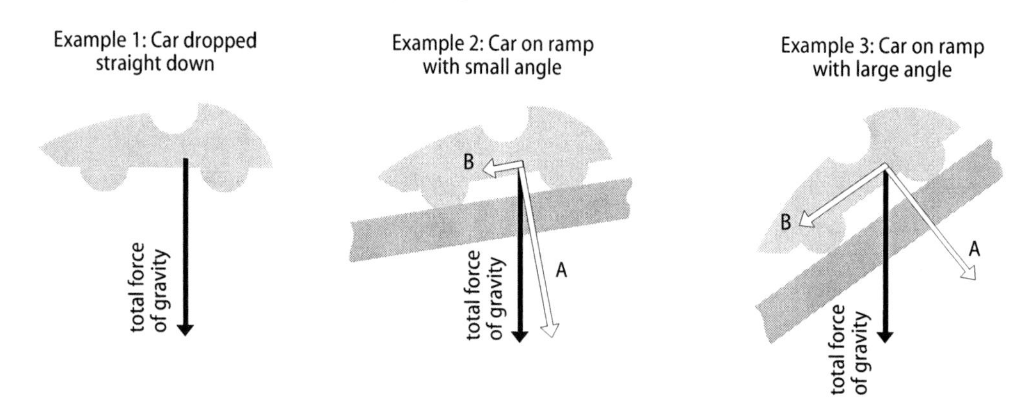

When the car is dropped, all components of the force of gravity act downward. On an incline, component A of the force of gravity holds the car against the surface of the ramp, and component B acts along the plane of the ramp to accelerate the car.

As the height of the ramp increases, the component of the force of gravity that acts parallel to the ramp increases and, thus, the car's acceleration increases. When the car reaches the end of the ramp, it is going faster than when the ramp height is lower. Even though the component of the force of gravity that acts parallel to the ramp increases as the ramp height increases, the total force of gravity on the car does not change. Thus, the component of the force of gravity that holds the car against the surface of the ramp must decrease as the ramp height increases. (See arrows labeled "A" in the examples on the previous page.)

Another way to think about this situation is to look at the total force on the car. When the car falls vertically, the only force on the car is gravity (ignoring the small force of friction with the air). When the car is on the track, in addition to gravity, there is the force of the track pushing up on the car. This track force is the same size as the component of gravity perpendicular to the track (A) and, thus, they cancel out. This leaves the parallel component of gravity (B) as the net force on the track. The lower the ramp, the smaller the net force and, thus, the acceleration.

Demonstration on an Incline

You can illustrate the role of gravity with a simple demonstration. Begin by attaching a rubber band to the back of a car. Hold the rubber band so that the car is suspended vertically. Note the length of the rubber band. Place the car on the incline and, while holding the rubber band in one hand, raise the incline with the other. Notice that the rubber band stretches more and more as you raise the ramp. (See figure at right.) This indicates that, as the height of the incline increases, the component of the force of gravity acting parallel to the ramp increases.

In terms of energy, raising the height of the ramp increases the gravitational potential energy of the car sitting at the top of the ramp. As the car rolls down the ramp, the gravitational potential energy is converted to energy of motion, kinetic energy. With more kinetic energy, the car attains a greater velocity. The faster the car is moving when it leaves the ramp, the more distance it can cover before friction slows and stops it.

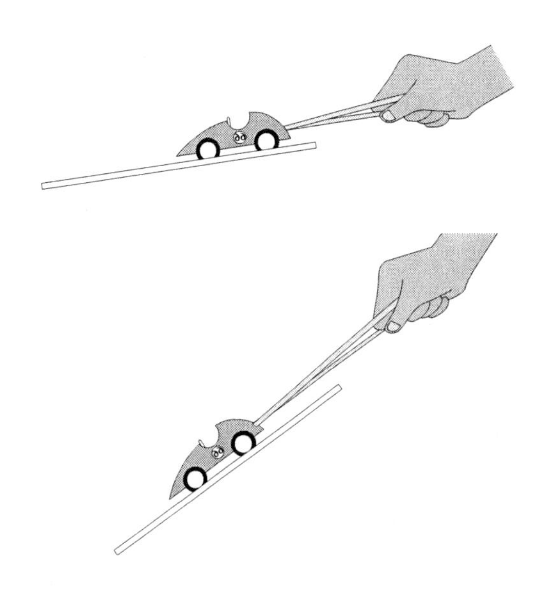

As the slope increases, so does the pull on the rubber band.

Cross-Curricular Integration

Language arts:

- Read aloud or suggest that students read the following book:
  - *The New Way Things Work,* by David Macaulay and Neil Ardley (young adult)
    This book explains the scientific principles behind hundreds of machines and devices, including digital machinery.

References

Activities Handbook for Teachers of Young Children, 5th ed.; Houghton-Mifflin: Boston, 1990.

Butzow, C. *Science through Children's Literature;* Teacher Idea: Englewood, CA, 1989.

Raise the Ramp

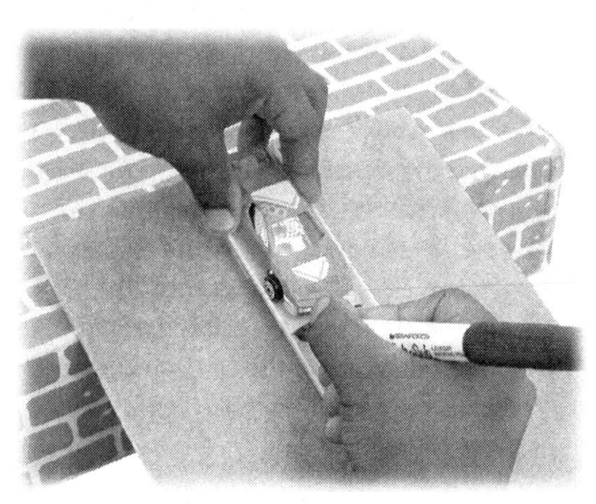

① Place one block under one end of the ramp and lay the track along the ramp. Hold your car at the top of the ramp (on the track) and mark a starting line just in front of it. All members of your group will use the same starting line and the same car to carry out the experiment.

② Have each group member mark where he or she predicts the front wheels of the car will stop after the car is released from the starting line. Measure the distance from the starting line to the place each group member predicts the car will stop.

Record each student's predicted travel distance in the data table.

Ramp Height of 1 Block			
Name of Student	**Predicted Travel Distance**	**Actual Travel Distance**	**Difference Between Predicted and Actual Travel Distances**
1.			
2.			
3.			
4.			
Average			

③ Have the first student place the car at the starting line and then release it. Be sure not to push the car; otherwise, the speed at which the car leaves the ramp may not be the same in each trial.

④ After the car comes to rest, mark the location of the front wheels. Measure the distance the car traveled.

Record the actual travel distance in the data table.

Calculate the difference between the first student's predicted distance and the actual distance. Record this number in the data table.

⑤ Give each group member a turn to do steps 3 and 4.

✎ Calculate the average actual travel distance for your group and record this number in the data table.

⑥ Raise the ramp by adding a second block. Repeat steps 2–5. This time create a data table on another piece of paper and record the results.

⑦ Raise the ramp again by adding two more blocks. Repeat steps 2–5. Create a data table on another piece of paper and record the results.

TIP: Notice that this time you are using four blocks, not three.

⑧ On graph paper, make a graph relating the number of blocks under the ramp to the average distances the car moved.

✎ Explain what this graph shows.

....

....

....

....

....

⑨ For each of the heights you tested in this experiment, look at your data tables for the differences between your predicted and actual travel distances.

✎ How did the accuracy of your predictions change as you went through the experiment?

....

....

....

....

....

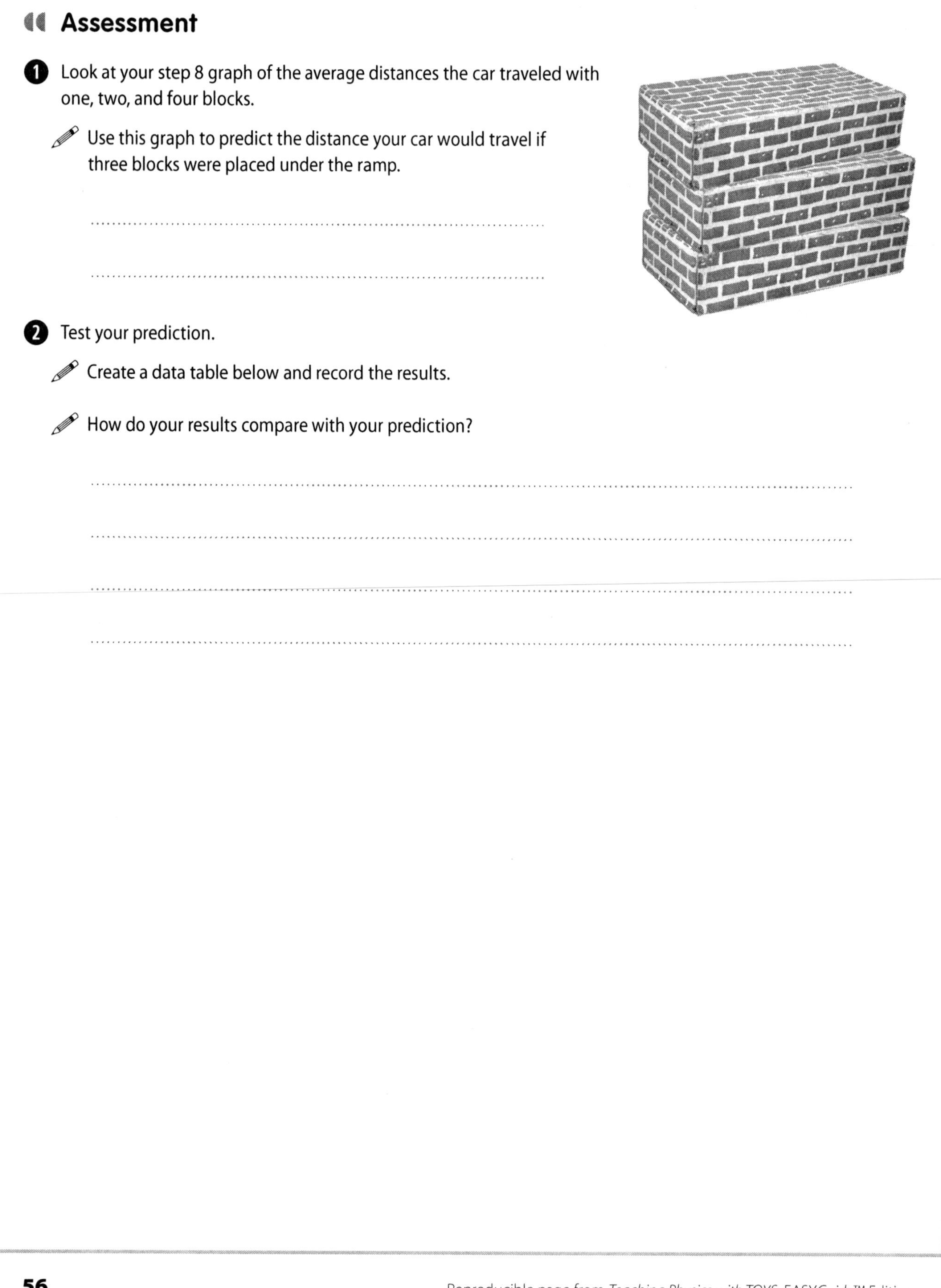

Assessment

1. Look at your step 8 graph of the average distances the car traveled with one, two, and four blocks.

   Use this graph to predict the distance your car would travel if three blocks were placed under the ramp.

2. Test your prediction.

   Create a data table below and record the results.

   How do your results compare with your prediction?

Understanding Speed

Students use the variables of time and distance traveled to determine which toy moves the fastest.

Grade Levels

Science activity appropriate for grades 4–7

Student Background

Students require no particular background preparation for this activity.

Time Required

Setup	5	minutes
Part A	45	minutes
Part B	45	minutes
Cleanup	5	minutes

Assessment time is not included.

Key Science Topics

- motion
- speed

National Science Education Standards Overview

See *www.terrificscience.org/physicsez/* for details of how these standards relate to the activity.

Science as Inquiry

Abilities Necessary to Do Scientific Inquiry

K–4 Plan and conduct a simple investigation.
K–4 Employ simple equipment and tools to gather data and extend the senses.
K–4 Use data to construct a reasonable explanation.
K–4 Communicate investigations and explanations.

5–8 Design and conduct a scientific investigation.
5–8 Use appropriate tools and techniques to gather, analyze, and interpret data.
5–8 Develop descriptions, explanations, predictions, and models using evidence.
5–8 Think critically and logically to make the relationships between evidence and explanations.
5–8 Communicate scientific procedures and explanations.
5–8 Use mathematics in all aspects of scientific inquiry.

Physical Science

K–4 Position and motion of objects

5–8 Motions and forces

Materials

For the Procedure

Part A, per group

- several toys (such as different-sized balls or a variety of toy cars)
- stiff cardboard or wooden ramp
- books for elevating ramp
- timepiece with second hand or stopwatch
- meterstick or measuring tape
- something to mark the toy's location (such as flat toothpicks or sticker dots)
- graph paper

Graph paper masters for copying are available at www.terrificscience.org/physicsez/.

Part B, per group

- wind-up walking toy
- timepiece with second hand or stopwatch
- meterstick or measuring tape
- graph paper
- (optional) 2 metersticks to keep toy walking straight
- (optional) masking tape to keep toy walking straight
- (optional) ink pad to stamp toy's feet
- (optional) string to help measure ink trail

For the Assessment

Per group

- toy car
- short ramp
- timepiece with second hand or stopwatch
- meterstick or measuring tape

Safety and Disposal

No special safety or disposal procedures are required.

Getting Ready

- Gather objects for groups to race in step 3 of Part A. These could be different-sized balls or a variety of toy cars. Pull-back cars are not suitable here, because they usually move too fast. Make sure the toys you select for each group travel at different speeds.
- For best results in Part B, test walking toys before using them to determine whether they walk in a reasonably straight line.
  - If the walking toys you have do not walk in sufficiently straight lines for students to be able to measure how far they walked, two metersticks parallel to one another can make a path.
  - Alternatively, several layers of masking tape will also make a sufficient barrier to keep the toys walking down the path.
  - Another option is to stamp the toys' feet on an ink pad. The toys will then leave a trail, which can be measured by laying a string along the trail and then measuring the appropriate length of string.

Procedure

This section provides teacher notes corresponding to each step of the student procedure. The procedure without teacher notes is included in the reproducible Student Notebook pages at the end of this activity and at www.terrificscience.org/physicsez/.

	Student Procedure	Teacher Notes
	Part A: Who's Fastest?	
①	**Look at this picture. All the runners started from the starting line at the same time. The cone on the left marks the finish line.** ✎ **Circle the fastest runner. What makes you think he is the fastest?**	Since the runners started at the same time and from the same starting line, the runner who crosses the finish line first is the fastest.
②	**This data table contains information for track stars from different states who have never raced each other.** ✎ **Which runner is the fastest? How do you know?** ✎ **Instead of having runners in a race go the same distance, what if they run for the same amount of time? How will you know who is the fastest?**	Students should conclude that Sally is the fastest runner, since she ran the 100 meters in the shortest time. If the distance of the race is the same, then the person with the shortest time is the fastest. If the time interval is the same, the fastest runner is the person who went the longest distance.
③	**Get some toys from your teacher. Work with your group to design an experiment to find out which toy is the fastest. Each toy must race for the same length of time. Plan to do at least three trials per toy.** ✎ **On another piece of paper, write a hypothesis and experimental design. Create a data table.**	Before students begin brainstorming, introduce or review the characteristics of good experiments. Depending on their experience, students may benefit from using the Experiment Planning Guide provided at *www.terrificscience.org/physicsez/.*
④	**Perform the experiment after your teacher reviews and approves your experimental design and data table.** ✎ **Record the results in your data table. Calculate the average for each toy.**	A reasonable plan is to give each object an equal start by releasing it from the top of a ramp (set at a fixed height) and measuring the distance each object goes in a given time, such as 3 seconds. One student can watch the timepiece and call out "stop" at the predetermined time interval. Another student can mark the spot where the object is when "stop" is called. Students will need to do at least three trials and repeat the same procedure for each object.
⑤	**On graph paper, make a graph of your results.** ✎ **Which of your toys is the fastest? How did you draw your conclusion?** ✎ **How do your results compare with your prediction?**	See Sample Answers for example data and a graph.

Student Procedure	Teacher Notes
⑥ Look at the distance and time data in the following table. ✎ Which toy is fastest? Explain how you decided.	Students should divide the distance by time to find out how far each toy traveled per second. (See Sample Answers.) The Darda® car is fastest.
Class Discussion	• Ask groups to share their explanations from step 6. Students should understand that, to figure out which toy is fastest, they need to figure out how far each toy went in just 1 second. Tell students that we use the term "speed" to indicate how far an object travels during some standard unit of time. • Ask students how they can figure out which toy from step 4 is the fastest in the class. Some students may realize that each group should calculate the speed of their fastest toy and then compare their results with the class.
Part B: Virtual Race	
① Work with your group to design an experiment to determine whether your walking toy is the fastest in the class without actually racing the toys against each other. ✎ On another piece of paper, write your plan and create a data table. Have your plan and data table approved by your teacher. TIP: Have your toy run three trials and make sure that your toy is wound up the same each time. ✎ Perform the experiment and record the results in your data table. Calculate your toy's speed in each trial. ✎ Find the average speed of your toy. Record your answer here and post it on the board. ✎ Which group's toy wins the race?	To determine speed, groups should measure the total distance their toy travels and the total travel time from the starting line to where the toy stops. Groups should run three trials. (See Sample Answers for example data.) Some students may design the experiment to get average speed in a different way (for example, over the first 10 seconds or the first 20 cm). Let students use different methods. When students compare their results, point out that they didn't all find the average for the same part of the motion and, thus, their decision on which toy is fastest may not be valid. So groups can more easily compare their results, make sure they post their toys' average speeds with the same units (such as cm/second).
② Think back to step 1. Did your toy move at the same speed during the entire distance it traveled? ✎ Write down what you remember about your toy's speed.	Students may notice the toy moves more quickly at first, or that the toy's speed becomes erratic toward the end of its walk.

	Student Procedure	Teacher Notes
③	Work with your group to design an experiment to test whether your walking toy changes speed during its operation. Plan to do at least three trials. ✎ On another piece of paper, write your plan and create a data table. Have your plan and data table approved by your teacher. ✎ Perform the experiment and record the results in your data table. Calculate an average for each interval and, on graph paper, make a graph of your results. ✎ Does your walking toy change speed during its operation? How did you draw your conclusion?	Procedures will vary. For example, students can start the toy and mark the distance it traveled at specific time intervals (such as every 3 seconds). As an alternative, students can mark a path for the toy with specific distance intervals and measure the time the toy takes to travel each interval. Students should take at least five measurements per trial and should make at least three trials. Some walking toys start out at a constant speed, and then slow down before coming to a stop. Other toys begin slowing down right away. (See Sample Answers for example data.)
④	Share and compare your results with the class.	Students should explain how their data supports their conclusion. • If they used constant time intervals, they might prove that their toy slowed down by noting that less distance was traveled in the later time intervals than in the first one. • If they used constant distance intervals, then they might conclude the toy slowed down because successive times got longer. • Some groups might choose to calculate the average speed for each interval and then compare each interval's speed. Have older students plot graphs of average speed versus time.
	Assessment	
❶	Your teacher will give your group a ramp and a toy car. Design an experiment to determine the average speed of your car over the distance it travels. Mark a starting line at the top of your ramp. Do not push your car when you release it from the starting line. ✎ Write your plan. ✎ Record your data in a data table and calculate your results.	Students should set up short ramps for the toy cars. Ramps make the motion more repeatable. Make sure that the cars are started from the same place each time. Students should measure the total time required for the car to travel its total travel distance, then calculate the car's average speed. (See Sample Answers for example data. See Explanation for a discussion of speed.)

Sample Answers

Where students follow the same procedure using the same materials, these answers are close to answers you can expect. Where students design their own experiment or model, students' results will vary.

Type of Ball	Distance Traveled in 3 Seconds			
	Trial 1	Trial 2	Trial 3	Average
tennis	1.6 m	2.0 m	2.2 m	1.9 m
Ping-Pong	2.1 m	2.1 m	2.1 m	2.1 m
golf	3.1 m	2.8 m	2.8 m	2.9 m

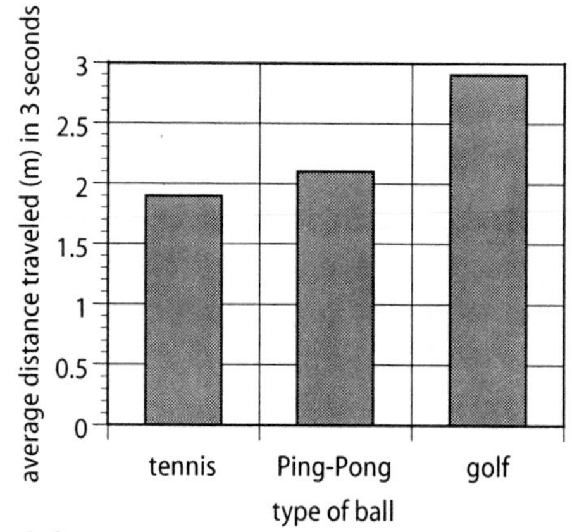

Example of data and graph for Part A

Trial	Total Distance Traveled	Total Travel Time	Speed
1	27.6 cm	15 seconds	1.8 cm/second
2	23.6 cm	12 seconds	2.0 cm/second
3	27.2 cm	15 seconds	1.8 cm/second

Example of virtual race data for Part B

Travel Time	Distance Traveled			
	Trial 1	Trial 2	Trial 3	Average
3 seconds	6.7 cm	6.0 cm	5.9 cm	6.2 cm
6 seconds	12.1 cm	12.2 cm	11.7 cm	12.0 cm
9 seconds	17.7 cm	18.0 cm	17.0 cm	17.6 cm
12 seconds	22.4 cm	23.6 cm	22.7 cm	22.8 cm
15 seconds	27.6 cm	28.0 cm	27.2 cm	27.6 cm
18 seconds	32.2 cm	32.2 cm	31.1 cm	31.8 cm

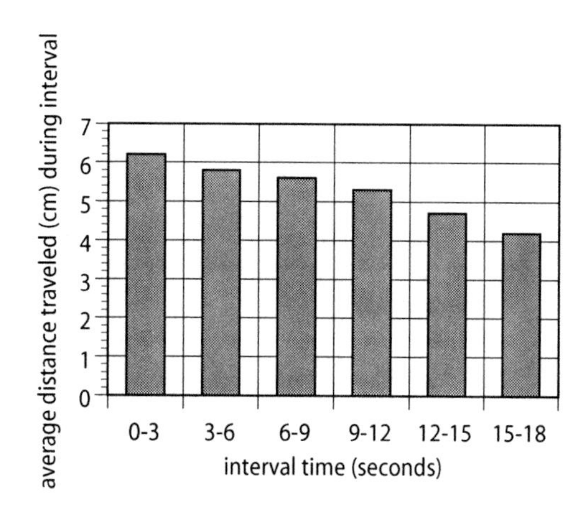

Example of walking toy speed test data and graph for Part B

Trial	Total Distance Traveled	Total Travel Time	Speed
1	188 cm	3 seconds	63 cm/second
2	196 cm	3 seconds	65 cm/second
3	198 cm	3 seconds	66 cm/second

Example of data for Assessment

Explanation

This section is intended for teachers. Modify the explanation for students as needed.

Speed and Velocity

As your students discover in this activity, an object's speed tells you how far the object will travel in some standard unit of time, such as 1 second or 1 hour. In other words, speed tells you how fast an object is moving. In order to calculate speed, both distance traveled and time required must be measured. An object's average speed can be calculated by dividing the total time required into the total distance traveled. Typical units of speed are m/second, feet/second, km/hour, and miles/hour.

average speed = distance traveled / time required

Sometimes when we use the word speed, we mean the average speed over some time interval and sometimes we mean instantaneous speed. A car's speedometer shows the car's instantaneous speed (how fast the car is going right now). One way of thinking about instantaneous speed is that it is average speed over a very short time interval.

Although we may use speed and velocity interchangeably in everyday language, in the language of physics, speed and velocity are not the same thing. Speed is a scalar quantity, meaning that it is fully described by magnitude alone. Like mass, it is a quantity that has no reference to direction. You might say that an object has a speed of 75 km/hour. Although velocity may seem similar to speed, velocity is a vector quantity, meaning that it must be described using both a magnitude and a direction. For example, you might describe an object's velocity as being 75 km/hour, eastward.

Changing Speed

The speed of an object can be constant during an entire motion, or the object may be speeding up or slowing down. The walking toys used in this activity differ in how constant their speed is before they eventually slow down and stop. Some will move with a fairly constant speed, slow down briefly, and then suddenly stop. Others will begin to slow down right away and keep slowing gradually until they stop.

To find out if the speed of their walking toys is constant, students may calculate the average speed in each time interval by dividing the distance traveled in each time interval by the number of seconds in each interval. Two sample graphs of speed versus time are provided below. The graph for the pig walking toy indicates that after slowing down over the first two time intervals, the speed is almost constant over the remaining three intervals. In other words, the pig covers approximately the same distance during each of the last three intervals. The graph for the dinosaur walking toy does not indicate a constant speed, showing that the toy is covering less distance in each time interval.

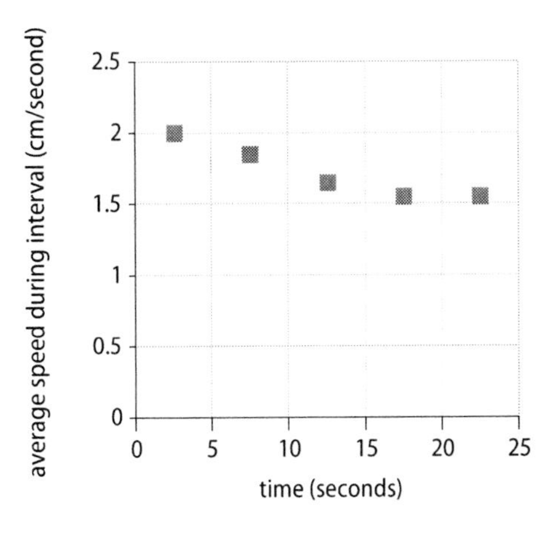

Graph showing the change in speed of the pig walking toy

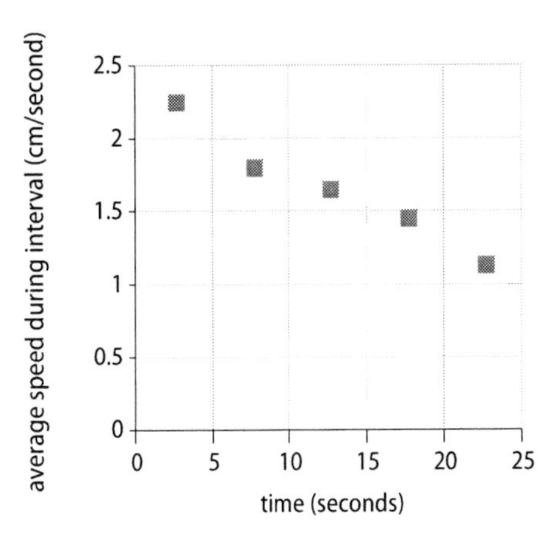

Graph showing the change in speed of the dinosaur walking toy

Cross-Curricular Integration

Earth science:

- Students can read about the speed at which the crustal plates move.

Language arts:

- Have students write about their recollections of races they have been in or seen.
- Have students write about the activity as if they were the walking toys.

Life science:

- Students can compare the speeds at which different organisms move.
- If parents have kept growth records, such as marks on a closet door or entries in a journal, students can calculate their growth "speed" in inches per year.

Math:

- Measurement and units, proportional reasoning, and averaging skills used in this activity can be connected to math lessons.

Physical education:

- Students can determine the speed of classmates walking, running, or jogging; then determine their heart rates after a certain amount of time.

Part A: Who's Fastest?

1. Look at this picture. All the runners started from the starting line at the same time. The cone on the left marks the finish line.

Circle the fastest runner. What makes you think he is the fastest?

....

....

2. This data table contains information for track stars from different states who have never raced each other.

Times for Women's 100-Meter Dash		
Runner's Name	**State**	**Time**
Nancy	WV	12.03 seconds
Sally	WY	11.47 seconds
Jennifer	TN	13.58 seconds
Jessica	CT	12.64 seconds

Which runner is the fastest? How do you know?

....

....

Instead of having runners in a race go the same distance, what if they run for the same amount of time? How will you know who is the fastest?

....

....

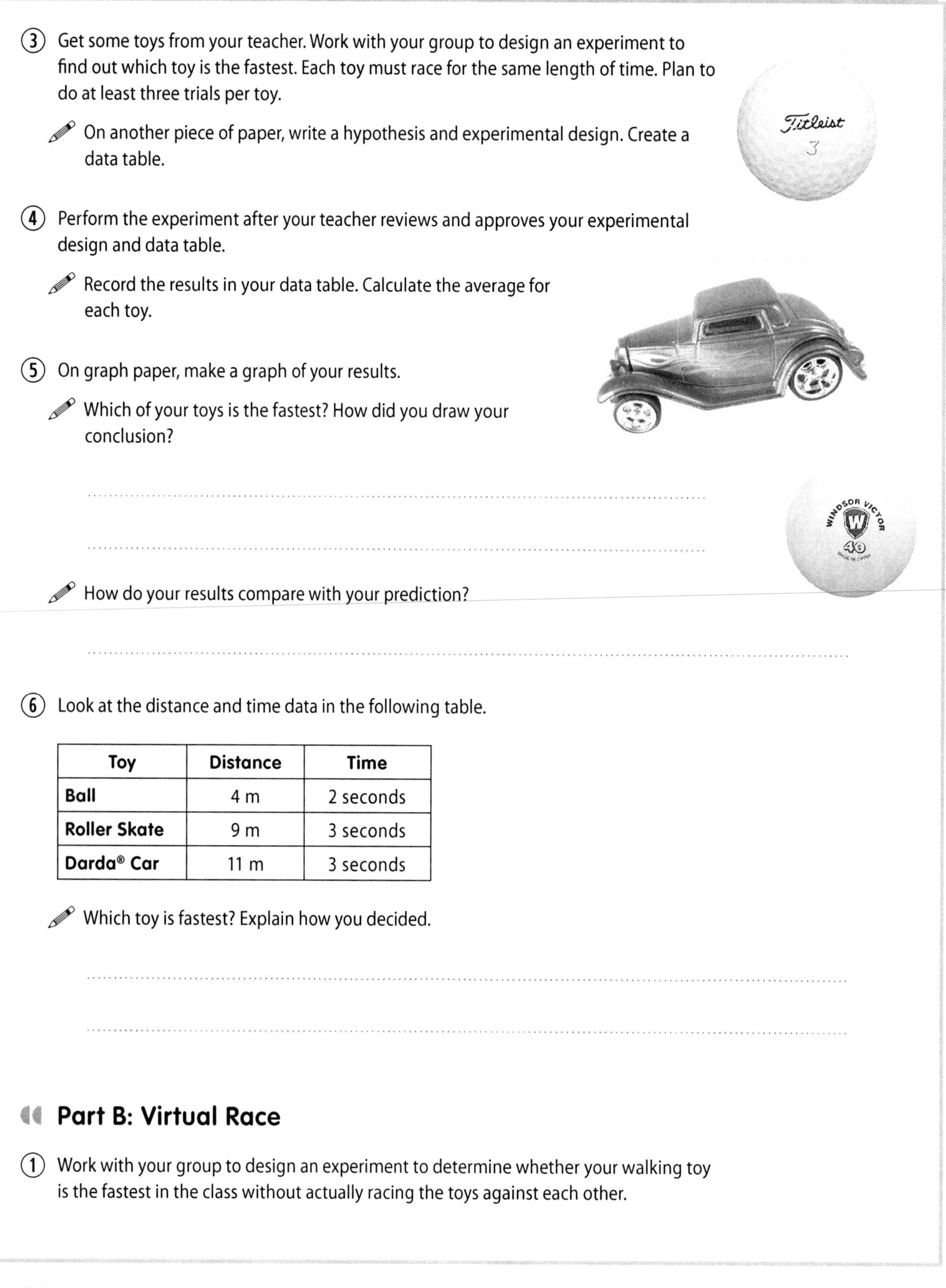

③ Get some toys from your teacher. Work with your group to design an experiment to find out which toy is the fastest. Each toy must race for the same length of time. Plan to do at least three trials per toy.

✎ On another piece of paper, write a hypothesis and experimental design. Create a data table.

④ Perform the experiment after your teacher reviews and approves your experimental design and data table.

✎ Record the results in your data table. Calculate the average for each toy.

⑤ On graph paper, make a graph of your results.

✎ Which of your toys is the fastest? How did you draw your conclusion?

....

....

✎ How do your results compare with your prediction?

....

⑥ Look at the distance and time data in the following table.

Toy	Distance	Time
Ball	4 m	2 seconds
Roller Skate	9 m	3 seconds
Darda® Car	11 m	3 seconds

✎ Which toy is fastest? Explain how you decided.

....

....

Part B: Virtual Race

① Work with your group to design an experiment to determine whether your walking toy is the fastest in the class without actually racing the toys against each other.

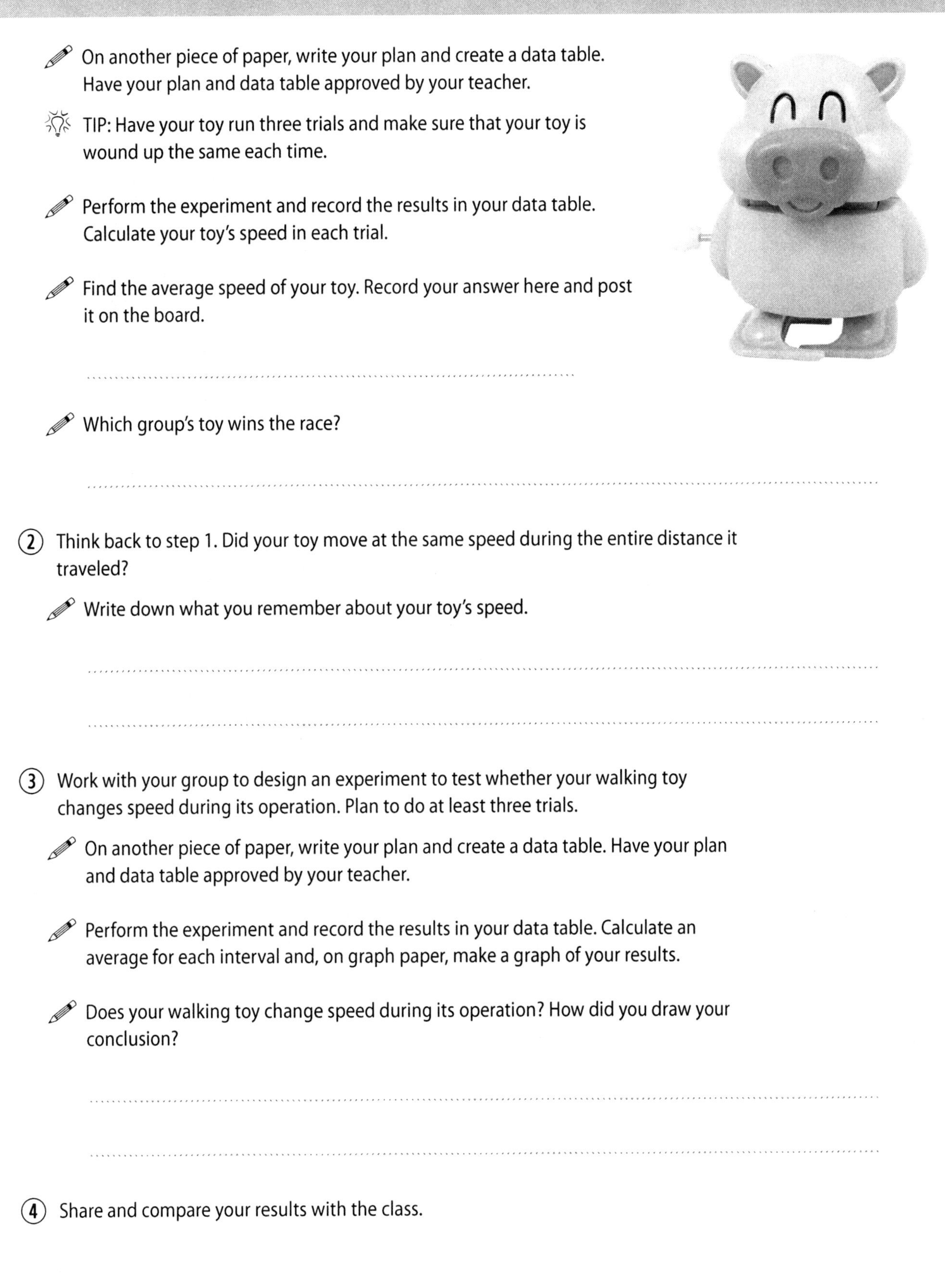

On another piece of paper, write your plan and create a data table. Have your plan and data table approved by your teacher.

TIP: Have your toy run three trials and make sure that your toy is wound up the same each time.

Perform the experiment and record the results in your data table. Calculate your toy's speed in each trial.

Find the average speed of your toy. Record your answer here and post it on the board.

....

Which group's toy wins the race?

....

(2) Think back to step 1. Did your toy move at the same speed during the entire distance it traveled?

Write down what you remember about your toy's speed.

....

....

(3) Work with your group to design an experiment to test whether your walking toy changes speed during its operation. Plan to do at least three trials.

On another piece of paper, write your plan and create a data table. Have your plan and data table approved by your teacher.

Perform the experiment and record the results in your data table. Calculate an average for each interval and, on graph paper, make a graph of your results.

Does your walking toy change speed during its operation? How did you draw your conclusion?

....

....

(4) Share and compare your results with the class.

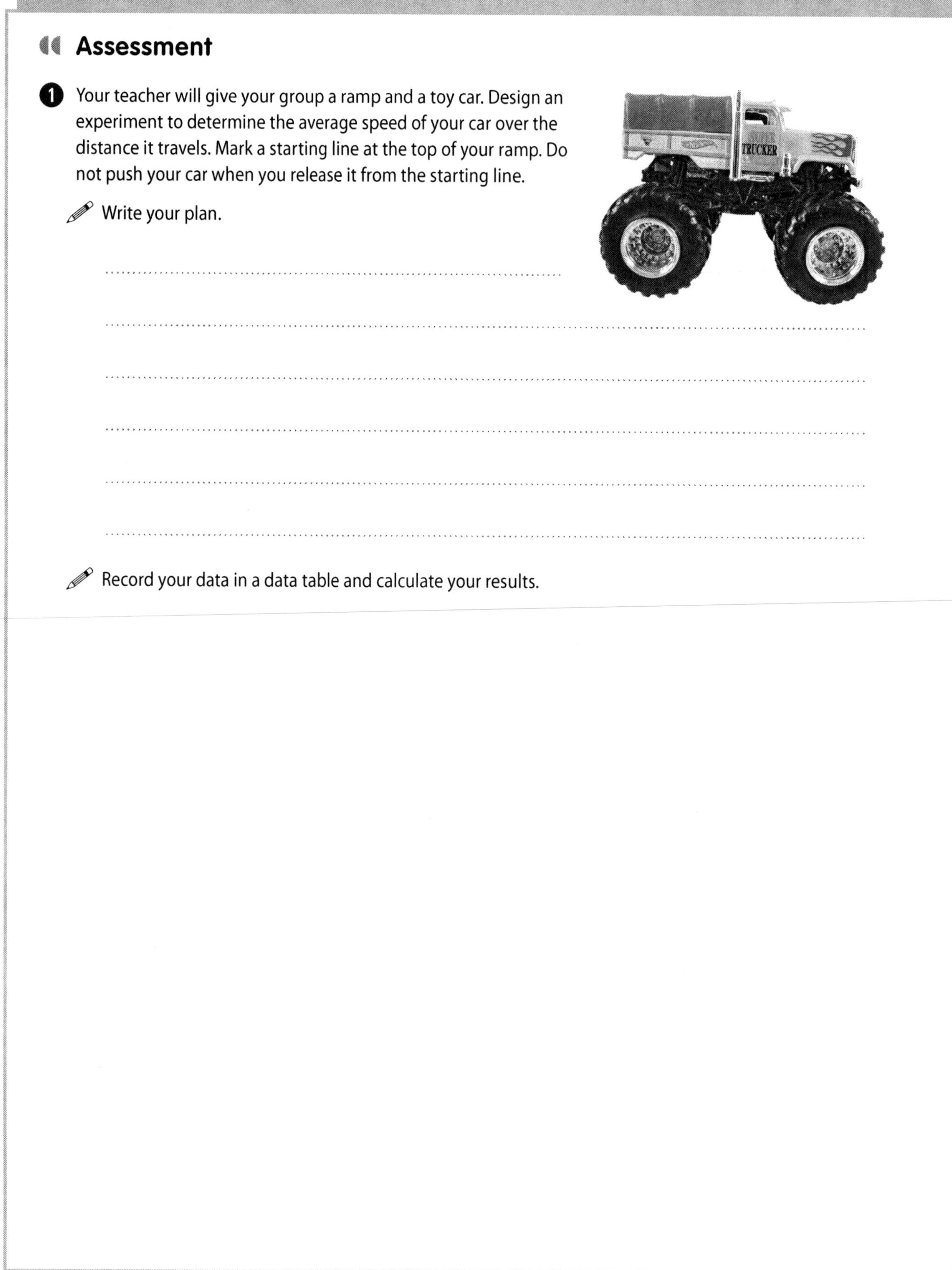

Assessment

1. Your teacher will give your group a ramp and a toy car. Design an experiment to determine the average speed of your car over the distance it travels. Mark a starting line at the top of your ramp. Do not push your car when you release it from the starting line.

Write your plan.

Record your data in a data table and calculate your results.

Car Coaster

Students use pull-back cars and looped track to explore potential energy, kinetic energy, energy transformations, and centripetal force.

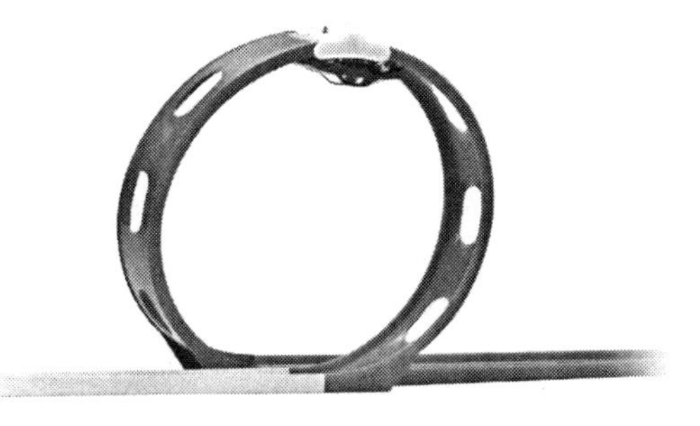

Grade Levels

Science activity appropriate for grades 5–9

Student Background

Students should be familiar with elastic potential energy, kinetic energy, and the idea that energy can be transformed from one type to another. If students have previously studied circular motion, the concept of centripetal force can be reviewed and reinforced with this lesson. If not, that concept may be omitted.

Time Required

Setup	10	minutes
Part A	15	minutes
Part B	15	minutes
Cleanup	5	minutes

Assessment time is not included.

Key Science Topics

- centripetal force
- energy transformations
- friction
- kinetic energy
- potential energy

National Science Education Standards Overview

See *www.terrificscience.org/physicsez/* for details of how these standards relate to the activity.

Science as Inquiry

Abilities Necessary to Do Scientific Inquiry

5–8 Develop descriptions, explanations, predictions, and models using evidence.
5–8 Think critically and logically to make the relationships between evidence and explanations.
5–8 Communicate scientific procedures and explanations.

9–12 Formulate and revise scientific explanations and models using logic and evidence.
9–12 Communicate and defend a scientific argument.

Physical Science

5–8 Motions and forces
5–8 Transfer of energy

9–12 Motions and forces
9–12 Conservation of energy and the increase in disorder

Materials

For the Procedure

Part A, per group

- pull-back car

Darda® cars made by Darda USA, Inc. work well.

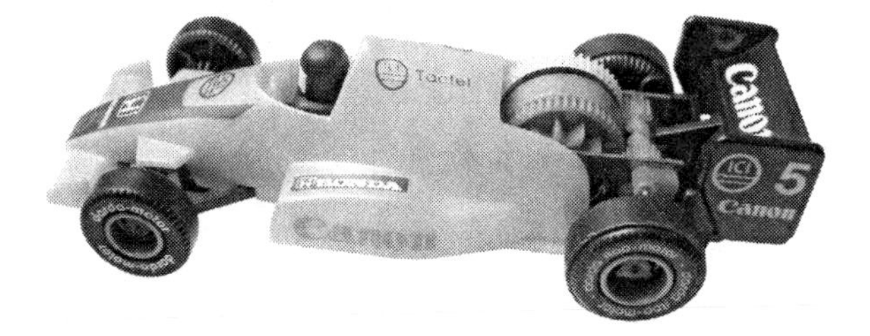

Part B, per group

- pull-back car
- track at least 1 m long with one loop

You may want to use cars and track from the same manufacturer to ensure that the cars fit on the track.

- meterstick or measuring tape
- (optional) track with a jump

For the Assessment

Assessment A, per group

- pop-up toy

For supply source suggestions, see www.terrificscience.org/supplies/.

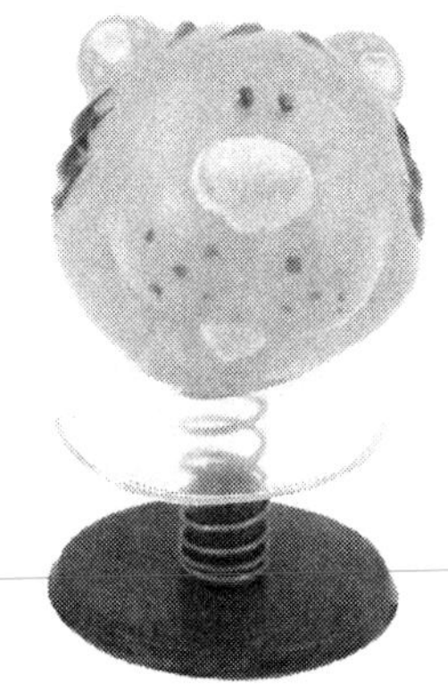

Assessment B, per group

All materials listed for the Procedure

Assessment C, per group

- 4–6 meters of track
- pull-back car
- stiff cardboard or wooden ramp
- several books or boxes
- meterstick or measuring tape

Safety and Disposal

No special safety or disposal procedures are required.

Getting Ready

- You may want to play with the pull-back cars enough to figure out how to work them consistently. With some brands of cars, you need to push down on the backs of the cars above the rear axle to make them wind to their full potential.
- If, after step 4 of Part B, you want to demonstrate how the car changes speed in the loop, you may want to practice getting the car to go down a ramp and move at speeds slow enough so that the car just barely makes it around the loop.

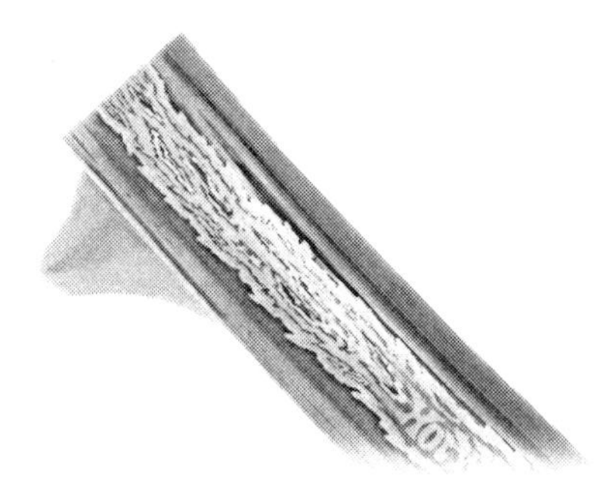

Procedure

This section provides teacher notes corresponding to each step of the student procedure. The procedure without teacher notes is included in the reproducible Student Notebook pages at the end of this activity and at www.terrificscience.org/physicsez/.

	Student Procedure	Teacher Notes
	Part A: Go the Distance	
1	Play with the pull-back car on the floor to get a general idea of how it works. TIP: You may need to push down on the back of the car above the rear axle to make it wind better. Record your observations relating how far you pull the car back to how far it travels after you let it go.	Generally, the cars travel farther when they are pulled back farther before being released. However, most cars do not perform consistently enough for students to record and graph data.
2	Think about the energy transformations taking place as you operate the pull-back car. Where does the car's energy come from first? What type of energy does the car store? What happens to this stored energy as the car moves forward?	The car's energy comes from the work done on the car as it is pulled back. The car stores potential energy. The car's potential energy is converted to kinetic energy when the car is released. Point out to students that the car has a spring that stores potential energy as elastic potential energy. You may want to take a car apart to show students that the spring is coiled like a roll of tape. Tell students that the car has the most kinetic energy just as the spring finishes unwinding.
	Part B: Ride the Loop	
1	Good scientists make careful observations and record them. Explore the pull-back car and the track with loop that your teacher has given you. Try these things while exploring: • Without winding the car, hold your car on the inside of the track at the top of the loop and let go. • Figure out what you need to do with your car to make it consistently go around the loop. • Try running your car in such a way that it will start going around the loop but will not complete the loop. Record what you observe while exploring the car and track.	This free exploration of the toy is an important part of the experiment. Allow enough time for the groups to do this.

	Student Procedure	Teacher Notes
②	Based on your observations, work with your group to determine what factors affect whether the car goes around the loop or not. ✎ Make a list of the factors that your group comes up with.	Students should list that the car's initial position on the track and how far the car is pulled back (in other words, the amount of energy initially stored) determine whether the car will go around the loop.
	Class Discussion	Ask groups to share the factors that they came up with in step 2. Make sure that these three factors are brought out: • The pull-back distance affects how much elastic potential energy is stored in the car and, therefore, how much kinetic energy the car will have just as the spring finishes unwinding. • The car has maximum kinetic energy just as the spring finishes unwinding. If the car reaches the loop before the spring finishes unwinding, some energy is still stored in the spring as elastic potential energy. The car may not have enough kinetic energy to make it around the loop. • The car loses kinetic energy to friction as it moves. If the distance to the loop is too long, the car may not have enough kinetic energy left to make it around the loop.
③	With your group, try to make your car complete a loop without pulling the car back to wind its spring and without pushing it. ✎ Record your alternate method for making the car complete the loop every time. ✎ Record the measurement your teacher has asked you to make.	If you need to provide a hint, tell the class to think of how a real roller coaster works. At the beginning of the ride, the cars are pulled up a high hill, and they gain speed as they go down the hill. If the coaster contains a loop, the speed of the cars going down the hill must be sufficient to provide the kinetic energy necessary to complete the loop. Students should figure out that they need to lift the end of the track to make a tall ramp. Once they have figured this out, tell groups to determine how high the beginning of the track must be in order for the car to make the loop. Have groups record this measurement.

Student Procedure	Teacher Notes
(4) As a group, decide why your alternate method works. ✏ Where does the car's energy come from first? ✏ How is this different from what happens when you pull the car back? ✏ How is this similar to what happens when you pull the car back?	Since the car moves when released at the top of the ramp, it must have energy. When the car is lifted onto the ramp of the track, work is being done and energy is being transferred to the car. This energy must be stored since the car does not move until you release it. This form of stored energy is called gravitational potential energy. Work is also being done when you pull back the car in Part A, but in that case the transferred energy is stored as elastic potential energy.
Class Discussion	• Point out that, regardless of whether it is elastic or gravitational, potential energy is converted to kinetic energy when the car is released. • Tell students that the car slows down when going up the loop and speeds up when coming down the other side, although this is very difficult to observe. (You may want to demonstrate by getting the car to go down the ramp and barely make it around the loop. Students may be able to observe the speed changes through the loop.) The reason the car slows down when going up the loop is that some of its kinetic energy is converted back to gravitational potential energy as it rises up the loop. The speed of the car increases coming down the other side of the loop as the car converts this gravitational potential energy back into kinetic energy. (See Explanation.) • You may want to discuss centripetal force with older students. (See Explanation.)
Optional	• Some brands of track have jumps in them. Demonstrate a car using the jump and discuss why the car is able to cross the jump. • Students could investigate what happens when the jump's gap is widened, what happens when one side of the jump is higher than the other, and whether it matters if the track on either side of the jump is level rather than slanted.

Student Procedure	Teacher Notes
Assessment A	
1 Play with a pop-up toy. ✎ Describe the energy conversions that take place. ✎ How do these energy conversions compare to those in the pull-back car?	When students push down on the pop-up toy, they do work. The kinetic energy used to do the work is converted to elastic potential energy stored in the toy's spring. The elastic potential energy is converted to kinetic energy when the toy pops up. The same energy conversions occur when the pull-back car is wound up and released. After the toy goes up and begins to fall, its energy conversions are comparable to those that occur when a ramp is used to start the motion of the pull-back car. Just as gravitational potential energy is converted to kinetic energy as the pop-up toy falls down, gravitational potential energy is converted to kinetic energy as the car goes down the ramp.
Assessment B	
1 Combine your track and loop with another group to make a double loop. Work with that group to design an experiment to decide what must be done to have a car successfully complete the two loops. ✎ On another piece of paper, write your hypothesis and experimental design. Record the reasons for your prediction.	Based on their experience, students should be able to predict that the car needs to start with more potential energy to complete two loops than it needs to complete one.
2 Perform the experiment after your teacher reviews and approves your experimental design. ✎ On another piece of paper, write a paragraph to explain what worked and how your results compared to your prediction.	A reasonable plan is to create a higher starting ramp or to pull the car back farther.
Assessment C	
1 Set up 4–6 meters of track by building a ramp and hill as shown. Use books or boxes to lift the ramp about 1 m high and to make a hill about one-half the height of the ramp. ✎ Measure the height of the hill from the floor. Record the data in the data table.	See Sample Answers for example data.

Student Procedure	Teacher Notes
2 Place the car on the ramp at a height equal to the height of the hill, but don't release it yet. ✎ What type of energy does the car possess sitting at rest on the ramp? ✎ Does the car have enough energy to make it over the hill?	Make sure that students do not pull the car back when placing it on the ramp. (You don't want any potential energy stored in the spring.) The car possesses gravitational potential energy when it is at rest on the ramp. Since the car will lose some of this energy to friction when it is released, there will not be enough kinetic energy when the car nears the top of the hill to make it over.
3 Without pulling back or pushing the car, let it go. ✎ Does the car have enough energy to make it over the hill? ✎ Where should you place the car on the ramp so that it will go over the hill?	The car does not have enough energy to make it over the hill. The car needs to begin somewhat higher on the ramp than the top of the hill in order to have enough energy to make it over.
4 Try to find the position on the ramp where the car, when released, just makes it over the hill. Run the car from this position several times to be sure it works all the time. ✎ Measure the height of this starting position from the floor and record the data in the data table above. ✎ Using the terms gravitational potential energy, kinetic energy, and energy lost due to friction, write a paragraph on another piece of paper explaining why the car must be placed at this position to make it over the hill.	See Sample Answers for example data. At rest on the ramp, the car possesses only gravitational potential energy. When the car is released, the gravitational potential energy is converted to kinetic energy as the car gains speed going down the ramp. When the car starts on the ramp at the same height as the hill, the car does not make it to the top of the hill because of friction between the wheels of the car and the track. Friction causes the car to lose some of its kinetic energy and not make it all the way up the hill. To compensate for friction, the car needs to start on the ramp at a height somewhat higher than the hill in order to make it over. At this greater starting height, the car possesses enough gravitational potential energy to provide the kinetic energy the car needs to go over the hill.

Sample Answers

Where students follow the same procedure using the same materials, these answers are close to answers you can expect. Where students design their own experiment or model, students' results will vary.

Car Just Makes It Over the Hill	
height of hill from floor	50 cm
car's starting height from floor	60 cm

Example of data for Assessment C

Explanation

This section is intended for teachers. Modify the explanation for students as needed.

Energy Conversions in the Pull-Back Car

In order to make the car move, you must give it energy. You can do this by pushing it, winding its spring, or lifting it (and the track) up. In all of these cases, you are exerting a force on the car and causing it to move over a distance—in other words, you are performing work on it. In doing work, you are transferring chemical energy from your body to the car.

The pull-back car contains a spring that tightens when the car is pulled back. (An enlarged view is shown below.) This spring converts the energy you transfer to the car through work into elastic potential energy.

When the car is lifted onto the ramp, the energy you transfer to the car through work is stored as gravitational potential energy. The higher up the car is, the more gravitational potential energy it has.

Regardless of whether the car stores elastic or gravitational potential energy, the energy is converted back into kinetic energy when the car is released.

Energy Conversions While in the Loop

As the car begins to go up the loop, it gains gravitational potential energy and therefore must lose kinetic energy (energy of motion). If the car makes it around the loop, it gains this kinetic energy back on the way down (except for a small amount used up as work done to overcome friction). If the car does not make it around the loop, it still has some kinetic energy left when it falls. It does not fall because it lost all of its energy.

Centripetal Force While in the Loop

So why does the car sometimes fall before it goes all the way around the loop? To understand why, we need to look at the two forces acting on the car as it goes around the loop—gravity and the force of the track against the wheels of the car. Together, these forces provide the centripetal force that keeps the car moving in a circle.

In the following paragraphs, we ignore friction between the track and the car, because it is small and because it would not significantly change the discussion.

Just before the car enters the loop, it is moving along a straight path. If no new force was exerted on the car, it would continue along this straight path forever. For the car to change direction and start moving in a circle, a force must act on it. As the car enters the loop, a force is exerted against its wheels by the track. (See figure below.)

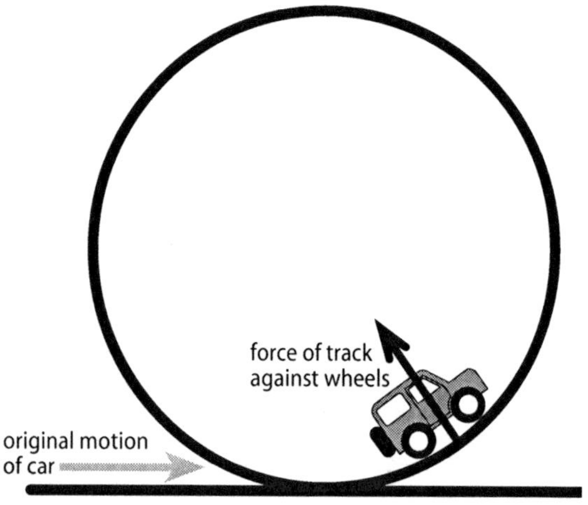

The track exerts a force that changes the direction of the car's motion.

The speed and mass of the car and the diameter of the loop determine the size of the force the track must exert to make the car change direction and start moving in a circle. The faster the car is moving, the greater the force of the track must be. As the car continues to move around the loop, it is continually changing direction and requires a continuous force towards the center of the circle (centripetal force) to make it do so.

You can visualize the role of centripetal force by imagining what would happen if, at any point in the car's path around the circle, a car-sized hole suddenly appeared in the track, right in front of the car. The car would shoot out of the hole, moving in a straight line tangent to the circle for that brief instant. (See figure below.) Then the car would follow a parabolic trajectory caused by gravitational force. If there is no force acting on the car, it continues to move at the same speed and in the same direction.

As the car goes around the track, two changes in its motion take place. As we just saw, the car is changing direction. On its way up the loop, the car is also slowing down as some of its kinetic energy is being converted into gravitational potential energy. Both these changes must be produced by a force. The direction change requires a force towards the center of the circle (perpendicular to the track) and the speed change requires a force directed opposite the direction of the motion (tangent to the track). Where do these forces come from? The force perpendicular to the track comes from two sources: One component is the force the track exerts on the car, which is at all points perpendicular to the track (just as the force a table exerts on a book which is laying on it is perpendicular to the table). Another component of the perpendicular force comes from the force of gravity. The force of gravity always acts vertically down on the car. To understand the effect of the gravity force, we need to think of it as consisting of two parts, one perpendicular to the track (p) and one tangent to the track (t), which add together to make the actual downward force (f). (See figure below.) Thus, the force of gravity also provides the force tangent to the track.

The relative sizes of the perpendicular and tangent parts of the gravity force change as the car moves around the circle. (See figure below.) Near the bottom or top of the track the tangent part is small and the perpendicular part is large. Near the middle of the track, the tangent part is large and the perpendicular part is small. As mentioned previously, the part of the gravity force tangent to the track is the force that changes the speed of the car. The part of the gravity force perpendicular to the track is added to the perpendicular force of the track pushing on the car and together they provide the force needed to make the car change direction. Any force that makes something move in a circle is called a centripetal force, so this combination of perpendicular forces is often referred to as the centripetal force even though it is not a single force.

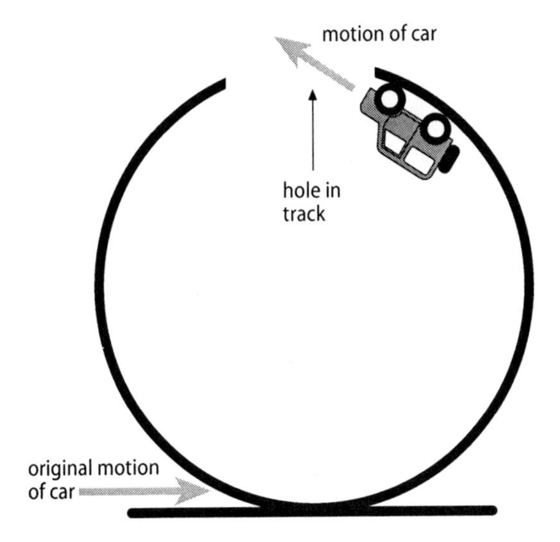

If a hole appeared in the track, the car would travel in a straight line out of the loop.

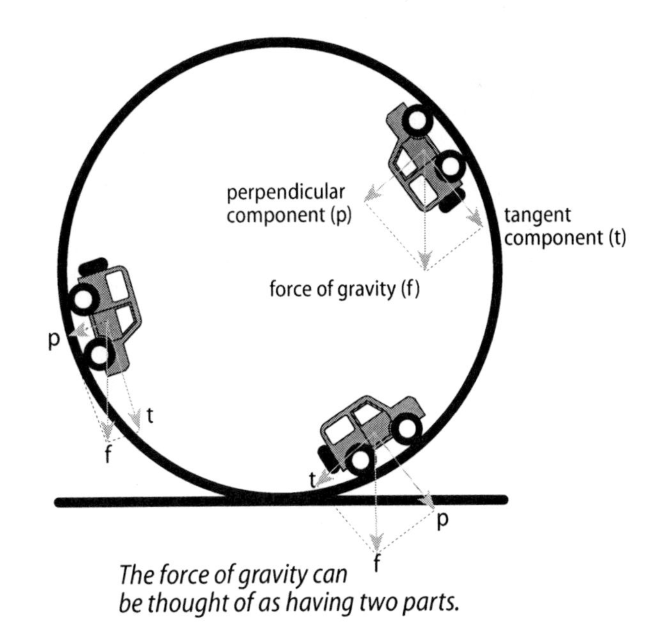

The force of gravity can be thought of as having two parts.

Remember that for the car to move in a circle, the net perpendicular force must be inward. When the car is in the bottom half of the loop, the part of the gravity force that is perpendicular to the track points away from the center of the circle. For the net perpendicular force to be inward, the track must exert a greater force than the perpendicular portion of the gravity force. When the car is in the top half of the loop, the perpendicular gravity force points toward the center of the circle. As a result, the track force can be smaller and still result in a net perpendicular force inward.

Now we are ready to answer the question with which we began: Why does the car sometimes fall before it goes all the way around the loop? The faster the car goes, the larger the size of the centripetal force needed to change the direction of the car's motion. If the car has a high speed, then the centripetal force needed is always larger than the perpendicular part of the gravity force and the track pushes inward to provide the additional force. Thus, the car stays in the loop. If the car is moving slowly, then the centripetal force needed may be smaller than the perpendicular part of the gravity force. The track does not need to push inward. In fact, part of the gravity force is "left over;" it isn't needed to move the car in a circle. The track can't grab the car and pull up to cancel the extra gravity force, so the car falls.

Cross-Curricular Integration

Language arts/Art:

- Have students make advertisements for the pull-back car and the track. Advertisements could be a billboard, magazine or newspaper ad, or TV commercial. Students should decide whether they are targeting the ad toward kids, teachers, or parents.
- Read aloud or suggest that students read the following books:
  - *Roller Coaster Science: 50 Wet, Wacky, Wild, Dizzy Experiments about Things Kids Like Best,* by Jim Wiese (grades 4–7)
    This book covers the science behind things found at amusement parks, playgrounds, and ball fields. Numerous easy experiments reinforce the scientific principles.
  - *Roller Coasters,* by Lynn M. Stone (grades 7 and up)
    This book covers the history and physics associated with different types of roller coasters.
  - *Roller Coasters: Or, I Had So Much Fun, I Almost Puked,* by Nick Cook (grades 3–6)
    This book describes the history, engineering, and physics behind roller coasters, including some specific modern examples.
  - *The Usborne Internet-Linked Library of Science: Energy, Forces, & Motion,* by Alastair Smith, Laura Howell, Corinne Henderson, and Judy Usborne Tatchell (grades 4–7)
    This book demonstrates the physics of everyday objects using photographs, detailed diagrams, and experiments. A list of related websites offers readers a way to extend the learning.

Social Studies:

- Research the history of roller coasters on the Internet or at the library. The following books could be helpful:
  - *American Roller Coaster,* by Scott Rutherford (adult)
    This book traces roller coaster history from its early roots in the mid-1800s through today, including coaster designers, manufacturers, and technologies.
  - *Rollercoasters: A Thrill Seeker's Guide to the Ultimate Scream Machines,* by Robert Coker (adult)
    This illustrated guide addresses the evolution of the roller coaster, covering 200 specific examples. The book offers details about coaster technology, design, and construction.

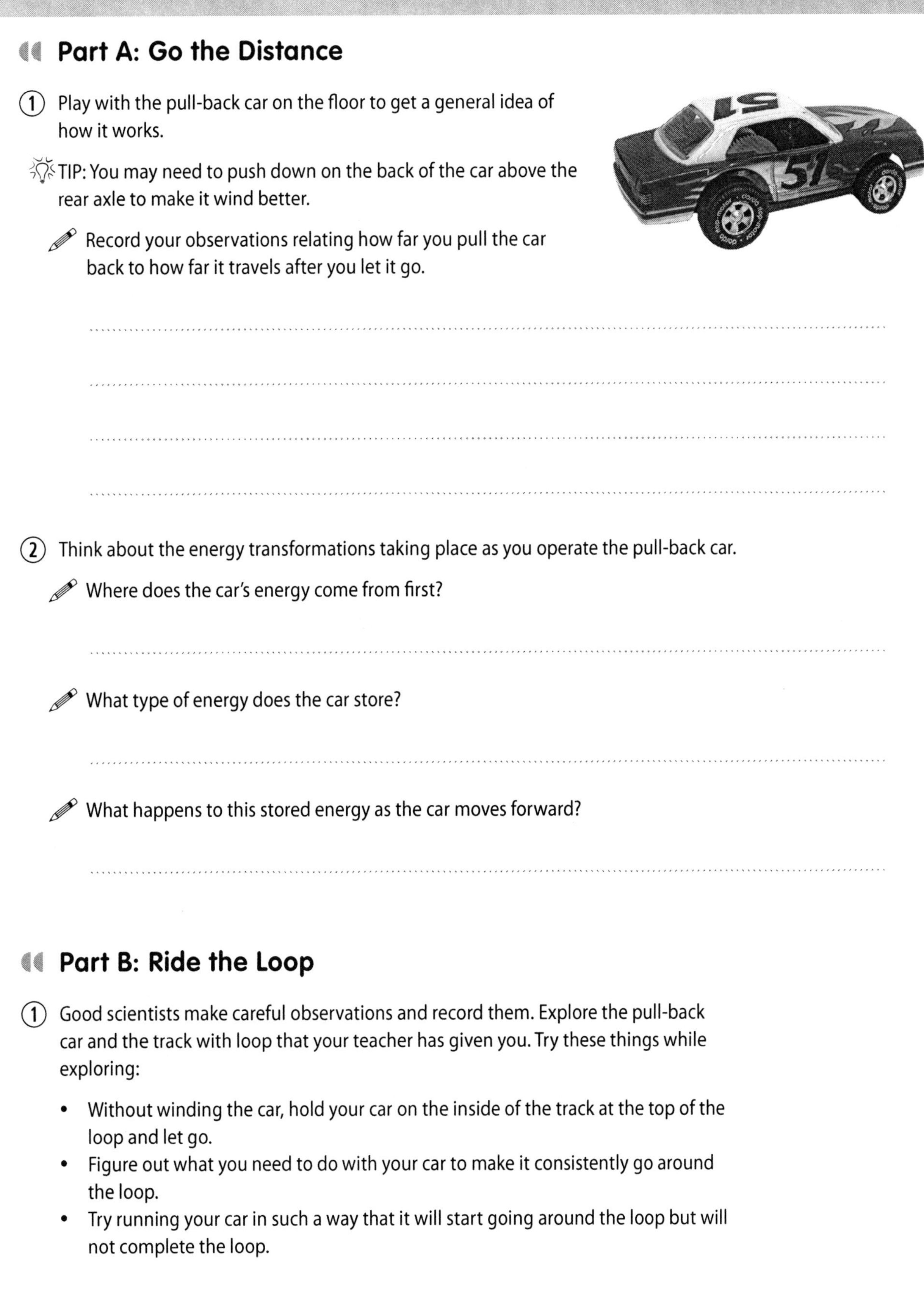

Part A: Go the Distance

① Play with the pull-back car on the floor to get a general idea of how it works.

TIP: You may need to push down on the back of the car above the rear axle to make it wind better.

✎ Record your observations relating how far you pull the car back to how far it travels after you let it go.

....

....

....

....

② Think about the energy transformations taking place as you operate the pull-back car.

✎ Where does the car's energy come from first?

....

✎ What type of energy does the car store?

....

✎ What happens to this stored energy as the car moves forward?

....

Part B: Ride the Loop

① Good scientists make careful observations and record them. Explore the pull-back car and the track with loop that your teacher has given you. Try these things while exploring:

- Without winding the car, hold your car on the inside of the track at the top of the loop and let go.
- Figure out what you need to do with your car to make it consistently go around the loop.
- Try running your car in such a way that it will start going around the loop but will not complete the loop.

Record what you observe while exploring the car and track.

....

....

....

② Based on your observations, work with your group to determine what factors affect whether the car goes around the loop or not.

Make a list of the factors that your group comes up with.

....

....

③ With your group, try to make your car complete a loop without pulling the car back to wind its spring and without pushing it.

Record your alternate method for making the car complete the loop every time.

....

....

Record the measurement your teacher has asked you to make.

....

④ As a group, decide why your alternate method works.

Where does the car's energy come from first?

....

How is this different from what happens when you pull the car back?

....

How is this similar to what happens when you pull the car back?

....

....

Assessment A

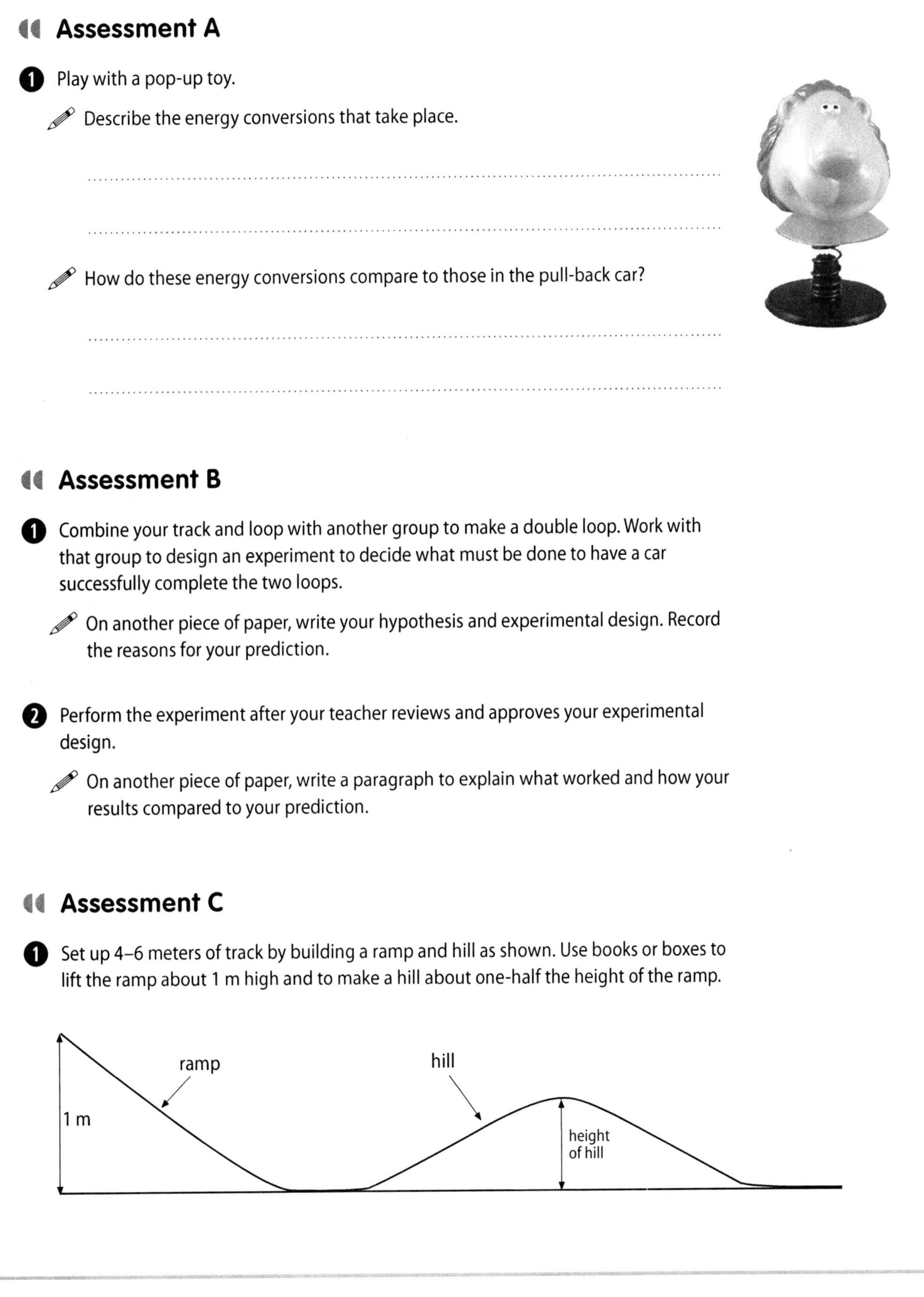

1. Play with a pop-up toy.

   ✎ Describe the energy conversions that take place.

   ....

   ....

   ✎ How do these energy conversions compare to those in the pull-back car?

   ....

   ....

Assessment B

1. Combine your track and loop with another group to make a double loop. Work with that group to design an experiment to decide what must be done to have a car successfully complete the two loops.

   ✎ On another piece of paper, write your hypothesis and experimental design. Record the reasons for your prediction.

2. Perform the experiment after your teacher reviews and approves your experimental design.

   ✎ On another piece of paper, write a paragraph to explain what worked and how your results compared to your prediction.

Assessment C

1. Set up 4–6 meters of track by building a ramp and hill as shown. Use books or boxes to lift the ramp about 1 m high and to make a hill about one-half the height of the ramp.

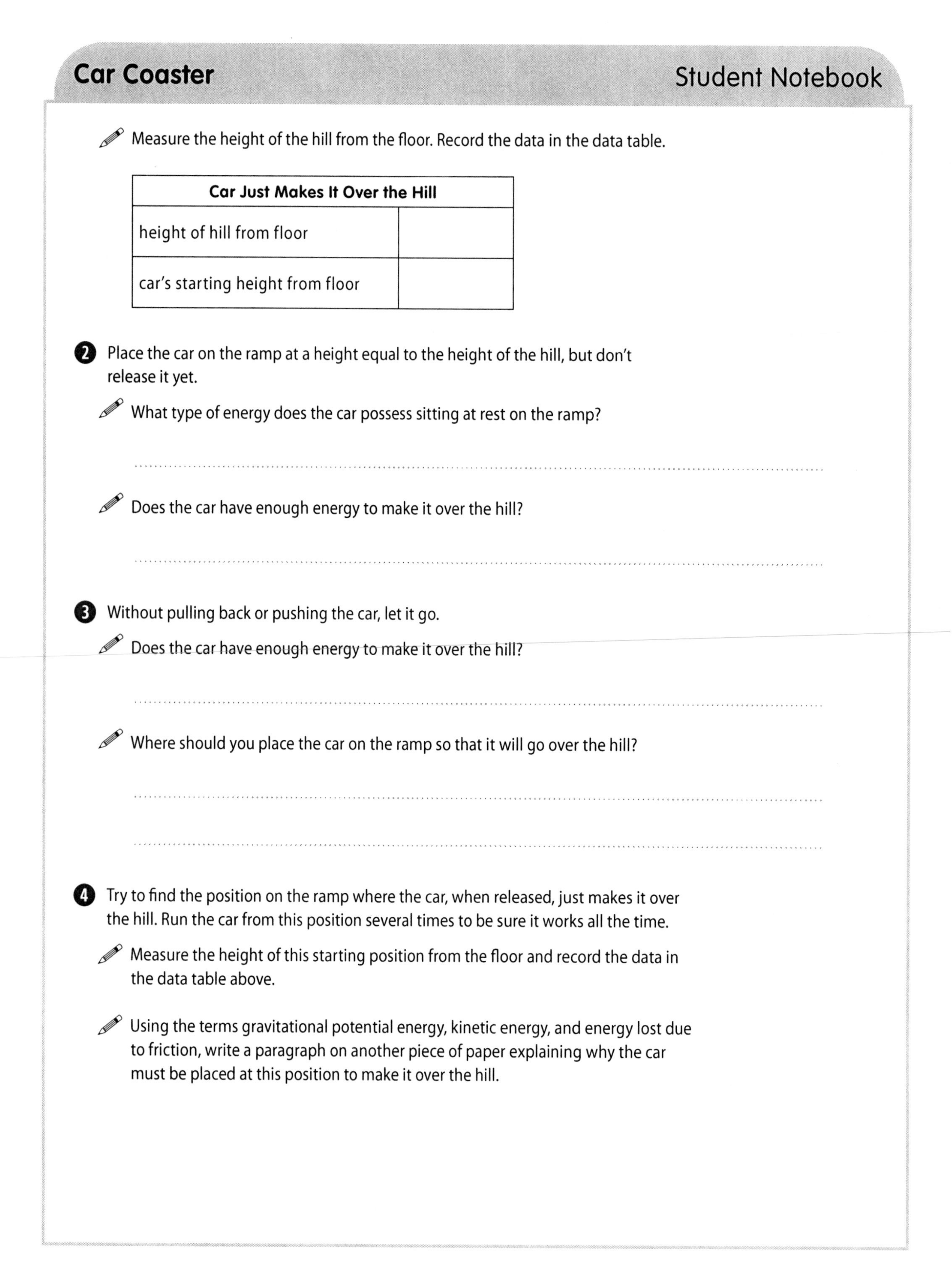

Measure the height of the hill from the floor. Record the data in the data table.

Car Just Makes It Over the Hill	
height of hill from floor	
car's starting height from floor	

2. Place the car on the ramp at a height equal to the height of the hill, but don't release it yet.

   What type of energy does the car possess sitting at rest on the ramp?

   ....

   Does the car have enough energy to make it over the hill?

   ....

3. Without pulling back or pushing the car, let it go.

   Does the car have enough energy to make it over the hill?

   ....

   Where should you place the car on the ramp so that it will go over the hill?

   ....

   ....

4. Try to find the position on the ramp where the car, when released, just makes it over the hill. Run the car from this position several times to be sure it works all the time.

   Measure the height of this starting position from the floor and record the data in the data table above.

   Using the terms gravitational potential energy, kinetic energy, and energy lost due to friction, write a paragraph on another piece of paper explaining why the car must be placed at this position to make it over the hill.

Push and Go

Students are introduced to the concepts of kinetic and potential energy and energy conversion, and they apply these concepts to a familiar toy.

Grade Levels

Science activity appropriate for grades 4–9

Student Background

This lesson introduces students to the ideas of potential and kinetic energy, so no previous background in energy is needed. Students should have some previous experience in connecting forces to the motion of objects and should have been introduced to the concept of work. At the upper levels, this activity is best used as a review of concepts learned in previous grades.

Time Required

Setup	5	minutes
Procedure	30	minutes
Cleanup	5	minutes

Assessment time is not included.

Key Science Topics

- energy conversion
- kinetic and potential energy
- work

National Science Education Standards Overview

See *www.terrificscience.org/physicsez/* for details of how these standards relate to the activity.

Science as Inquiry

Abilities Necessary to Do Scientific Inquiry

K–4 *Communicate investigations and explanations.*

5–8 *Develop descriptions, explanations, predictions, and models using evidence.*
5–8 *Communicate scientific procedures and explanations.*

9–12 *Formulate and revise scientific explanations and models using logic and evidence.*

Physical Science

K–4 *Position and motion of objects*

5–8 *Motions and forces*
5–8 *Transfer of energy*

9–12 *Motions and forces*

Science and Technology

Abilities of Technological Design

K–4 *Evaluate a product or design.*

5–8 *Evaluate completed technological designs or products.*

Materials

For Getting Ready

- (optional) Push 'n' Go™ toy

For supply source suggestions, see www.terrificscience.org/supplies/.

For the Procedure

Per class

- ball

Per group

- toy car
- Push 'n' Go toy
- screwdriver
- small cup

For the Assessment

Per group

- newspaper and magazine advertisements
- scissors
- tape or stapler
- (optional) toy from home

Safety and Disposal

Depending on the age of your students, you may not want them to handle the screwdrivers and take apart the toys. If this is the case, disassemble one Push 'n' Go toy beforehand and pass it around for step 4. No special safety or disposal procedures are required.

Getting Ready

- Some Push 'n' Go toys are hard to put back together correctly. Rather than have students take apart the Push 'n' Go they use in steps 1–3, you may want to have an extra toy available to take apart ahead of time and pass around for step 4. (Use a rubber band to hold the toy together.)
- If each group will be taking apart a toy, you may want to loosen all the screws to make sure that none are stuck and then gently retighten them.

Procedure

This section provides teacher notes corresponding to each step of the student procedure. The procedure without teacher notes is included in the reproducible Student Notebook pages at the end of this activity and at www.terrificscience.org/physicsez/.

	Student Procedure	Teacher Notes
	Activity Introduction	Conduct a class demonstration as follows: • Gently roll a ball away from you. Have a student roll it back. • Listen to students' observations and help them to conclude that you pushed away and the ball moved away. • Review the rule that when a force acts on a simple stationary object, the object moves in the direction of the force. Also review the idea that work is done on the ball during the time and over the distance the ball is being pushed.
	Make the Push 'n' Go Go	
1	Gently roll a car away from you. On another piece of paper, draw the car and an arrow that shows the direction of the force you applied. Label this arrow "force." Draw another arrow that shows the direction of the car's motion. Label this arrow "motion."	Students will observe that when the car is pushed away, the car moves away. (See Sample Answers for student drawing example.)
2	Explore the Push 'n' Go toy. Make the toy run by pushing down on the driver and releasing it. On another piece of paper, draw the toy. Draw an arrow showing the direction of the applied force. Label this arrow "force." Draw another arrow that shows the direction of the toy's motion. Label this arrow "motion." How is what happens to the Push 'n' Go different than what happens to the car in step 1?	Students will observe that, unlike the car in step 1, the force is downward and the movement is forward. They should be able to conclude that the object does not move in the direction of the force. (See Sample Answers for student drawing example.) Students may also notice a timing difference. With the car, motion begins as soon as the force begins. With the Push 'n' Go, motion does not begin until the force is no longer being applied.
3	As a group, discuss what you think is happening inside the toy. Record your idea. Be sure to include something about force and motion and how these relate to the internal parts of the toy.	

	Student Procedure	Teacher Notes
④	Take the toy apart to check out your hypothesis. Use the screwdriver to remove the screws from the bottom of the Push 'n' Go. Put the screws in a small cup. Operate the toy, make observations, and figure out how the toy works. ✎ What part of the toy stores the energy? ✎ Explain how the direction of the force is changed. ✎ How are the gears prevented from moving while the user applies a force to the driver?	Some students will recognize that the toy stores energy in its spring. A series of gears changes the direction of the force. Since the middle gear disengages from the rest of the gears when the driver is pressed downward, the wheels don't move and the toy doesn't move forward until the driver is released.
⑤	With your group, discuss the energy conversions that make the toy work. ✎ Record your conclusions. Be sure to discuss the following in your answer: • What type of energy does the toy gain as you push down on the driver? • What happens to the toy's energy when you release the driver?	If this is your students' first encounter with energy concepts, you will need to define the terms kinetic energy and potential energy before they can answer the questions. See Explanation for a discussion of the energy conversions that make the toy work.
	Class Discussion	Through discussion, bring the class to consensus on how the toy stores energy, changes the direction of the force, and keeps the gears from turning while the driver is being pushed down. (See Explanation.)
	Assessment	
❶	Search through newspaper and magazine advertisements to find a picture of a toy (besides Push 'n' Go) in which energy is changed from one kind to another. Attach the picture to this page or draw the toy below. If you have such a toy at home, bring it in to show the class. ✎ Describe the conversion from potential energy to kinetic energy that takes place in this toy. Explain the energy conversion of the toy to other groups in your class or to family and friends.	Many toys illustrate energy conversion. • All toys using springs, whether compressed or wound, can be analyzed in a manner similar to the Push 'n' Go. • Battery-operated toys convert chemical potential energy to kinetic energy. (See "From Magnets to Motors" in this book.) • Even simpler toys, such as paddleballs and bouncing balls, can be used to illustrate energy conversion. (See "Bounceability" in this book.)

Sample Answers

Where students follow the same procedure using the same materials, these answers are close to answers you can expect. Where students design their own experiment or model, students' results will vary.

Example of student drawing for step 1

Example of student drawing for step 2

Explanation

This section is intended for teachers. Modify the explanation for students as needed.

When you operate the Push 'n' Go, you apply a downward force to the driver and the driver moves downward as a result. (See figure below.) Thus, an applied force has moved through a distance and work has been done. As you push, the driver pushes back on your finger, showing that a force is being exerted. Clearly, the movement of the driver downward occurs while the force is being applied.

Underneath the driver is a rather stiff spring. As the driver moves down, the spring compresses, storing energy. Once the driver goes as far down as possible, nothing happens until you remove your finger.

If you are demonstrating and explaining at the same time, this is a good time to hold your finger still and mention that your work was converted into elastic potential energy that is now stored in the spring. Your body's energy was transferred, through the process of doing work, to the spring.

When the driver is released, the spring extends, and it pushes the driver back up. As the driver rises, a rack

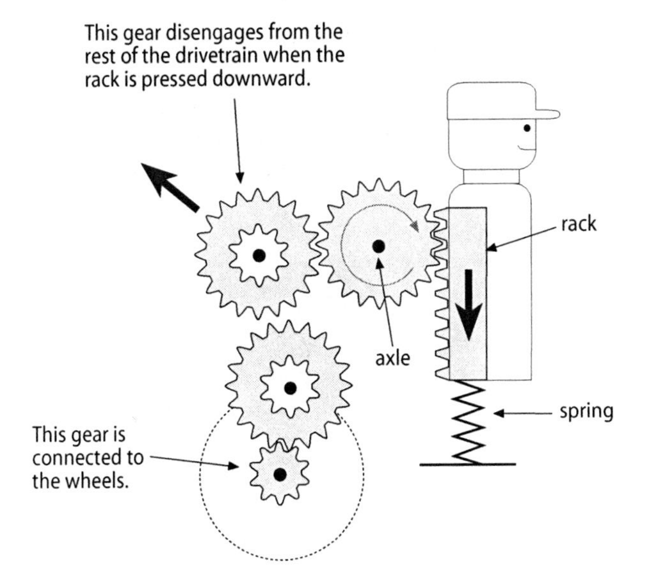

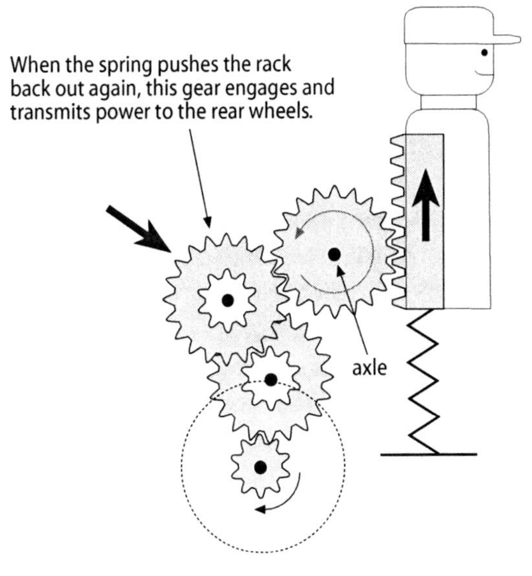

When the driver is pushed down, the spring compresses, storing energy. When the driver rises, a rack on the driver causes the gears to turn.

(series of teeth or notches) on the driver exerts a force to turn a pinion (gear) that is mounted on the axle. The pinion transfers this force through all the other gears to the wheels, causing the Push 'n' Go to roll forward. You can turn the axle by hand to see the gears move in slow motion. The removal of one additional screw on the side of the gear housing detaches it from the base of the vehicle, making the lower two gears more easily visible.

The process can be described in energy terms as follows: the elastic potential energy of the spring is converted into kinetic (or motion) energy of the gears and the toy as a whole. A small part of the energy is converted to thermal energy by friction between the parts.

When the Push 'n' Go stops rolling, you might think the energy has disappeared. However, the energy is not lost. Rather, it has changed into thermal energy because of the friction between the wheels and the floor. On a carpeted floor, where there is a greater amount of friction, the Push 'n' Go will not travel as far before all the kinetic energy has been changed to thermal energy. This concept is explored further in the activity "Exploring Friction" in this book.

Since the rack causes the axle to turn while the driver is being pushed up by the spring, why doesn't the same thing happen when the driver is being pushed down by the user? The motion of the axle while the spring is being compressed is prevented by an interesting feature of the middle gear. This gear is mounted in a slot that allows about one half centimeter of vertical travel. When the spring is extending, the pinion pushes down on the middle gear, causing it to turn and to make contact with the gear below it. When the spring is being compressed, the pinion turns in the opposite direction and lifts up on the middle gear, causing it to move forward. Thus, no contact is made between the middle gear and the gear below, and the axle does not turn.

Cross-Curricular Integration

Language arts:

- Have students imagine the following scenario. Push 'n' Go toys have been popular for a number of years because of their unique design and durability, but suppose sales have recently slumped. You have the following assignment as an advertising agent: Describe your designs for the new and improved Push 'n' Go toy line. Include in your discussion 1) why you think this design will sell, 2) who your targeted buyers are, and 3) a possible price for the product. Finally, make up an advertisement for the new design and include illustrations.
- Read aloud or suggest that students read the following book:
    - *Hands-On Science™: Get in Gear*, by Sholly Fisch (grades K–3)
      A robot named Doc Sprocket guides students through a series of wacky gear configurations that they build directly on the pages in this hands-on book that includes actual gears and a working motor.

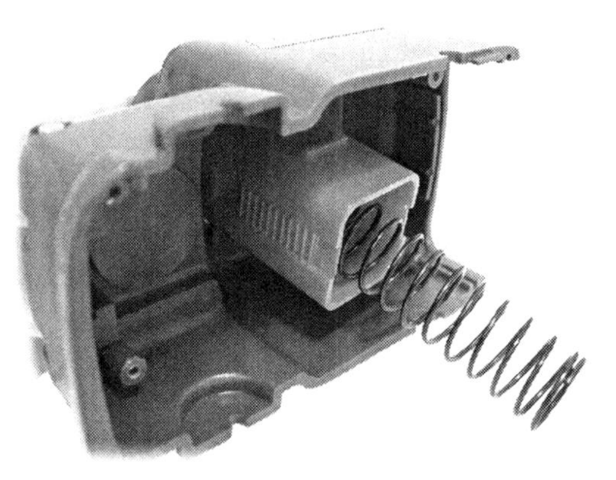

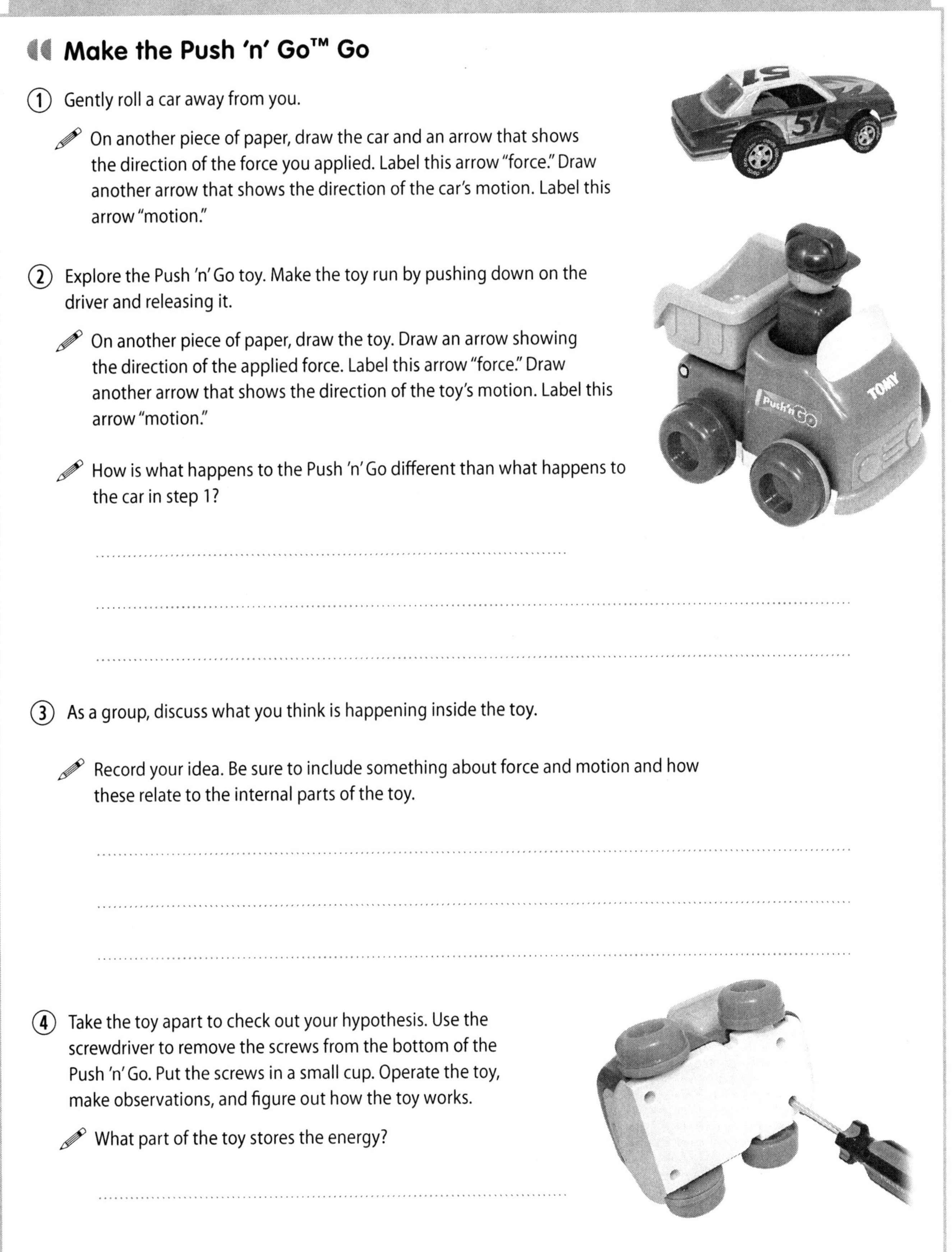

Make the Push 'n' Go™ Go

① Gently roll a car away from you.

On another piece of paper, draw the car and an arrow that shows the direction of the force you applied. Label this arrow "force." Draw another arrow that shows the direction of the car's motion. Label this arrow "motion."

② Explore the Push 'n' Go toy. Make the toy run by pushing down on the driver and releasing it.

On another piece of paper, draw the toy. Draw an arrow showing the direction of the applied force. Label this arrow "force." Draw another arrow that shows the direction of the toy's motion. Label this arrow "motion."

How is what happens to the Push 'n' Go different than what happens to the car in step 1?

....

....

....

③ As a group, discuss what you think is happening inside the toy.

Record your idea. Be sure to include something about force and motion and how these relate to the internal parts of the toy.

....

....

....

④ Take the toy apart to check out your hypothesis. Use the screwdriver to remove the screws from the bottom of the Push 'n' Go. Put the screws in a small cup. Operate the toy, make observations, and figure out how the toy works.

What part of the toy stores the energy?

....

Explain how the direction of force is changed.

How are the gears prevented from moving while the user applies a force to the driver? Why is this necessary?

⑤ With your group, discuss the energy conversions that make the toy work.

Record your conclusions. Be sure to discuss the following in your answer:

- What type of energy does the toy gain as you push down on the driver?
- What happens to the toy's energy when you release the driver?

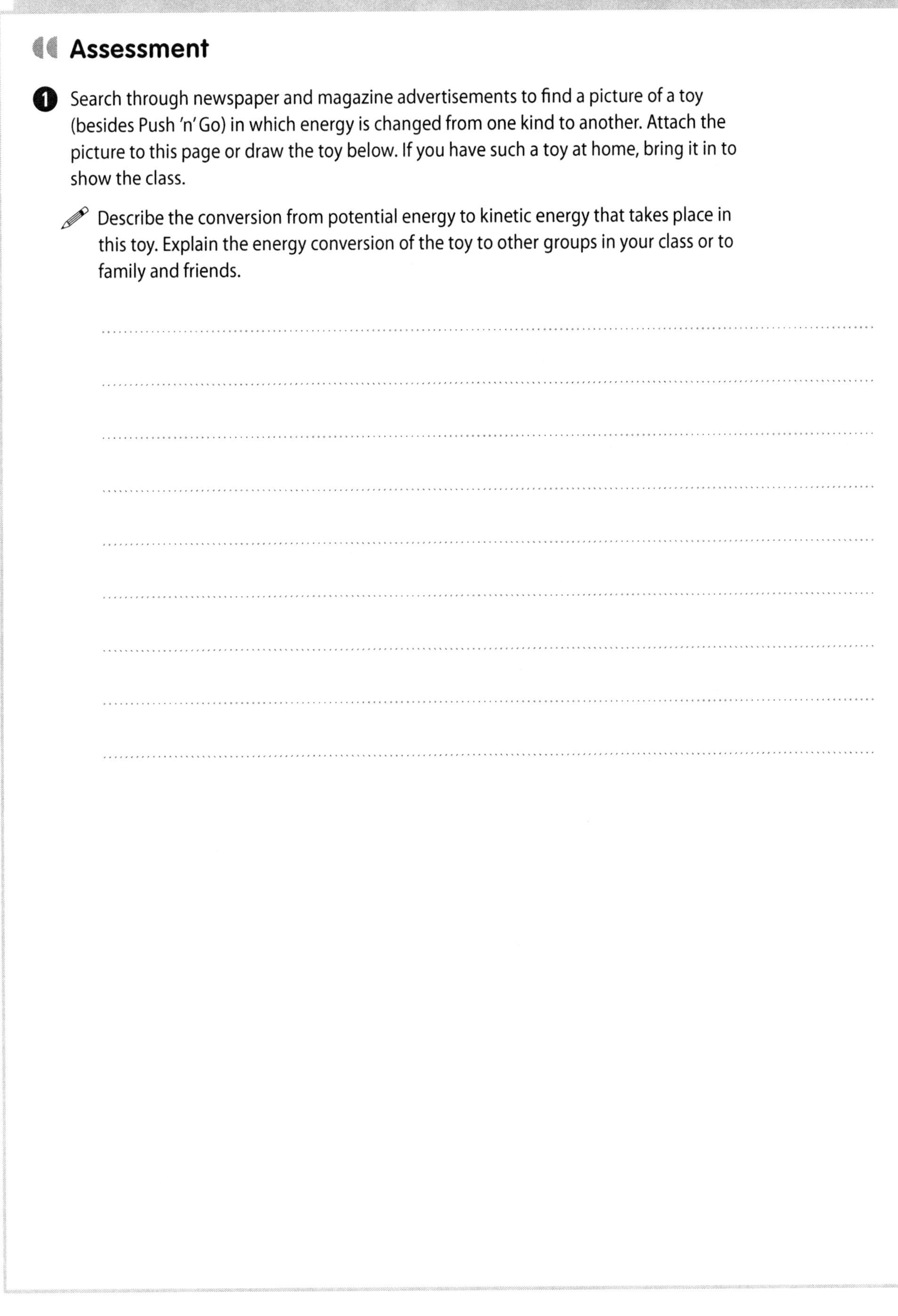

Assessment

1. Search through newspaper and magazine advertisements to find a picture of a toy (besides Push 'n' Go) in which energy is changed from one kind to another. Attach the picture to this page or draw the toy below. If you have such a toy at home, bring it in to show the class.

   Describe the conversion from potential energy to kinetic energy that takes place in this toy. Explain the energy conversion of the toy to other groups in your class or to family and friends.

Exploring Friction

Students are introduced to friction by observing and measuring how far a Push 'n' Go™ toy goes on various surfaces.

Grade Levels

Science activity appropriate for grades 4–6

Student Background

Students should have prior knowledge of potential and kinetic energy. This activity is a good follow-up to the activity "Push and Go" in this book.

Time Required

Setup		negligible
Procedure	30	minutes
Cleanup	5	minutes

Assessment time is not included.

Key Science Topics

- energy conversion
- force
- friction
- kinetic energy

National Science Education Standards Overview

See *www.terrificscience.org/physicsez/* for how these standards relate to the activity.

Science as Inquiry

Abilities Necessary to Do Scientific Inquiry

K–4 *Plan and conduct a simple investigation.*
K–4 *Employ simple equipment and tools to gather data and extend the senses.*
K–4 *Use data to construct a reasonable explanation.*

5–8 *Design and conduct a scientific investigation.*
5–8 *Use appropriate tools and techniques to gather, analyze, and interpret data.*
5–8 *Develop descriptions, explanations, predictions, and models using evidence.*
5–8 *Think critically and logically to make the relationships between evidence and explanations.*

Physical Science

K–4 *Position and motion of objects*

5–8 *Motions and forces*
5–8 *Transfer of energy*

Science and Technology

Abilities of Technological Design

K–4 *Communicate a problem, design, and solution.*

5–8 *Communicate the process of technological design.*

Materials

For the Procedure

Per group

- Push 'n' Go toy

For supply source suggestions, see www.terrificscience.org/supplies/.

- several types of horizontal surfaces (such as vinyl, linoleum, carpet, wood, foam rubber pad, asphalt, sandpaper, and concrete)
- meterstick or measuring tape
- graph paper

Graph paper masters for copying are available at www.terrificscience.org/physicsez/.

- (optional) objects that fit in or on the Push 'n' Go toy

Safety and Disposal

No special safety or disposal procedures are required.

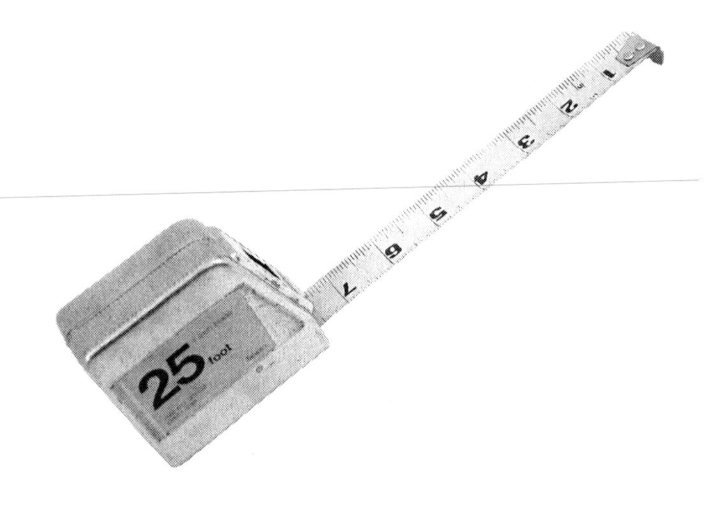

Getting Ready

Collect or identify the surfaces you wish to offer the class. Test the Push 'n' Go to make sure it runs on these surfaces within the surface lengths that are available.

Procedure

This section provides teacher notes corresponding to each step of the student procedure. The procedure without teacher notes is included in the reproducible Student Notebook pages at the end of this activity and at www.terrificscience.org/physicsez/.

	Student Procedure	Teacher Notes
	Push 'n' Go on Different Surfaces	
①	Play with the Push 'n' Go. Think about what causes the Push 'n' Go to eventually stop moving. ✎ Write your hypothesis.	Provide time for free exploration to allow the students to examine and play with the Push 'n' Go. Through discussion, help students realize that the toy stops moving because its kinetic energy is converted to another form of energy. They should also conclude that friction between the toy and the floor has something to do with this conversion of kinetic energy.
②	Feel the rubber on the Push 'n' Go wheels and notice the temperature. Working with a partner, send the Push 'n' Go rapidly back and forth a few times. Feel the rubber on the wheels again. ✎ What do you observe? ✎ What can you infer from this observation about what happened to the toy's kinetic energy?	Students may be able to feel that the rubber on the wheels becomes warmer after the Push 'n' Go has traveled back and forth a few times. They should conclude that the kinetic energy was changed into thermal energy. If students are not able to feel warmth on the wheels, have them slide their hands along a carpeted surface a few times. Explain that the same thing is happening to the toy's wheels. Students could predict that the toy would not travel as far if the friction between the wheels and the surface increased. Conversely, it would travel farther if the friction force were reduced.
③	Work with your group to design an experiment to determine which of several surfaces exerts the largest friction force on the toy. Make sure to plan for at least three trials per surface. ✎ On another piece of paper, write your hypothesis and experimental design. Create a data table.	Before students begin brainstorming, introduce or review the characteristics of good experiments. Depending on their experience, students may benefit from using the Experiment Planning Guide provided at *www.terrificscience.org/physicsez/.*
④	Perform the experiment after your teacher reviews and approves your experimental design and data table. ✎ Record the results in your data table.	A reasonable plan would include running the toy on several different surfaces and measuring the distance the toy travels on each surface. Because some Push 'n' Go vehicles may run better than others, use the same Push 'n' Go toy for all the work by each group. Data tables should include space for at least three trials per surface and for trial averages.

	Student Procedure	Teacher Notes
⑤	Average the results of your trials. ✎ Record these averages in your data table.	See Sample Answers for example data.
⑥	On graph paper, make a graph of your results. ✎ Which of your surfaces exerts the largest amount of friction? How did you draw your conclusion? ✎ How do your results compare with your prediction?	So groups can more easily compare their results, you may want to decide on a common format for the graphs. For example, ask students to plot the independent variable (different surfaces) on the horizontal axis (x) and the dependant variable (travel distance) on the vertical axis (y). (See Sample Answers for an example graph.) Students may conclude that the toy travels the shortest distance on the surface with the greatest friction force.
⑦	Share and compare your results with the class.	The results of different groups are unlikely to be identical. In some cases, a group's results may deviate greatly from the rest of the class. Have students discuss possible reasons for these variations. Reasons may include: • **poor experimental design** (for example, too many variables) • **faulty materials** (for example, surface not level) • **faulty execution of procedure** (for example, not completely pushing down the driver of the Push 'n' Go toy each time) • **errors in collecting and recording data** (for example, misreading the meterstick)
	Class Discussion	• Ask students to look carefully at each of the surfaces. What physical differences are there between high friction ones and low friction ones? • Ask students which of their surfaces had the least amount of friction and how they drew that conclusion. Make sure students understand the relationship between distance traveled and the size of the friction force exerted by each surface.

Student Procedure	Teacher Notes
Optional	You may want to have older students design and conduct an experiment to determine the performance of the Push 'n' Go on a single surface while carrying different loads. (See Sample Answers for results using quarters and Explanation for a discussion.)
Assessment	
1 Suppose that a group of students wants to determine which surface exerts the largest friction force on a toy. Their plan is to use three different toys to test three different surfaces. They plan to use a Push 'n' Go on the classroom carpet, a wind-up car on the wooden gymnasium floor, and a pull-back train on the sidewalk. What is wrong with their plan? How could you improve it?	The plan changes two variables: the toy and the surface. In order to know what is causing the variation in results, students should use the same toy for all surfaces.

Sample Answers

Where students follow the same procedure using the same materials, these answers are close to answers you can expect. Where students design their own experiment or model, students' results will vary.

Surface	Distance Toy Traveled					
	Trial 1	Trial 2	Trial 3	Trial 4	Trial 5	Average
hardwood floor	491 cm	534 cm	480 cm	462 cm	508 cm	495 cm
vinyl floor	457 cm	490 cm	452 cm	439 cm	457 cm	459 cm
concrete floor	312 cm	325 cm	299 cm	310 cm	304 cm	310 cm
low-pile carpet	178 cm	203 cm	196 cm	213 cm	175 cm	193 cm

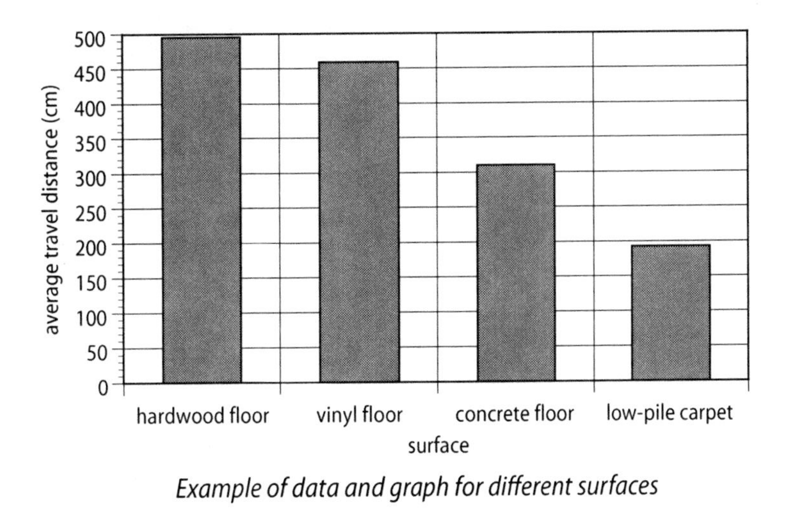

Example of data and graph for different surfaces

Number of Quarters	Distance Toy Traveled					
	Trial 1	Trial 2	Trial 3	Trial 4	Trial 5	Average
0	427 cm	439 cm	445 cm	437 cm	442 cm	438 cm
8	394 cm	394 cm	399 cm	396 cm	401 cm	397 cm
16	378 cm	376 cm	371 cm	358 cm	363 cm	369 cm

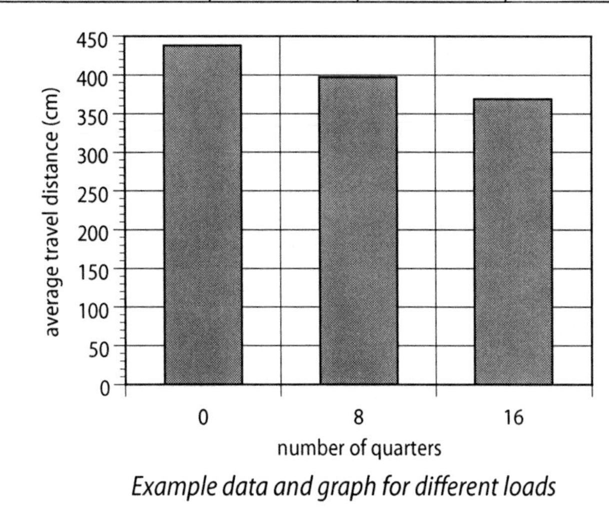

Example data and graph for different loads

Explanation

This section is intended for teachers. Modify the explanation for students as needed.

Friction and Different Surfaces

When the Push 'n' Go stops rolling, its kinetic energy has been changed into thermal energy because of the friction between the wheels and the floor. (See the activity "Push and Go" in this book for a discussion of energy conversions in the Push 'n' Go.) Students may have felt this warmth on the tires after rolling the Push 'n' Go back and forth a few times. Or, they felt their hand become warmer after moving it across a carpeted surface a few times. On surfaces with greater friction, the kinetic energy is converted to thermal energy more rapidly, and the Push 'n' Go does not travel as far. (See Sample Answers for data table and graph examples.)

Why does the amount of friction change on different surfaces? To answer, let's consider the factors affecting the frictional force exerted by one surface upon another.

- One factor is the perpendicular force one surface exerts on the other. This perpendicular force is called the normal force. In the case of the Push 'n' Go moving on a level surface, this normal force is simply the weight of the Push 'n' Go. This force does not change when the Push 'n' Go is on different surfaces. If you test the performance of the Push 'n' Go on a single surface while carrying increasing loads, the distance traveled decreases as the weight increases. This happens because the increasing weight increases the normal force and, thus, the friction force. The toy gets the same kinetic energy each time you push it. The friction force does work, which converts the kinetic energy to thermal energy. Since *work* = *force* × *distance*, when the friction force is larger, it takes less distance to convert the kinetic energy into thermal energy.
- The other factor affecting the frictional force is the nature or texture of the surfaces that are in contact. As each point on the Push 'n' Go's rotating wheels touches the surface the wheels are rolling over, the irregularities in the surface structures allow the rubber and plastic wheel material and the surface material to "grab" one another. The force of this grab is the friction between the wheels and the surface, and it changes depending on the surface that the wheels are rolling over.

Three Types of Friction

Three general types of friction occur: static, sliding, and rolling.

- Static friction between two surfaces occurs when the two surfaces are at rest relative to one another, such as a book lying on a table. If you gradually raised one end of the table, the book would remain in place until the component of the gravitational force parallel to the table becomes larger than the force of static friction. (See "Ramps and Cars" in this book for further explanation.)
- Sliding friction between two surfaces occurs when one of the surfaces moves over the other one while they are in contact, such as a book being pushed or pulled across a table.
- Rolling friction occurs when at least one surface is round and rolls relative to the other, such as a wheel rolling on a surface.

In the case of the Push 'n' Go, rolling friction exists between the wheels of the car and the surface over which it is rolling. A rolling wheel helps reduce friction between two surfaces because, for the most part, the two surfaces are not moving relative to one another when they make contact. In other words, the wheel is not sliding across the surface. As the wheel turns, every point on the wheel just touches the surface and then moves off of it. As each point of the wheel touches, static friction exists between the wheel and the surface. When you push a wheel, its static friction at the point of contact with the surface causes it to "tip over" and start rolling. This static friction gives the wheel traction.

The path of a point on a moving, rotating wheel is called a cycloid. The following figure provides an idea of this motion. Search the Internet using the terms "cycloid animation" or "cycloid movie," and you will be likely to find several moving versions to view.

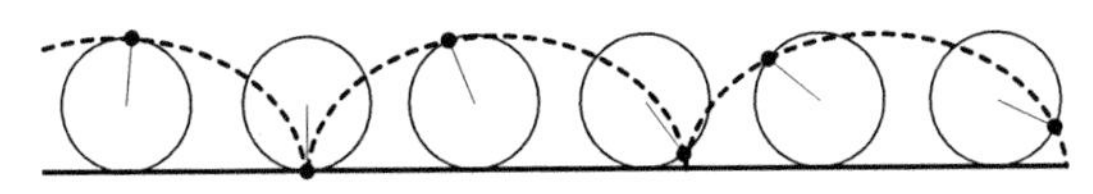

A point on a rotating wheel moves forward and down, touches the surface for an instant, then continues forward and up.

Wheels and Newton's Third Law

If we were studying wooden blocks sliding across different surfaces after being pushed or pulled to get them started, our explanation could stop here. However, the behavior of the Push 'n' Go is a little more complex. Because the Push 'n' Go exerts a force on the surface to get itself started in motion, we must also consider Newton's third law. As it turns, the wheel exerts a backward force on the surface and the surface exerts a forward force on the wheel. (See figure below.)

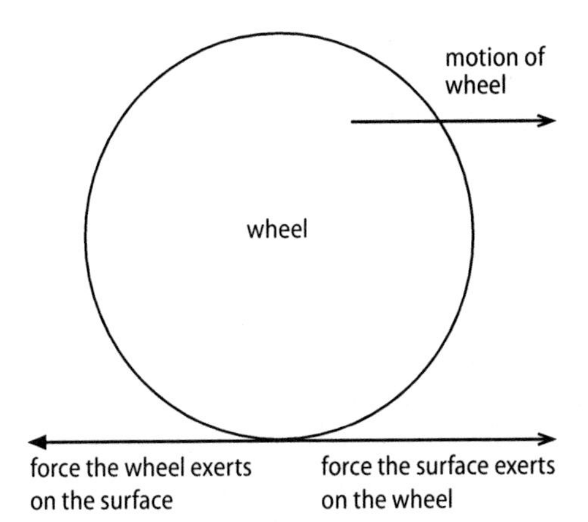

If the surface is not flexible, the force the wheel exerts on the surface is returned to the wheel, maximizing its motion forward. If the surface is flexible (such as carpet), it moves as the wheel exerts a force on it, so the wheel is not able to exert a very large force on the carpet. (Think about trying to push on something floating in a swimming pool. You can't push very hard because, as soon as you start to push, the object moves away from you.) The Push 'n' Go doesn't travel as far on the carpet, partially because the carpet exerts a greater friction force than the tile floor, but also partially because its initial velocity is lower due to its inability to exert as large a force on the carpet.

Cross-Curricular Integration

Math:

- Review with students when bar graphs should be used to present data and when line graphs are more appropriate.

Push 'n' Go™ on Different Surfaces

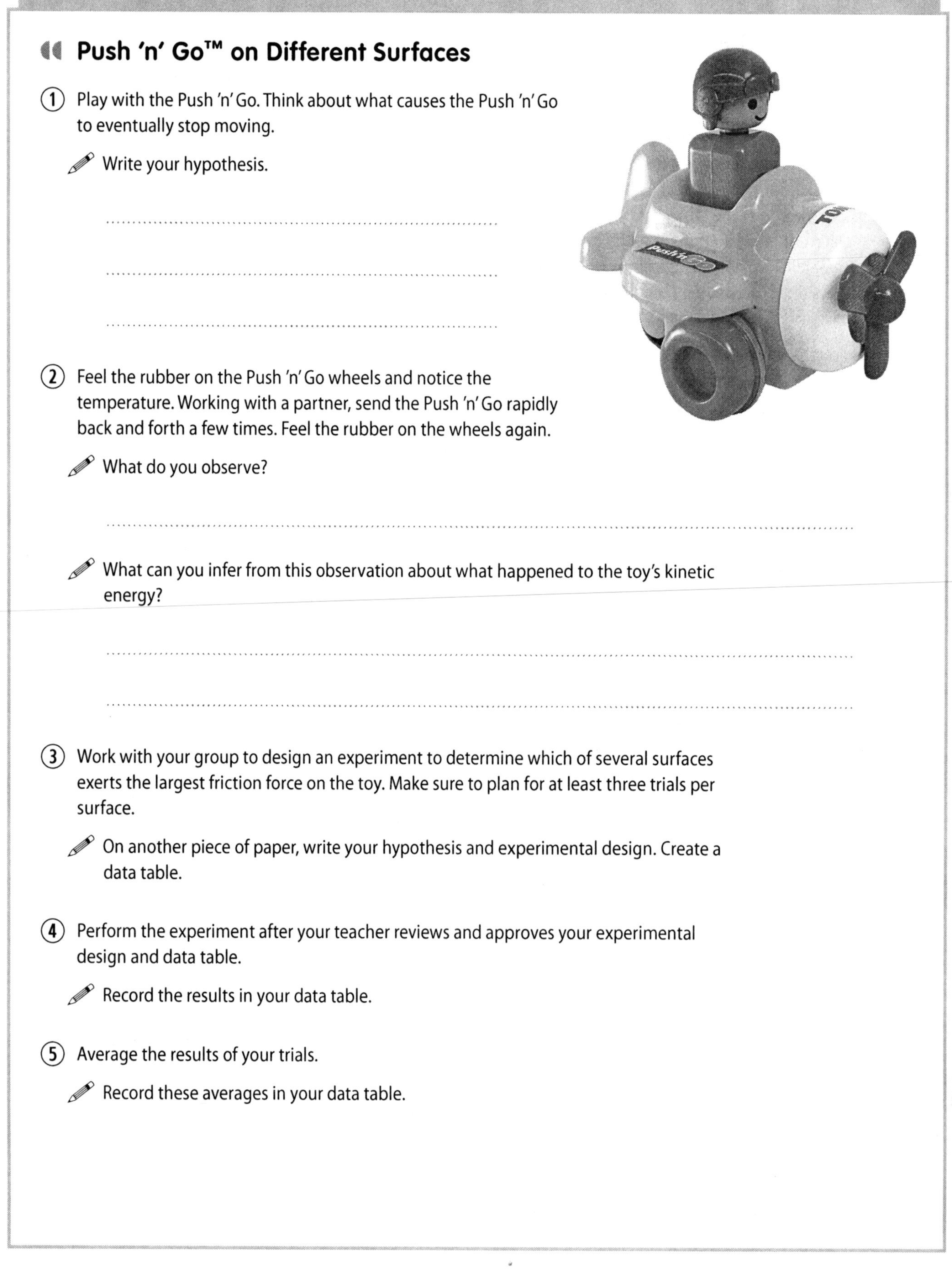

① Play with the Push 'n' Go. Think about what causes the Push 'n' Go to eventually stop moving.

Write your hypothesis.

....

....

....

② Feel the rubber on the Push 'n' Go wheels and notice the temperature. Working with a partner, send the Push 'n' Go rapidly back and forth a few times. Feel the rubber on the wheels again.

What do you observe?

....

What can you infer from this observation about what happened to the toy's kinetic energy?

....

....

③ Work with your group to design an experiment to determine which of several surfaces exerts the largest friction force on the toy. Make sure to plan for at least three trials per surface.

On another piece of paper, write your hypothesis and experimental design. Create a data table.

④ Perform the experiment after your teacher reviews and approves your experimental design and data table.

Record the results in your data table.

⑤ Average the results of your trials.

Record these averages in your data table.

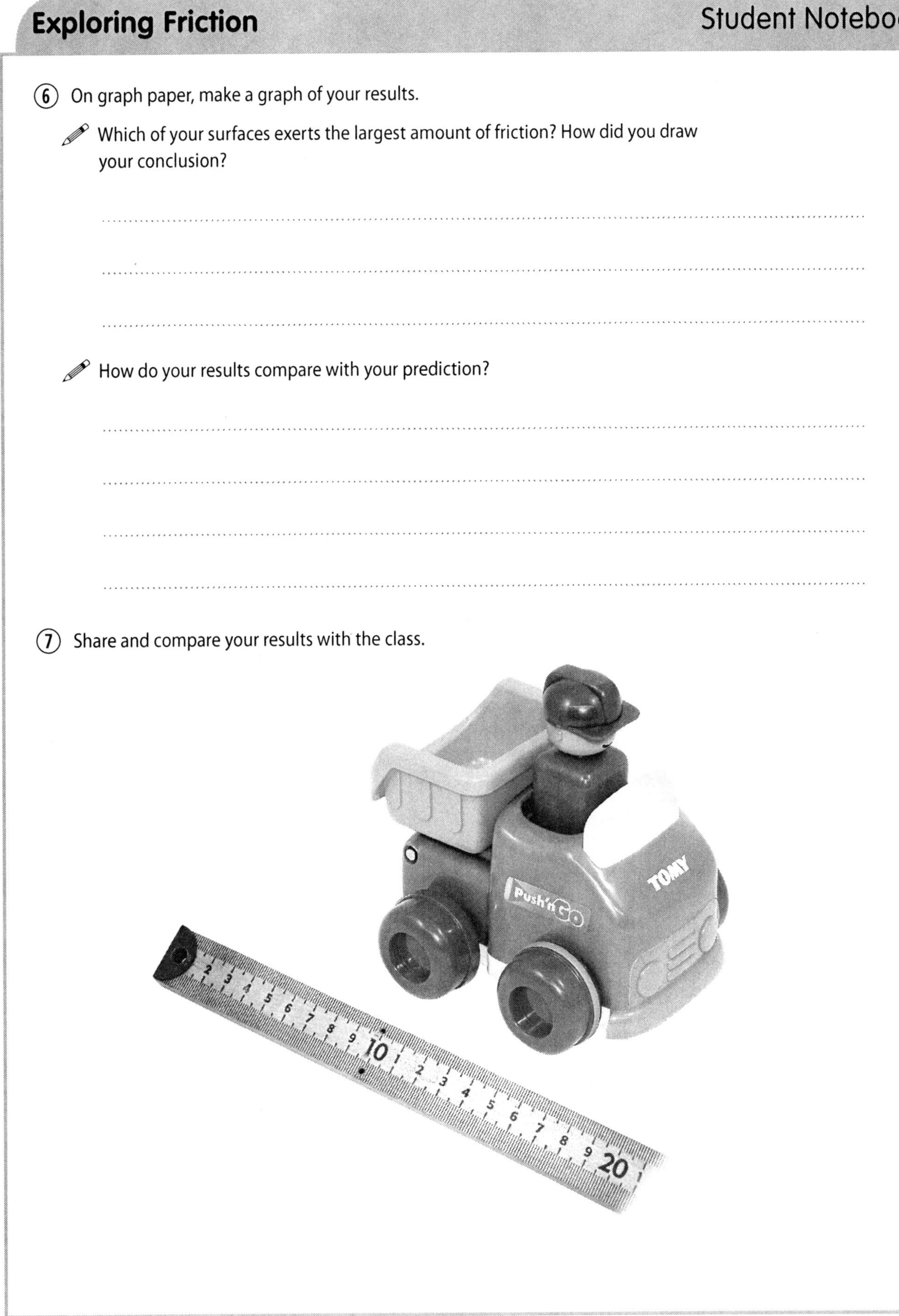

⑥ On graph paper, make a graph of your results.

✎ Which of your surfaces exerts the largest amount of friction? How did you draw your conclusion?

....

....

....

✎ How do your results compare with your prediction?

....

....

....

....

⑦ Share and compare your results with the class.

Assessment

1. Suppose that a group of students wants to determine which surface exerts the largest friction force on a toy. Their plan is to use three different toys to test three different surfaces. They plan to use a Push 'n' Go on the classroom carpet, a wind-up car on the wooden gymnasium floor, and a pull-back train on the sidewalk.

   What is wrong with their plan?

   ....

   ....

   ....

   How could you improve it?

   ....

   ....

   ....

   ....

   ....

Toy That Returns

Explore the concepts of kinetic energy and elastic potential energy using rubber bands to store energy.

Grade Levels

Science activity appropriate for grades 3–8

Student Background

Students should have already been introduced to the concept of kinetic energy and elastic potential (or stored) energy. This activity is a good follow-up to the activity "Push and Go" in this book.

Time Required

Setup	25	minutes
Part A	15	minutes
Part B	30	minutes
Cleanup	5	minutes

Assessment time is not included.

Key Science Topics

- elastic potential energy
- inertia
- kinetic energy
- motion

National Science Education Standards Overview

See *www.terrificscience.org/physicsez/* for details of how these standards relate to the activity.

Science as Inquiry

Abilities Necessary to Do Scientific Inquiry

K–4 *Employ simple equipment and tools to gather data and extend the senses.*
K–4 *Communicate investigations and explanations.*

5–8 *Use appropriate tools and techniques to gather, analyze, and interpret data.*
5–8 *Develop descriptions, explanations, predictions, and models using evidence.*
5–8 *Think critically and logically to make the relationships between evidence and explanations.*
5–8 *Communicate scientific procedures and explanations.*

Physical Science

K–4 *Position and motion of objects*

5–8 *Motions and forces*
5–8 *Transfer of energy*

Science and Technology

Abilities of Technological Design

K–4 *Evaluate a product or design.*

5–8 *Implement a proposed design.*
5–8 *Evaluate completed technological designs or products.*

Materials

For Getting Ready

Per group

- (optional) material to prepare coffee can
  - empty and clean small (1 pound) coffee can
  - *Larger cans can be used but will require larger rubber bands.*
  - can opener (for teacher use only)
  - electrical or masking tape
- (optional) material to prepare soft-drink bottles
  - empty and clean plastic 2-L soft-drink bottle
  - drill with ⅝-inch bit (for teacher use only)
  - box for support during drilling
  - scissors
  - electrical or masking tape
  - unwanted CDs (2 for each bottle)
  - hot-melt glue gun (for teacher use only)

For the Procedure

Activity Introduction, per class

- (optional) rubber-band-powered toy airplane with a visible rubber band

Part A, per class

- one of the following toys:
  - coffee can come-back toy made in Getting Ready
  - commercial come-back toy
  - *For supply source suggestions, see www.terrificscience.org/supplies/.*

Part B, per group

- material to make one of three styles of come-back toys (K'NEX®, soft-drink bottle, or coffee can):
  - K'NEX come-back toy
    - assorted K'NEX rods and connectors
    - *For specifics, see K'NEX Parts Inventory at www.terrificscience.org/physicsez/ or www.terrificscience.org/supplies/.*
    - K'NEX assembly diagrams (Print in color from *www.terrificscience.org/physicsez/*.)
    - 2 unwanted CDs
    - 2 Styrofoam® plates
    - 2 size 33 rubber bands (about 8–10 cm long)
    - steel ball bearing about ¾ inch in diameter
  - soft-drink bottle come-back toy
    - 3 large paper clips
    - 2 size 33 rubber bands (about 8–10 cm long)
    - piece of wire or twist tie about 10 cm long
    - weight (See coffee can come-back toy.)
    - soft-drink bottle prepared in Getting Ready
  - coffee can come-back toy
    - 2 plastic lids for the coffee can
    - scissors
    - size 33 rubber band (about 8–10 cm long)
    - 2 paper clips
    - piece of wire or twist tie about 10 cm long
    - weight (such as bolt and several nuts, large fishing sinker, or pennies tied in cloth)
    - coffee can prepared in Getting Ready
- masking tape
- meterstick or measuring tape
- graph paper
- *Graph paper masters for copying are available at www.terrificscience.org/physicsez/.*

For the Assessment

Per class

- assorted materials including rubber bands

Safety and Disposal

Cover sharp edges on the cans and bottles with electrical or masking tape. No special disposal procedures are required.

Getting Ready

- Make a coffee can come-back toy for Part A if you don't have a commercial toy. (See How to Make a Come-Back Toy in the Student Notebook.)
- If making coffee can toys, students can collect cans and plastic lids in advance. Cut out both metal ends of the can. Cover the cut edges with tape.
- If making soft-drink bottle toys, prepare each 2-L bottle as follows: drill a ⅝-inch hole in the center bottom (hold the bottle against the corner of a box for support); use scissors to cut a window about 5 cm x 8 cm in the side; cover the cut edges with tape; and hot glue a CD to each end, making sure not to fill the holes in the bottle and the CDs.

Procedure

This section provides teacher notes corresponding to each step of the student procedure. The procedure without teacher notes is included in the reproducible Student Notebook pages at the end of this activity and at www.terrificscience.org/physicsez/.

	Student Procedure	Teacher Notes
	Activity Introduction	Have students review their ideas about kinetic energy and potential energy. You may want to facilitate the discussion by using a rubber-band-powered airplane to model their ideas.
	Part A: How Does the Toy Work?	
1	Watch your teacher demonstrate a toy, then answer these questions. What kind of energy does the toy have while it is moving? Where does the energy come from? Why do you think the toy returns? Why does the toy stop?	Demonstrate the motion of a come-back toy by placing it on the floor and giving it a push. (Use the coffee can come-back toy or a commercial come-back toy so students cannot see the toy's internal mechanism.) Some students will realize that your hand provides the energy to make the toy move, the toy has kinetic energy while moving, stored energy causes the toy to return, and friction causes the toy to eventually stop.
2	Think about what could be inside the toy to store energy. On another piece of paper, draw a diagram of what you think might be inside the toy. Label the parts.	
3	Examine the toy to find out how it stores energy. On another piece of paper, draw and label what is inside the toy.	Remove the lids from the toy to reveal the mechanism to the class. Students should draw and label the rubber band and the attached weight. (See Sample Answers for student drawing example.)
	Class Discussion	• Explain how the mechanism of the toy works. Turn the lids to illustrate how the rubber band winds up. • Discuss the toy's energy conversions. (See Explanation.)

Student Procedure	Teacher Notes
Part B: Make Your Own Toy	
① Make a come-back toy by following one of the methods on the How to Make a Come-Back Toy instruction sheets.	Select one or more of the three come-back toys shown in How to Make a Come-Back Toy and make the materials available to the class. If using K'NEX, students will also need color K'NEX assembly diagrams to build the model. (See *www.terrificscience.org/physicsez/*.) You could have different groups make different types of come-back toys and compare their results in step 7.
② Try out the toy by placing it on its side on a flat surface and giving it a push to start it rolling. If it does not return, try making minor adjustments such as adding more weight, tightening the rubber band, or adjusting the weight so it doesn't touch the side of the toy.	Some toys may not work on the first try. Troubleshooting toys that don't work is an important part of the students' learning process.
③ Observe the motion of your toy. ✎ Explain how kinetic energy is put into your toy, where potential energy is stored, and what happens when the potential energy goes back to kinetic energy.	Students should be able to discuss kinetic energy, elastic potential energy, and friction.
④ Put a strip of masking tape (about 5 m long) on the floor in a straight line.	Make sure that the masking tape is placed as straight as possible on the floor. If the floor is tiled, students can follow a seam between tiles.
⑤ Beginning at about the middle of the tape, push the come-back toy so that the toy travels along the length of the tape. If the toy goes past the end of the tape, start over and push more gently. Mark the tape at each point the toy changes direction until the toy stops. Label the point of the first direction change as D1, the point of the second direction change as D2, and so on. ✎ Measure the distance between each point and record this information in the data table. ✎ On graph paper, draw a graph showing the movement of your come-back toy.	Come-back toys that have been pushed too hard will go past the end of the tape. In this case, students can't mark the toy's movement. The data table provides space for eight measurements. However, each toy will vary and students should adjust the table as needed. Help students understand how to measure and graph their results. (See Sample Answers.)

Student Procedure	Teacher Notes
⑥ Work with your group to answer the following questions. ✎ Where does the kinetic energy come from to cause the come-back toy to travel from the start to D1? ✎ Why does the toy stop at D1? ✎ Why does the toy begin moving again from D1 to D2? ✎ Why doesn't the toy stop when it gets back to the starting location? ✎ Why does the toy stop at D2? ✎ How does the distance from D1 to D2 compare to the distance from D2 to D3? Why does this happen? Does the pattern continue for the other distances?	• The kinetic energy to travel from the start to D1 comes from the work you do on the toy when you push it with your hand. • The toy stops at D1 because all of the kinetic energy has been transformed to either elastic potential energy (stored in the rubber band) or thermal energy (from friction with the floor). • The toy begins moving again from D1 to D2 because the elastic potential energy is being transformed into kinetic energy. • The toy continues to roll back past the starting location (even after the rubber band is completely unwound) because of the toy's inertia. (You may want to review Newton's first law.) This motion winds the rubber band in the opposite direction. • The toy stops at D2 because, again, all of the kinetic energy has been transformed to either elastic potential energy (stored in the rubber band) or thermal energy (from friction with the floor). • The distance from D1 to D2 and all subsequent distances will continue to get smaller until the toy stops. This happens because some kinetic energy is converted to thermal energy on each trip.
⑦ Share and discuss your results with the class.	Have students share their results and their explanations for what they see on the graph. Through discussion, help students see how the data they have collected and graphed corresponds to the answers to the questions in step 6. Discuss the energy conversions that occur when operating the toy.
Assessment	
❶ Design and make a toy or other object that uses a rubber band to store energy. ✎ How is kinetic energy put into your object? ✎ Draw your object below. Show where potential energy is stored and what happens when the potential energy goes back to kinetic energy. You may need to draw more than one picture.	Students can make any object that has a rubber band that stores energy. Students should be able to explain how their object works. (See Sample Answers for design examples.)

Sample Answers

Where students follow the same procedure using the same materials, these answers are close to answers you can expect. Where students design their own experiment or model, students' results will vary.

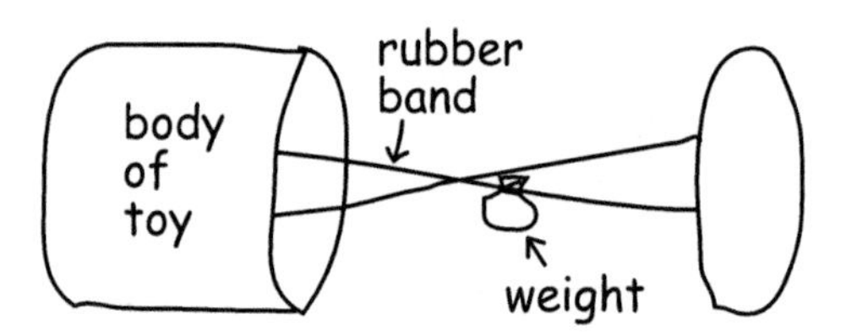

Example of student drawing for Part A

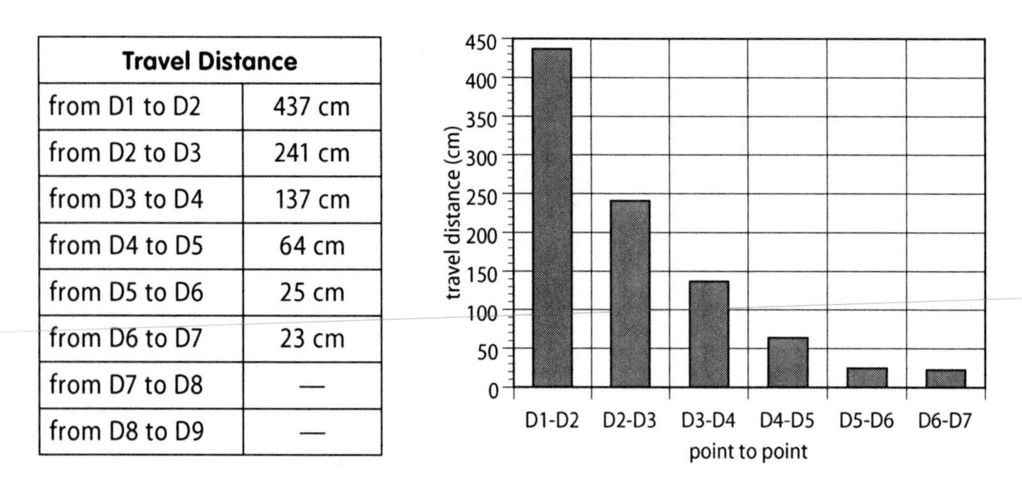

Travel Distance	
from D1 to D2	437 cm
from D2 to D3	241 cm
from D3 to D4	137 cm
from D4 to D5	64 cm
from D5 to D6	25 cm
from D6 to D7	23 cm
from D7 to D8	—
from D8 to D9	—

Sample data and graph for K'NEX come-back toy

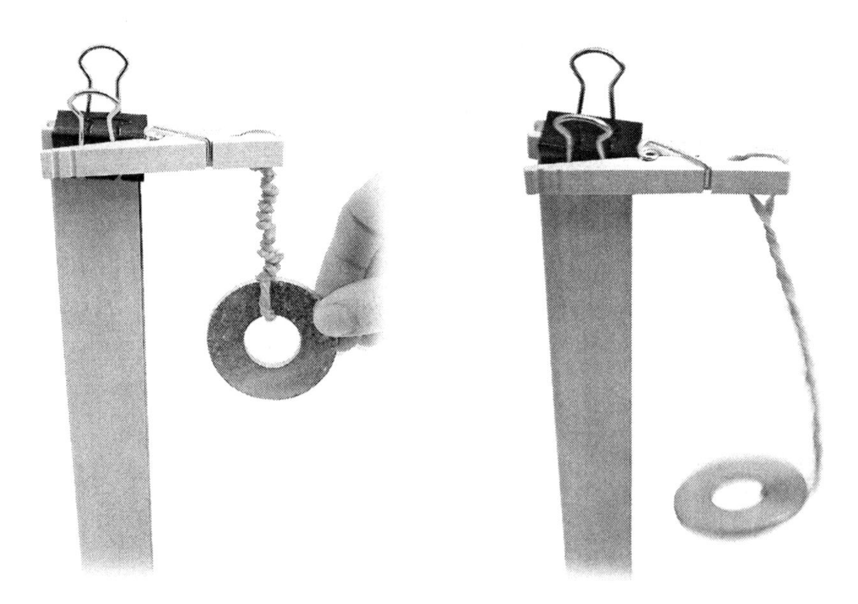

Example of student design for Assessment

Explanation

This section is intended for teachers. Modify the explanation for students as needed.

The come-back toy starts moving because your hand applies a force to the toy. You do work with this force. For the simple case of a constant force applied in the direction of motion, work is defined as the product of the force applied to an object and the distance the object moves while the force is applied. The work done on the come-back toy starts the toy moving in two ways: First, the toy moves in a straight line. Second, it rotates. The work done gives the toy both linear kinetic energy and rotational kinetic energy. (Kinetic energy is energy of motion.)

The suspended weight is the only item in the toy that does not turn when the toy moves. Because the weight does not turn with the toy, the rubber band winds up. Rubber bands store energy when they are stretched or twisted and release the energy later. The energy stored in the twisted rubber band is called elastic potential energy. The toy slows down as it moves away from you because the kinetic energy is being converted to elastic potential energy. At some point the toy stops and all the energy is stored as elastic potential energy. This elastic potential energy is released and returns to linear and rotational kinetic energy as the toy moves back to you.

The toy continues to roll back past the start even after the rubber band is completely unwound. This is due to the toy's inertia. Newton's first law, also called the law of inertia, says an object will continue to move with a constant velocity unless a force acts to change its motion. This motion winds the rubber band in the opposite direction, again storing elastic potential energy until the toy stops again and rolls forward. This process continues until friction converts all the kinetic energy into thermal energy and the toy stops.

Cross-Curricular Integration

Language arts:

- Have students make a foldable concept map as shown in the figure at right. Have students record examples of potential and kinetic energy on the inside of the concept map. (For each type of energy, have students record one example from the activity and one other example.) Students can then write one paragraph explaining kinetic energy and one paragraph explaining potential energy on the outside of the concept map. Each paragraph should define the type of energy and cite the examples.

Fold paper along its width, leaving a 1-inch margin at the top of the page.

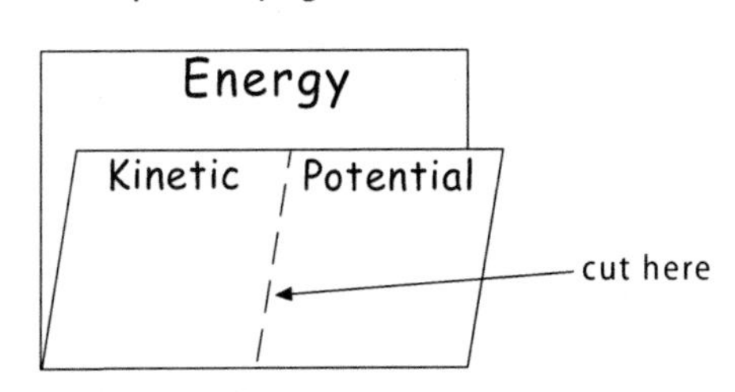

Cut just the front layer in half to the fold. Label the sections as shown.

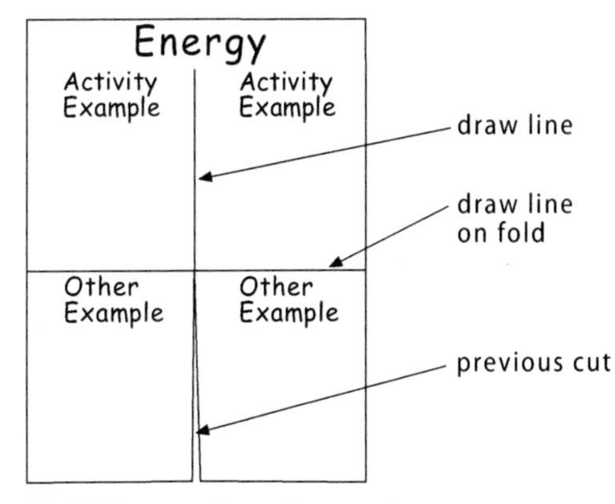

Unfold paper. Draw lines and label sections as shown.

Math:

- Have students use the travel data from their come-back toys to calculate the percent decrease between D1-D2 and D2-D3, between D2-D3 and D3-D4, and so on. Are these percentages always the same?

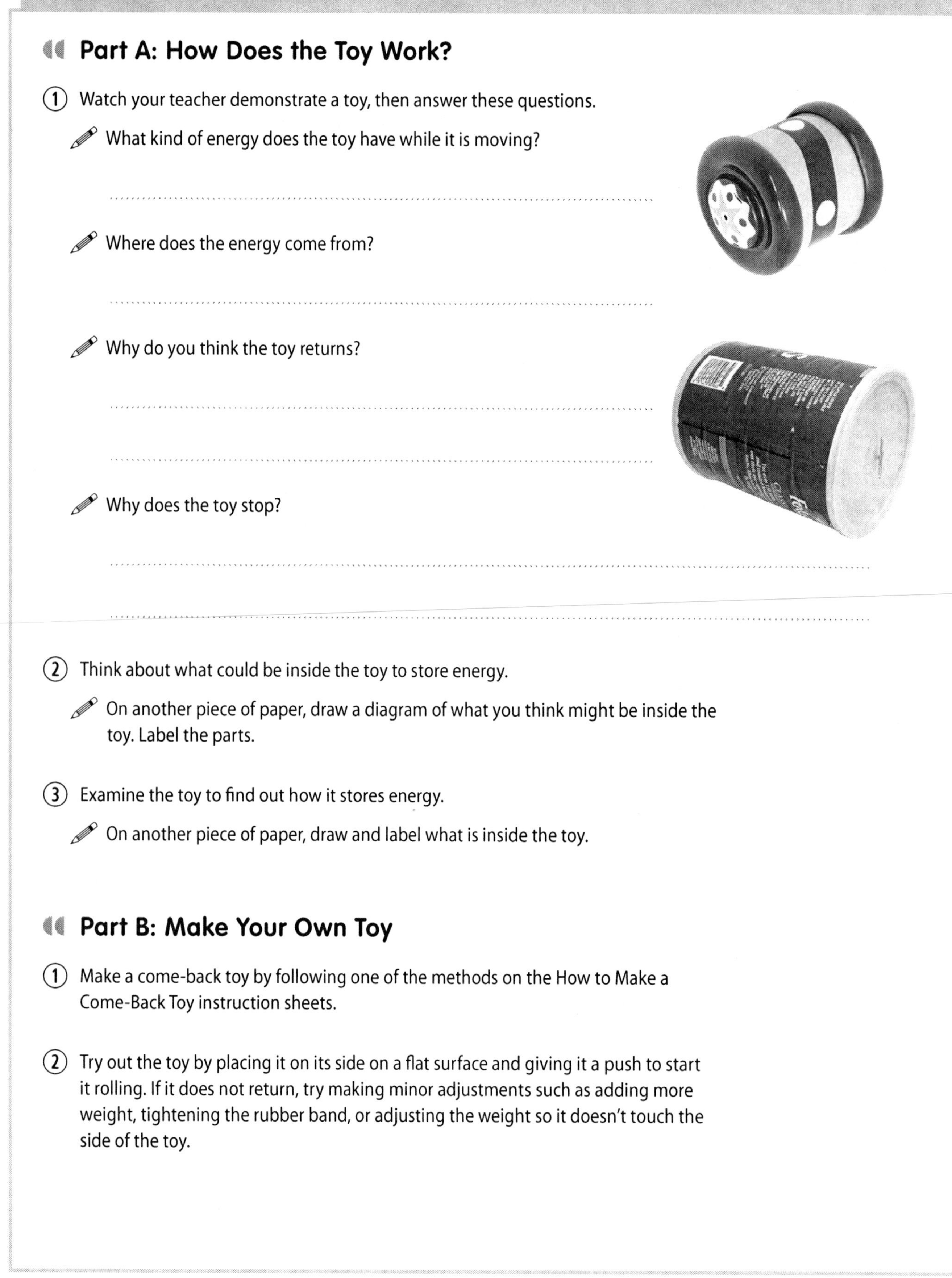

Part A: How Does the Toy Work?

① Watch your teacher demonstrate a toy, then answer these questions.

What kind of energy does the toy have while it is moving?

....

Where does the energy come from?

....

Why do you think the toy returns?

....

....

Why does the toy stop?

....

....

② Think about what could be inside the toy to store energy.

On another piece of paper, draw a diagram of what you think might be inside the toy. Label the parts.

③ Examine the toy to find out how it stores energy.

On another piece of paper, draw and label what is inside the toy.

Part B: Make Your Own Toy

① Make a come-back toy by following one of the methods on the How to Make a Come-Back Toy instruction sheets.

② Try out the toy by placing it on its side on a flat surface and giving it a push to start it rolling. If it does not return, try making minor adjustments such as adding more weight, tightening the rubber band, or adjusting the weight so it doesn't touch the side of the toy.

③ Observe the motion of your toy.

✎ Explain how kinetic energy is put into your toy, where potential energy is stored, and what happens when the potential energy goes back to kinetic energy.

....

....

④ Put a strip of masking tape (about 5 m long) on the floor in a straight line.

⑤ Beginning at about the middle of the tape, push the come-back toy so that the toy travels along the length of the tape. If the toy goes past the end of the tape, start over and push more gently. Mark the tape at each point the toy changes direction until the toy stops. Label the point of the first direction change as D1, the point of the second direction change as D2, and so on.

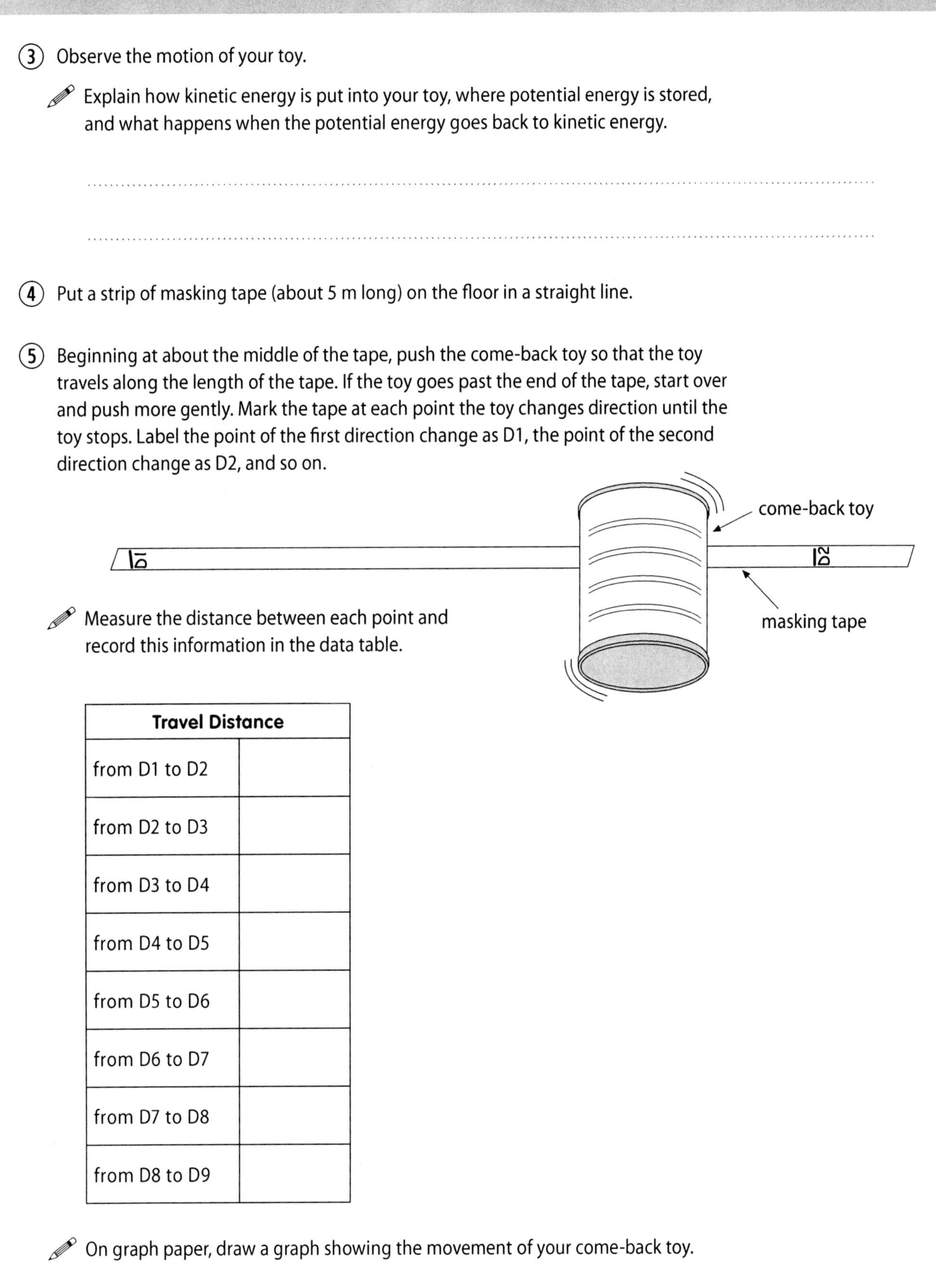

✎ Measure the distance between each point and record this information in the data table.

Travel Distance	
from D1 to D2	
from D2 to D3	
from D3 to D4	
from D4 to D5	
from D5 to D6	
from D6 to D7	
from D7 to D8	
from D8 to D9	

✎ On graph paper, draw a graph showing the movement of your come-back toy.

6. Work with your group to answer the following questions.

   Where does the kinetic energy come from to cause the come-back toy to travel from the start to D1?

   Why does the toy stop at D1?

   Why does the toy begin moving again from D1 to D2?

   Why doesn't the toy stop when it gets back to the starting location?

   Why does the toy stop at D2?

   How does the distance from D1 to D2 compare to the distance from D2 to D3? Why does this happen? Does the pattern continue for the other distances?

7. Share and discuss your results with the class.

Assessment

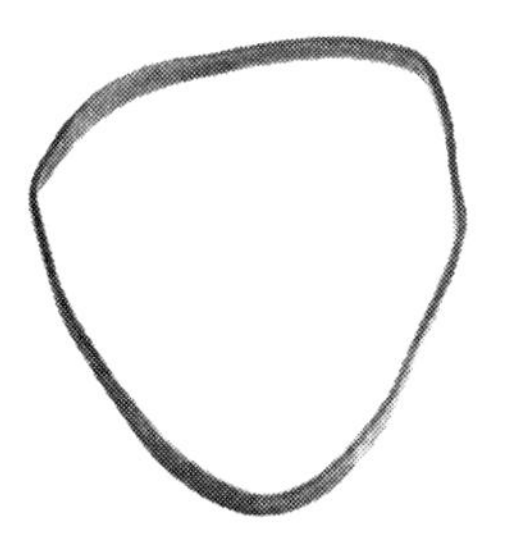

1. Design and make a toy or other object that uses a rubber band to store energy.

   How is kinetic energy put into your object?

   ....

   ....

   Draw your object below. Show where potential energy is stored and what happens when the potential energy goes back to kinetic energy. You may need to draw more than one picture.

How to Make a Come-Back Toy

A **K'NEX® Come-Back Toy: Use K'NEX pieces, unwanted CDs, Styrofoam plates, a steel ball bearing, and rubber bands. (This page provides an assembly overview. Your teacher will provide detailed K'NEX assembly diagrams.)**

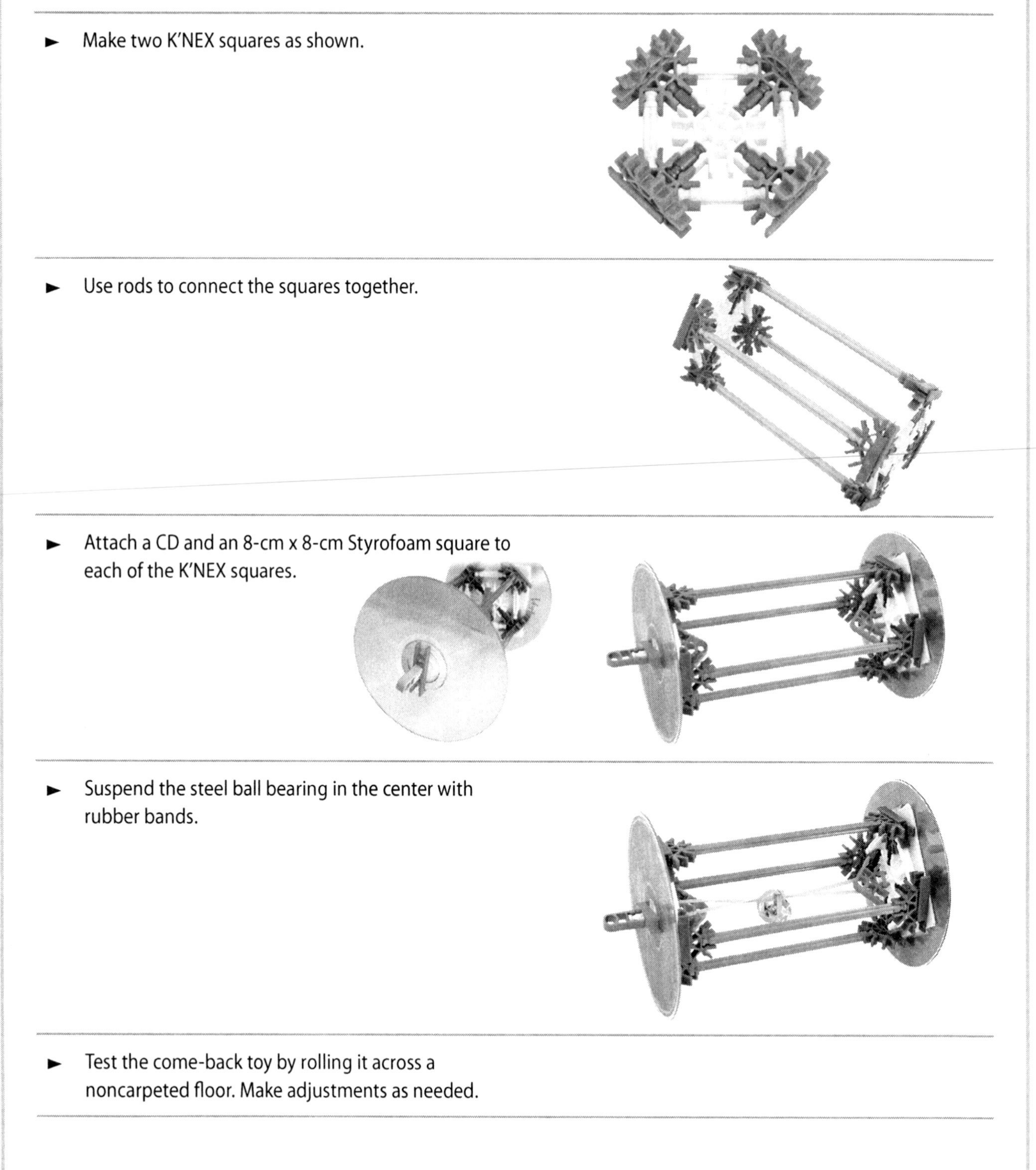

- Make two K'NEX squares as shown.
- Use rods to connect the squares together.
- Attach a CD and an 8-cm x 8-cm Styrofoam square to each of the K'NEX squares.
- Suspend the steel ball bearing in the center with rubber bands.
- Test the come-back toy by rolling it across a noncarpeted floor. Make adjustments as needed.

B **Soft-Drink Bottle Come-Back Toy: Use a prepared 2-L plastic soft-drink bottle (with hole drilled, window cut, and CDs attached), weight, wire or twist tie, paper clips, and rubber bands.**

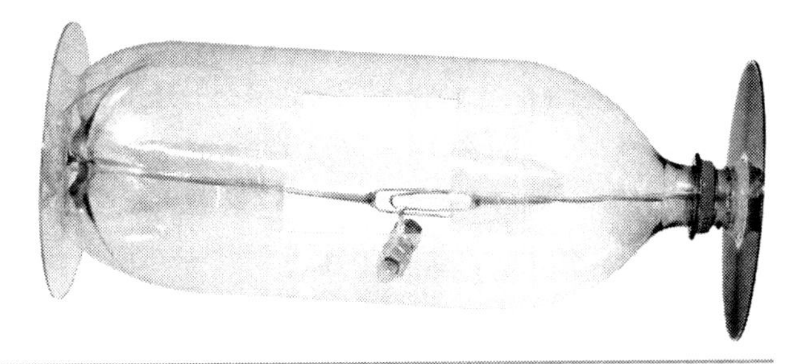

- Assemble the paper clips and rubber bands as shown. Fasten one end of the twist tie to the center of the weight as shown. Fasten the other end of the twist tie to the center paper clip. The length of the twist tie from the clip to the weight should be no more than about 1 cm.

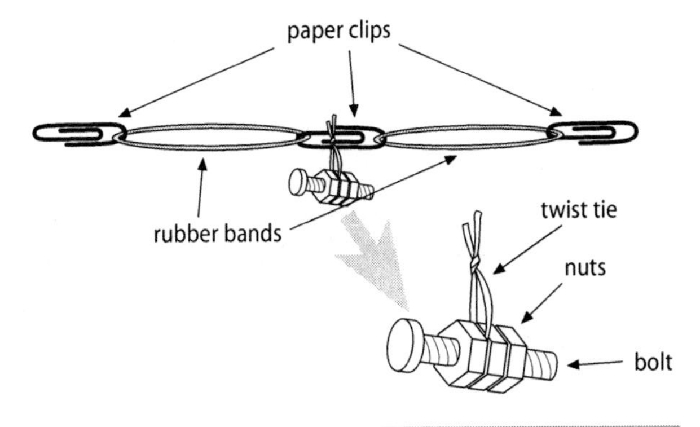

- Place the paper clip, rubber band, and weight assembly into the bottle through the window in the side of the bottle. Reach through the window, feed one end of the assembly through the mouth of the bottle, and turn the paper clip across the hole as shown. Feed the other end of the assembly through the hole in the bottom of the bottle and turn the paper clip across the hole.

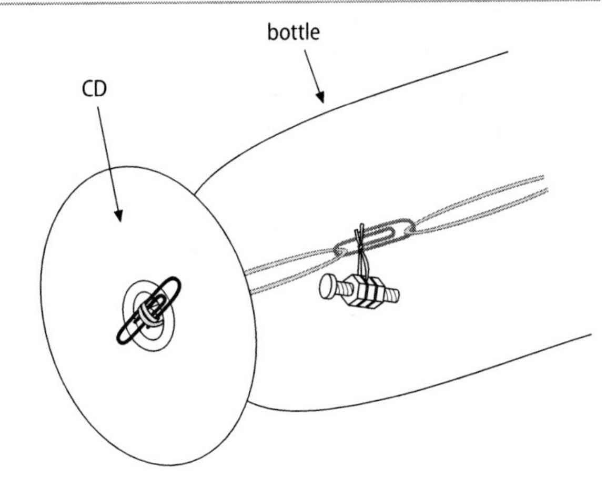

- The assembly should now be stretched through the bottle and the weight suspended in the middle or near the middle of the bottle as shown. Turn the bottle on its side. If the weight is touching the side of the bottle, tighten the rubber bands by winding them around the paper clips.

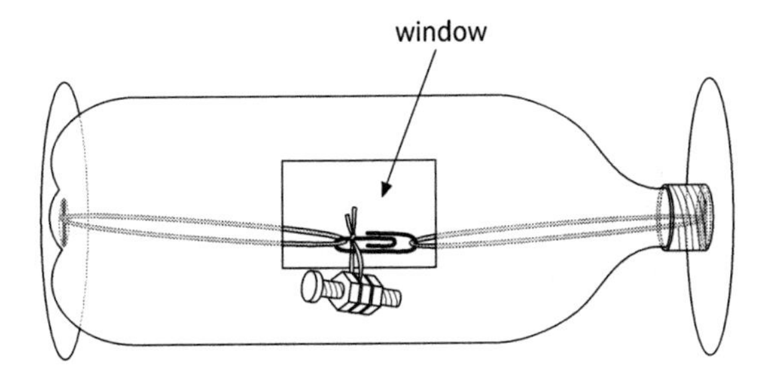

- Test the come-back toy by rolling it across a noncarpeted floor. Make adjustments as needed.

C Coffee Can Come-Back Toy: Use a prepared coffee can (with both metal ends cut out), plastic lids, weight, wire or twist tie, paper clips, and a rubber band.

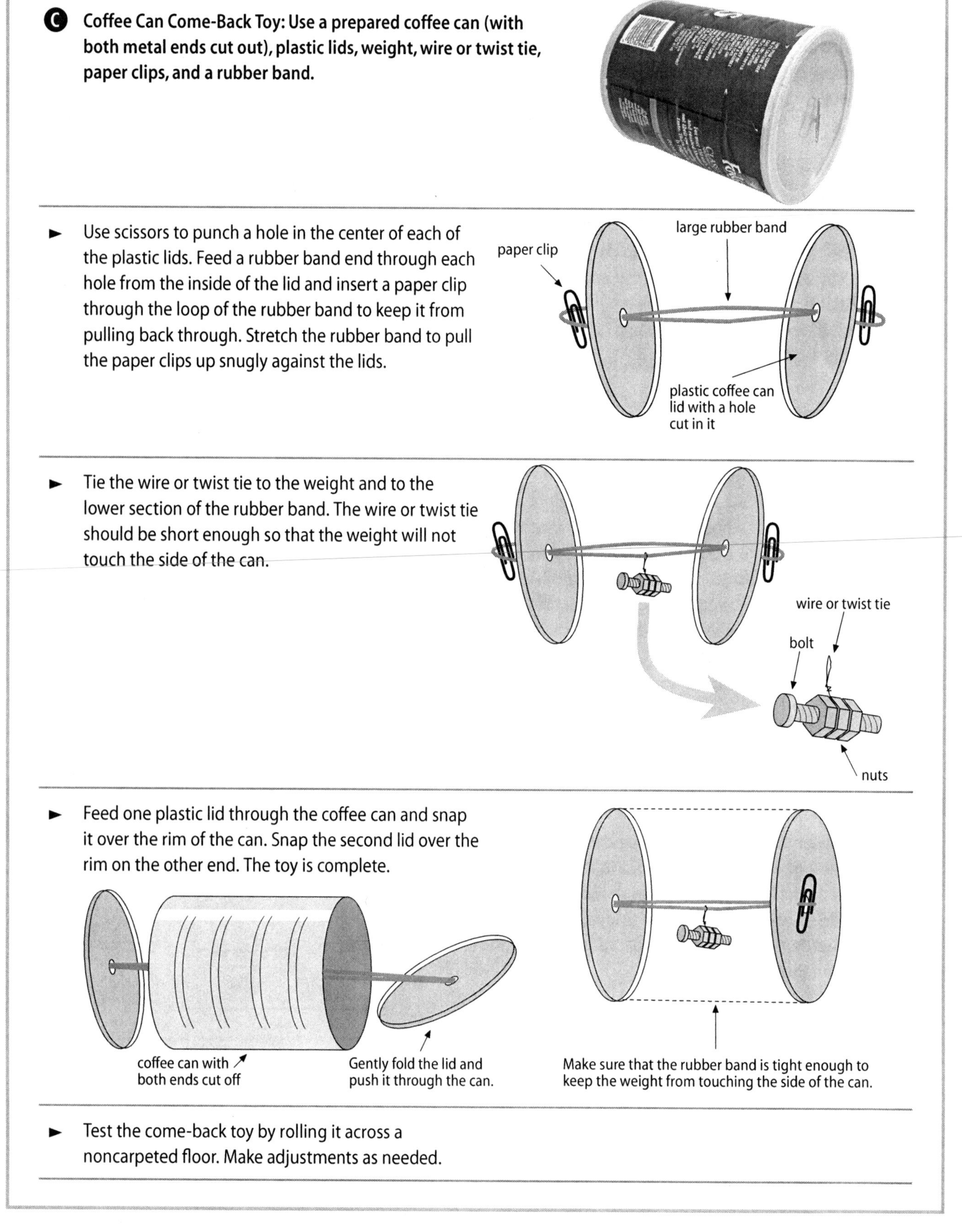

- Use scissors to punch a hole in the center of each of the plastic lids. Feed a rubber band end through each hole from the inside of the lid and insert a paper clip through the loop of the rubber band to keep it from pulling back through. Stretch the rubber band to pull the paper clips up snugly against the lids.

- Tie the wire or twist tie to the weight and to the lower section of the rubber band. The wire or twist tie should be short enough so that the weight will not touch the side of the can.

- Feed one plastic lid through the coffee can and snap it over the rim of the can. Snap the second lid over the rim on the other end. The toy is complete.

- Test the come-back toy by rolling it across a noncarpeted floor. Make adjustments as needed.

Bounceability

Students investigate bouncing balls to determine why some are better bouncers than others.

Grade Levels

Science activity appropriate for grades 3–7

Student Background

Students should already be familiar with toys that store elastic potential energy (in springs and rubber bands). The activities "Push and Go" and "Toy That Returns" in this book cover this concept.

Time Required

Setup	5 minutes
Part A	45 minutes
Part B	45 minutes
Cleanup	5 minutes

Assessment time is not included.

Key Science Topics

- elastic potential energy
- gravitational potential energy
- kinetic energy

National Science Education Standards Overview

See *www.terrificscience.org/physicsez/* for details of how these standards relate to the activity.

Science as Inquiry

Abilities Necessary to Do Scientific Inquiry

K–4 *Plan and conduct a simple investigation.*
K–4 *Employ simple equipment and tools to gather data and extend the senses.*
K–4 *Use data to construct a reasonable explanation.*
K–4 *Communicate investigations and explanations.*

5–8 *Design and conduct a scientific investigation.*
5–8 *Use appropriate tools and techniques to gather, analyze, and interpret data.*
5–8 *Develop descriptions, explanations, predictions, and models using evidence.*
5–8 *Think critically and logically to make the relationships between evidence and explanations.*
5–8 *Communicate scientific procedures and explanations.*
5–8 *Use mathematics in all aspects of scientific inquiry.*

Physical Science

K–4 *Properties of objects and materials*
K–4 *Position and motion of objects*

5–8 *Motions and forces*
5–8 *Transfer of energy*

Materials

For the Procedure

Activity Introduction, per class

- one or more stored energy toys (such as pop-up toy or pull-back car)

For supply source suggestions, see www.terrificscience.org/supplies/.

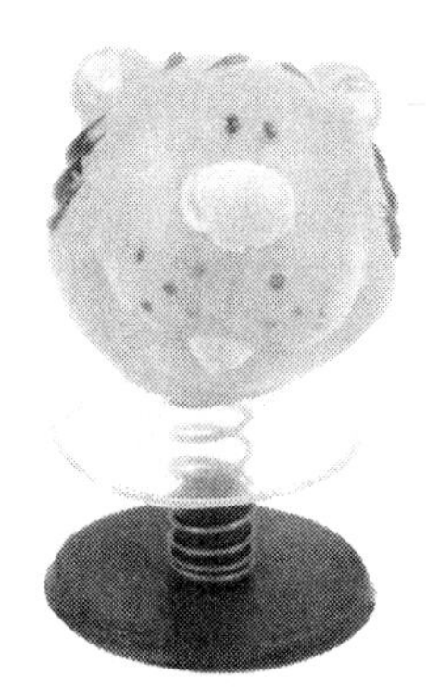

Part A, per group

- different kinds of balls (such as tennis, golf, rubber, Styrofoam®, hi-bounce, hard plastic, and racquetball)

The exact number of balls is not important.

- meterstick
- graph paper

Graph paper masters for copying are available at www.terrificscience.org/physicsez/.

- small ball of clay

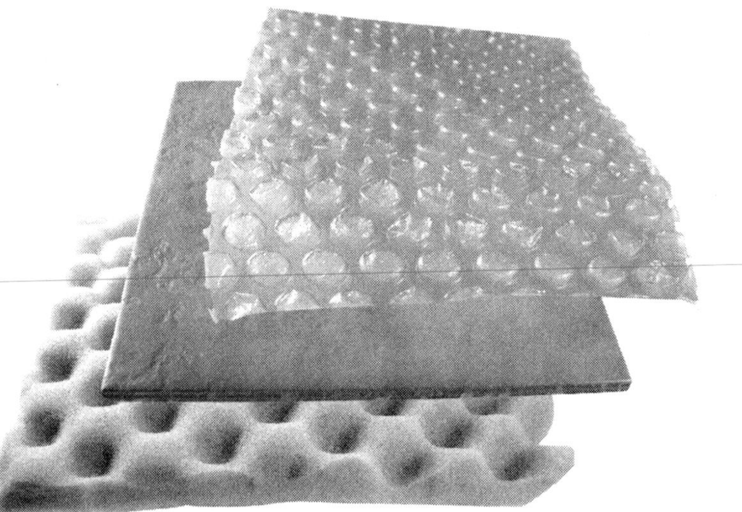

Part B, per group

All materials listed for Part A, plus

- different surfaces to bounce the balls on (such as carpet, grass, floor tile, ceiling tile, metal plate, corrugated cardboard, cork, foam pad, and Styrofoam)

The exact number of surfaces is not important.

For the Assessment

Assessment A, per group

- soft, pliable ball (such as polyfoam stress ball, squeezable therapeutic ball, or Koosh® ball)

For supply source suggestions, see www.terrificscience.org/supplies/.

- meterstick

Safety and Disposal

No special safety or disposal procedures are required.

Getting Ready

You may want to ask students to bring balls from home for use in this activity.

Procedure

This section provides teacher notes corresponding to each step of the student procedure. The procedure without teacher notes is included in the reproducible Student Notebook pages at the end of this activity and at www.terrificscience.org/physicsez/.

Student Procedure	Teacher Notes
Activity Introduction	• Review the concepts of kinetic and potential energy using one or more stored-energy toys. For instance, demonstrate a pop-up toy and have students describe the energy transformations that take place. • Tell students that a ball seems to be the simplest possible stored-energy toy; however, it is really a lot more complicated than they might think. Through discussion, review the energy changes that a bouncing ball goes through. (See Explanation.) ∘ Before the ball is dropped, it has an amount of gravitational potential energy that depends on its height above the floor. ∘ As the ball falls, this potential energy is gradually turned into kinetic energy. ∘ On the way back up, the kinetic energy is being turned back into gravitational potential energy.
Part A: Bouncing Balls	
(1) **Your teacher has given you a set of different types of balls. Look at and feel the balls, but don't bounce them yet.** **What variables do you think might affect how high a ball bounces?** **Which ball do you think is the best bouncer and which ball is the worst bouncer? Explain your choices.**	Have students share their ideas. Depending on your students' experience, you may want to define the term variable. Possible variables include the material the ball is made of, its temperature, whether it is dropped or thrown, the height from which it is dropped, and the surface that it hits.

	Student Procedure	Teacher Notes
②	Work with your group to design an experiment to determine which ball is the best bouncer. Measure only the first bounce and include at least three trials per ball. ✎ On another piece of paper, write your hypothesis and experimental design. Create a data table.	Before students begin brainstorming, introduce or review the characteristics of good experiments. Depending on their experience, students may benefit from using the Experiment Planning Guide provided at *www.terrificscience.org/physicsez/*.
③	Perform the experiment after your teacher reviews and approves your experimental design and data table. ✎ Record the results in your data table. Calculate the average bounce height for each ball.	A reasonable plan is to hold a meterstick upright with its zero end resting on the floor. Drop each ball in front of the meterstick from a standard height (such as 1 m) measured from a standard position on the ball (say, the bottom) onto a hard surface, and then "eyeball" the rebound height when the ball bounces. Students should do at least three trials per ball and calculate average values. (See Sample Answers for example data.)
④	On graph paper, make a graph of your results. ✎ Which ball is the best bouncer? How did you draw your conclusion? ✎ How do your results compare with your prediction of which ball is the best bouncer?	So groups can more easily compare their results in step 5, you may want to decide on a common format for the graphs, including the order in which the different balls are listed. (See Sample Answers for an example graph.)
⑤	Share and compare your results with the class.	The results of different groups are unlikely to be identical. In some cases, a group's results may deviate greatly from the rest of the class. Have students discuss possible reasons for these variations. Reasons may include: • **poor experimental design** (for example, too many variables) • **faulty materials** (for example, lopsided balls) • **faulty execution of procedure** (for example, throwing the balls down rather than dropping them) • **errors in collecting and recording data** (for example, misreading the meterstick)

	Student Procedure	Teacher Notes
6	After discussing class results, answer the following questions. ✎ How do your results compare to the results of other groups? ✎ Do any of the balls bounce back to their original heights? Why or why not? ✎ Why do you think some balls bounce better than others?	Help students realize that a ball can only return to its original height if it regains all of the potential energy it starts with. If a ball does not return to its original height, some of the energy must have gone elsewhere. (See Explanation.) Students should be able to conclude that some balls bounce better than others because they recover more potential energy after the bounce.
7	Making a model will help you learn more about why balls bounce. ✎ Roll some clay into a ball. Draw the ball on another piece of paper. ✎ Drop the ball of clay onto a hard surface. Draw what the ball looks like now. ✎ Drop one of the other balls. Does the ball look any different after you've dropped it? Why or why not?	See Sample Answers for examples of student drawings. If students could see a slow motion movie of a rubber ball dropping, they would see it gets a bit flatter just as it hits the surface, much like the ball of clay. Then, after the ball bounces back up, it regains its original shape. It does not stay flattened as the clay does. Explain that much of the difference in how balls bounce is a result of how much they flatten, or deform, upon hitting the ground and, even more importantly, how fast they regain their original shape. You may want to discuss the energy changes associated with this change in shape. (See Explanation.) You may want to have students try using a digital camera to take pictures at the exact moment a ball hits the floor in order to see the ball flatten out. This will be challenging, but easier if the camera has a setting for capturing fast action. Also, a softer ball will be easier to capture.
	Part B: Effect of Different Surfaces	
1	Your teacher has given you some different surface materials and a ball. Look at and feel the surfaces, but don't bounce the ball on them yet. ✎ On which surface do you think your ball will bounce the best? Explain your answer.	Have students use the balls from Part A, but assign a different ball to each group.

Student Procedure	Teacher Notes
② Work with your group to design an experiment to determine how different surfaces affect how high your ball bounces. Measure only the first bounce and include at least three trials per surface. ✎ On another piece of paper, write your experimental design. Create a data table.	
③ Perform the experiment after your teacher reviews and approves your experimental design and data table. ✎ Record the results in your data table. Calculate the average bounce height for each surface.	This experiment can be set up similarly to the experiment designed in Part A. (See Sample Answers for example data.)
④ On graph paper, make a graph of your results. ✎ How do different surfaces affect how high your ball bounces? How did you draw your conclusion? ✎ How do your results compare with your prediction of which surface would produce the best bounce?	So groups can more easily compare their results in step 5, you may want to decide on a common format for the graphs, including the order in which the different surfaces are listed. You may want to have groups make large graphs on chart paper and post them around the room to facilitate comparisons in the next step. (See Sample Answers for an example graph.)
⑤ Share and compare your results with the class; then, answer the following question. ✎ Why do balls bounce better on some surfaces than others?	Students should be able to conclude that a ball bounces better on some surfaces than others because the ball recovers more potential energy on some surfaces after the bounce.
Class Discussion	Ask students to discuss their observations with the class. Some of the following differences and similarities among groups' results may emerge: • The best bouncer on one surface may not be the best bouncer on every surface. • Students will see little variation from one ball to another on the foam. You can explain that the foam pad deforms easily, so the stiffness of the ball doesn't matter much. • Students should observe the biggest variation on the metal plate because the metal deforms very little.

Student Procedure	Teacher Notes
Assessment A	
1 Your teacher has given you a ball to experiment with. ✎ Based on your results with the other types of balls, predict how high this ball will bounce on a hard surface. ✎ Explain the reasons for your prediction.	Students should be able to predict that this ball will not bounce as high as all or most of the balls they've observed because it will not regain its original shape very well. Therefore, it will not recover as much potential energy.
2 Following the same experimental procedure that you used with the different types of balls in Part A, test the bounce height of the new ball. Observe the ball carefully when it hits the surface. ✎ How high does the ball bounce? ✎ Describe how the ball looks when it hits the surface. ✎ How does your result compare with your prediction?	Soft, pliable balls don't bounce as high as other balls and may not bounce at all. When dropped, they flatten on impact and regain their shape slowly or not at all. Therefore, little or no kinetic energy is returned to the ball after it hits the surface.
Assessment B	
1 The graphs show the bounce heights for two groups of balls. ✎ What question do you think the scientist was trying to answer by doing this experiment? ✎ What did the scientist find out? ✎ Explain the results of this experiment.	Students may realize that the experiment was done to determine the effect of temperature on bounce heights. The graphs show that the tennis and Styrofoam balls did not bounce as high at the colder temperature. (See Explanation.)
Assessment C	
1 Look at the graph below. ✎ What happens to the bounce heights as the number of bounces increases? ✎ Explain why the bounce heights change.	Students may realize that the bounce heights decrease because part of the balls' energy is lost during each bounce. Depending on their experience, some students may say this energy is lost because of friction, sound waves, and vibrational kinetic energy (thermal energy).

Sample Answers

Where students follow the same procedure using the same materials, these answers are close to answers you can expect. Where students design their own experiment or model, students' results will vary.

Type of Ball	First Bounce Height on Table			
	Trial 1	Trial 2	Trial 3	Average
tennis	55 cm	49 cm	51 cm	52 cm
golf	77 cm	74 cm	73 cm	75 cm
Styrofoam	30 cm	30 cm	36 cm	32 cm

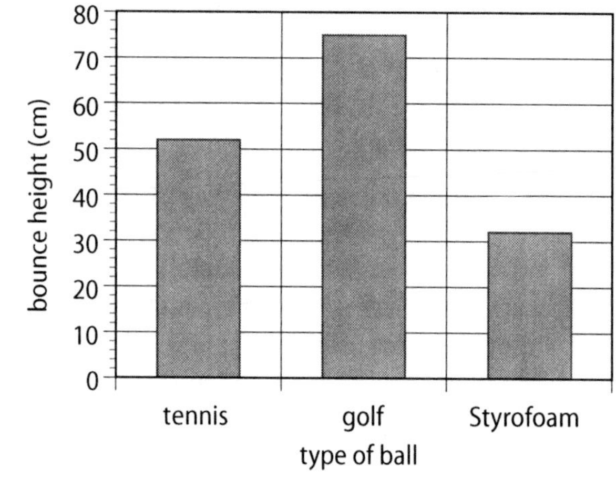

Examples of data and graph for Part A

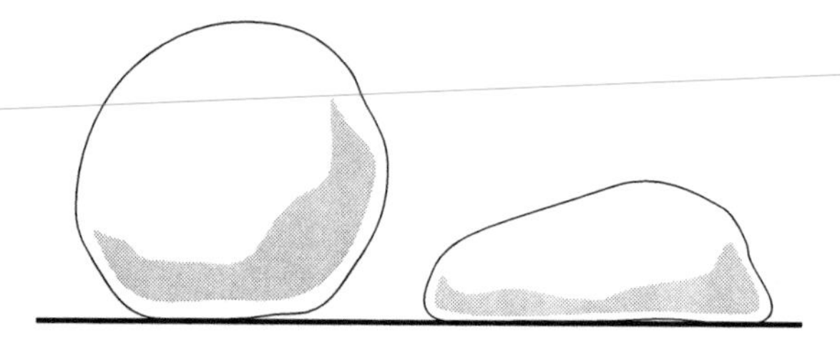

Examples of student drawings for Part A

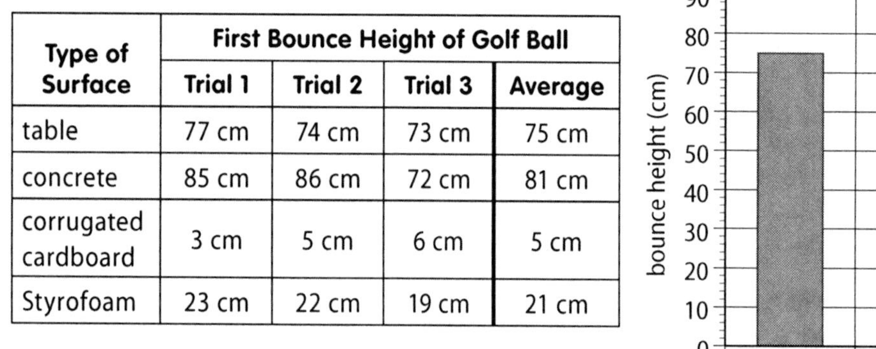

Type of Surface	First Bounce Height of Golf Ball			
	Trial 1	Trial 2	Trial 3	Average
table	77 cm	74 cm	73 cm	75 cm
concrete	85 cm	86 cm	72 cm	81 cm
corrugated cardboard	3 cm	5 cm	6 cm	5 cm
Styrofoam	23 cm	22 cm	19 cm	21 cm

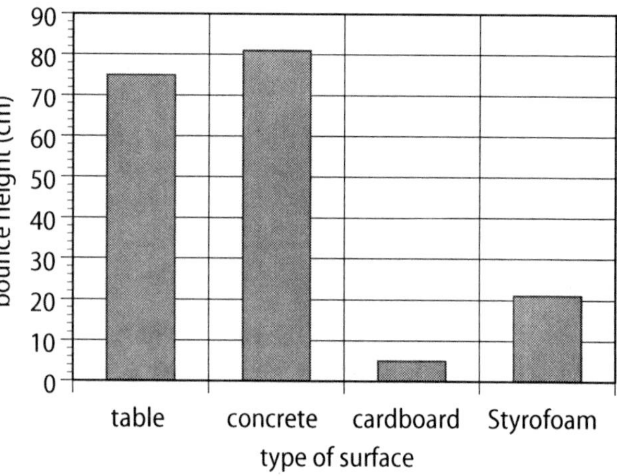

Examples of data and graph for Part B

Explanation

This section is intended for teachers. Modify the explanation for students as needed.

Energy Conversions

Objects can store energy in many ways. A stretched or twisted rubber band stores elastic potential energy. The stored chemical energy in foods and fuels is a form of potential energy. In this activity, students explore forms of stored energy called gravitational potential energy and elastic potential energy.

Imagine lifting a ball off a surface and holding it some distance above that surface. The product of the force used to lift the ball and the lift distance is the work done on the ball. Once you stop lifting the ball and hold it still, the energy your work added to the ball is now gravitational potential energy stored in the ball.

Before a ball is dropped, it has an amount of gravitational potential energy that depends on the ball's weight and its height above the surface. As the ball falls, this gravitational potential energy is gradually turned into kinetic energy. The greater the ball's speed as it falls, the greater the kinetic energy. When the ball collides with the surface, some of this kinetic energy is stored as elastic potential energy in the ball and the surface. The particles in the ball and the surface stretch and squeeze together like tiny springs. On the way back up, the kinetic energy is being turned back into gravitational potential energy. At the top of the bounce, the ball's energy is once again all gravitational potential energy. Since the ball does not bounce back to its starting height, it has less gravitational potential energy than it had before being dropped.

Since energy cannot be "lost," the missing energy must go into some other form of energy. In our dropped ball example, the energy transfer happens during the interaction between the ball and the surface. Part of the energy goes into sound waves that are produced when the ball hits the surface. Part of the energy is converted to vibrational kinetic energy that may be felt as an increase in the temperature of the ball and the surface. If you throw down a ball of clay several times in rapid succession, the clay may begin to feel warmer.

Bounce Height

What determines how high different balls bounce on the same surface? Much of the difference is a result of how much the balls deform and, even more importantly, how fast they regain their original shape. During the bounce, the shape of the ball changes. This shape change takes energy, just as stretching a rubber band does. Flattening the ball is similar to compressing a spring. The energy of compression (elastic potential energy) converts to kinetic energy as the shape goes back to normal. The clay doesn't bounce well because it stays deformed. If the ball is still partially deformed after it leaves the surface, the energy that was stored in that deformation does not return to kinetic energy of the ball even though the ball later returns to its original shape.

What determines how high the balls bounce on different surfaces? During the bounce, both the shape of the ball and the shape of the surface are deformed. The height of the bounce is determined by how much energy of compression is returned as the shape of both the ball and the surface go back to normal. Each ball type and surface type interact differently, producing a unique result. Even so, some surfaces produce fairly consistent results with all types of balls. For example, all of the balls bounce on the foam pad because the foam deforms rather than the ball, acting much like a trampoline. In contrast, if the surface stays deformed as the Styrofoam surface may, then the energy that went into causing the deformation does not return to the ball.

Elastic and Inelastic Collisions

In physics, we talk about two basic types of collisions: elastic and inelastic. An elastic collision is one in which two objects collide and no energy is transformed into permanent deformation of the objects and no thermal energy is produced. An example of a perfectly elastic collision would be a dropped ball rebounding to the same height from which it was dropped. In reality, there are no perfectly elastic collisions. A Super Ball® that rebounds to over 90% of its original height comes fairly close, and billiard balls colliding with one another come even closer.

An inelastic collision is one in which some of the initial kinetic energy is converted into thermal energy. Some examples are balls bouncing or being kicked, cars colliding at low speeds, and golf balls being struck by golf clubs. Most collisions are inelastic.

Sometimes a third type of collision is discussed in physics books. This type is really just a subcategory of inelastic collisions called perfectly (or completely) inelastic. In a perfectly inelastic collision, the objects stick together after the collision. Examples include cars that collide and become entangled, a football player who tackles another player, and a bullet that hits and stays inside a target.

Elasticity can arise from both the springiness of air inside the ball and the springiness of the ball itself. Examples of air-filled balls include tennis balls, volleyballs, and footballs. Solid balls, such as baseballs and golf balls, rely solely on the ball's material for elasticity.

Temperature has an effect on a ball's elasticity. In general, cold balls bounce less than warm ones. For example, cold rubber is usually less flexible than warm rubber. This lack of flexibility causes more of the energy to go into making the molecules vibrate and less into elastic potential energy.

Cross-Curricular Integration

Language arts:

- Read aloud or suggest that students read the following book:
  - *Sports Science Projects: The Physics of Balls in Motion*, by Madeline Goodstein (grades 4–7)
    This book encourages exploration as readers learn about the design and performance of different balls used in sports.

Physical education:

- Discuss why different kinds of balls are used in different sports.

Reference

Exploratorium Website. That's the Way the Ball Bounces. http://www.exploratorium.edu (accessed February 22, 2005).

Part A: Bouncing Balls

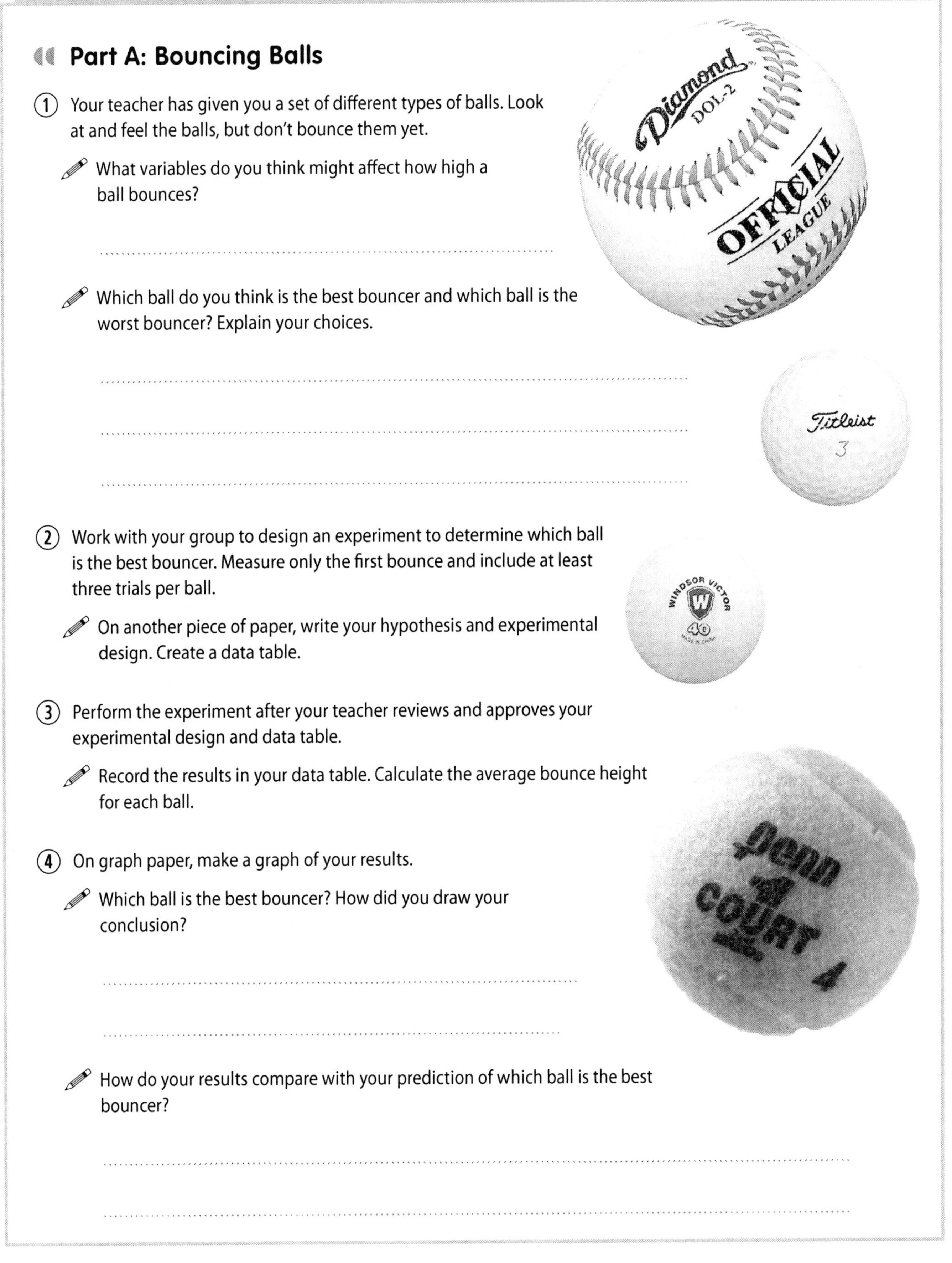

① Your teacher has given you a set of different types of balls. Look at and feel the balls, but don't bounce them yet.

✎ What variables do you think might affect how high a ball bounces?

....

✎ Which ball do you think is the best bouncer and which ball is the worst bouncer? Explain your choices.

....

....

....

② Work with your group to design an experiment to determine which ball is the best bouncer. Measure only the first bounce and include at least three trials per ball.

✎ On another piece of paper, write your hypothesis and experimental design. Create a data table.

③ Perform the experiment after your teacher reviews and approves your experimental design and data table.

✎ Record the results in your data table. Calculate the average bounce height for each ball.

④ On graph paper, make a graph of your results.

✎ Which ball is the best bouncer? How did you draw your conclusion?

....

....

✎ How do your results compare with your prediction of which ball is the best bouncer?

....

....

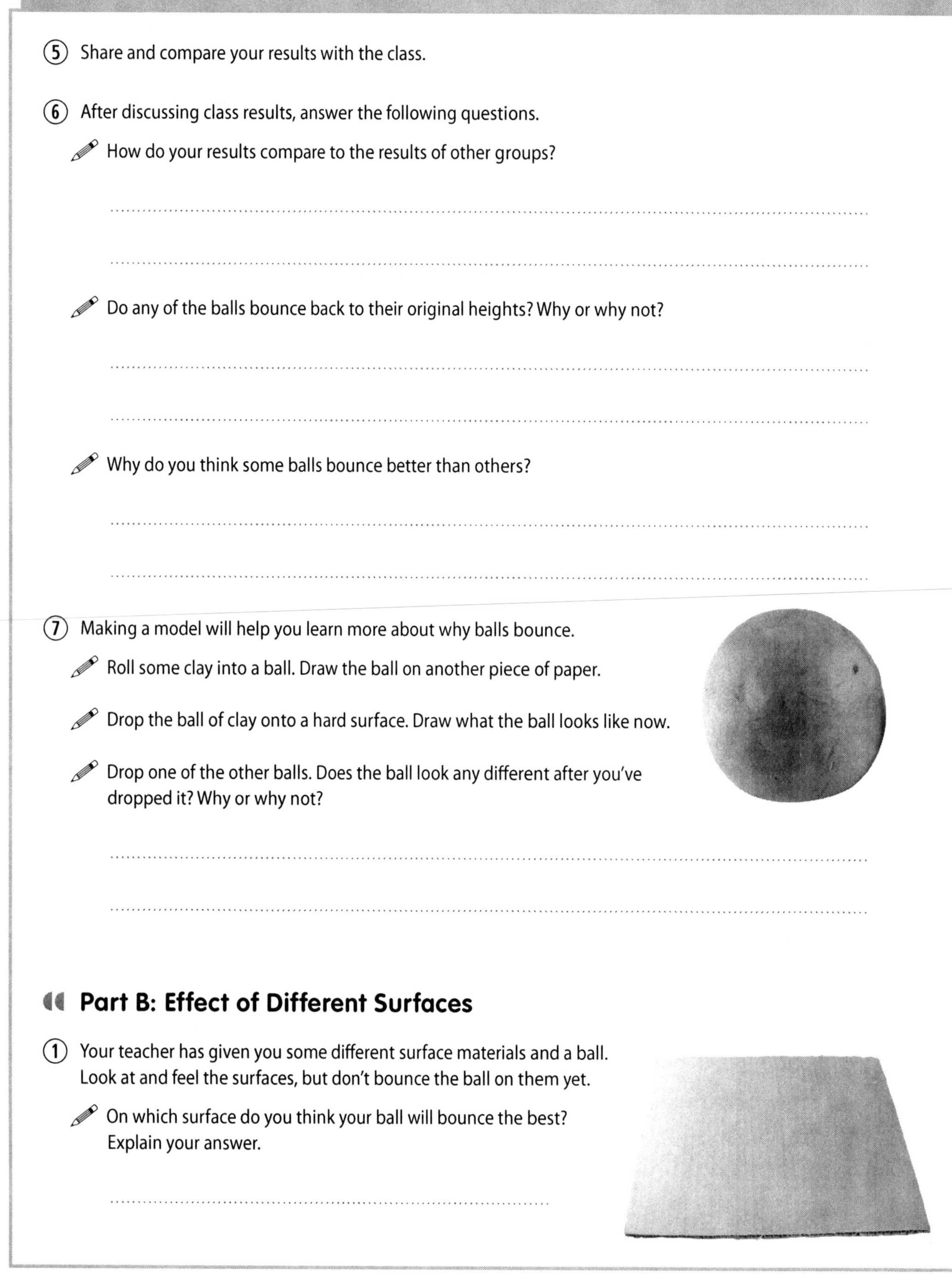

⑤ Share and compare your results with the class.

⑥ After discussing class results, answer the following questions.

✎ How do your results compare to the results of other groups?

....

....

✎ Do any of the balls bounce back to their original heights? Why or why not?

....

....

✎ Why do you think some balls bounce better than others?

....

....

⑦ Making a model will help you learn more about why balls bounce.

✎ Roll some clay into a ball. Draw the ball on another piece of paper.

✎ Drop the ball of clay onto a hard surface. Draw what the ball looks like now.

✎ Drop one of the other balls. Does the ball look any different after you've dropped it? Why or why not?

....

....

Part B: Effect of Different Surfaces

① Your teacher has given you some different surface materials and a ball. Look at and feel the surfaces, but don't bounce the ball on them yet.

✎ On which surface do you think your ball will bounce the best? Explain your answer.

....

② Work with your group to design an experiment to determine how different surfaces affect how high your ball bounces. Measure only the first bounce and include at least three trials per surface.

✎ On another piece of paper, write your experimental design. Create a data table.

③ Perform the experiment after your teacher reviews and approves your experimental design and data table.

✎ Record the results in your data table. Calculate the average bounce height for each surface.

④ On graph paper, make a graph of your results.

✎ How do different surfaces affect how high your ball bounces? How did you draw your conclusion?

....

....

....

....

✎ How do your results compare with your prediction of which surface would produce the best bounce?

....

....

⑤ Share and compare your results with the class; then, answer the following question.

✎ Why do balls bounce better on some surfaces than others?

....

....

....

....

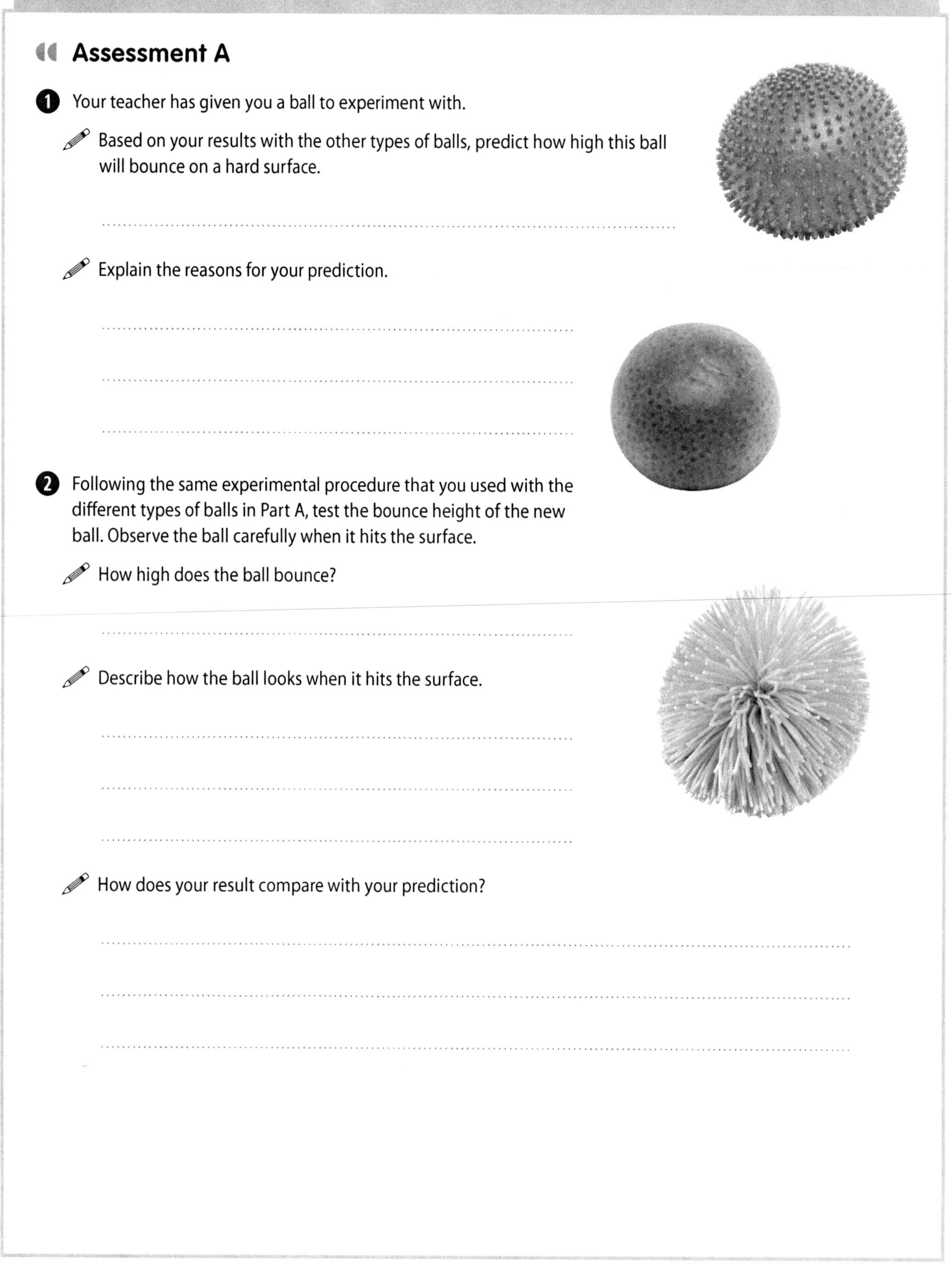

Assessment A

1 Your teacher has given you a ball to experiment with.

Based on your results with the other types of balls, predict how high this ball will bounce on a hard surface.

....

Explain the reasons for your prediction.

....

....

....

2 Following the same experimental procedure that you used with the different types of balls in Part A, test the bounce height of the new ball. Observe the ball carefully when it hits the surface.

How high does the ball bounce?

....

Describe how the ball looks when it hits the surface.

....

....

....

How does your result compare with your prediction?

....

....

....

Assessment B

1 The graphs show the bounce heights for two groups of balls.

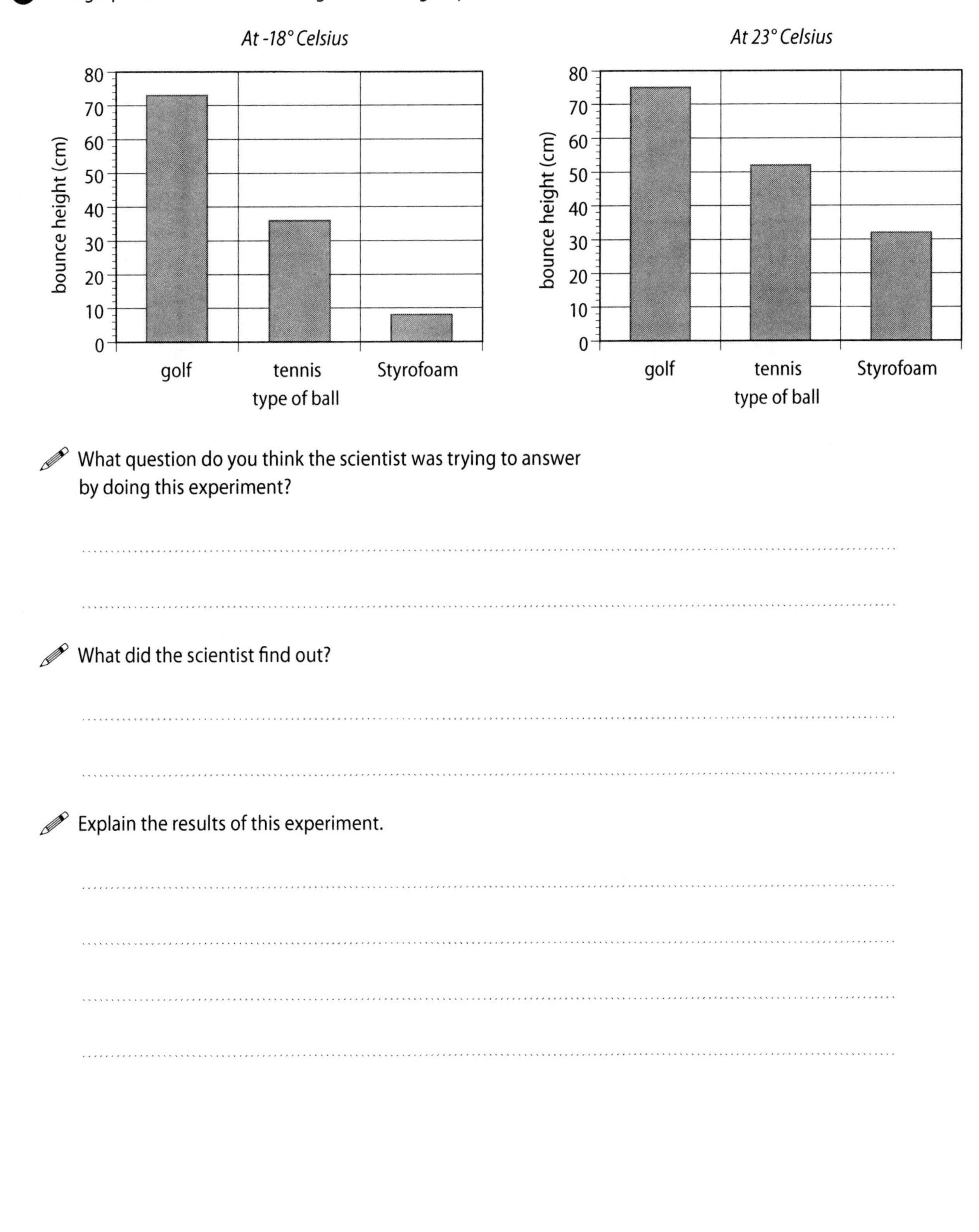

What question do you think the scientist was trying to answer by doing this experiment?

....

....

What did the scientist find out?

....

....

Explain the results of this experiment.

....

....

....

....

Assessment C

❶ Look at the graph below.

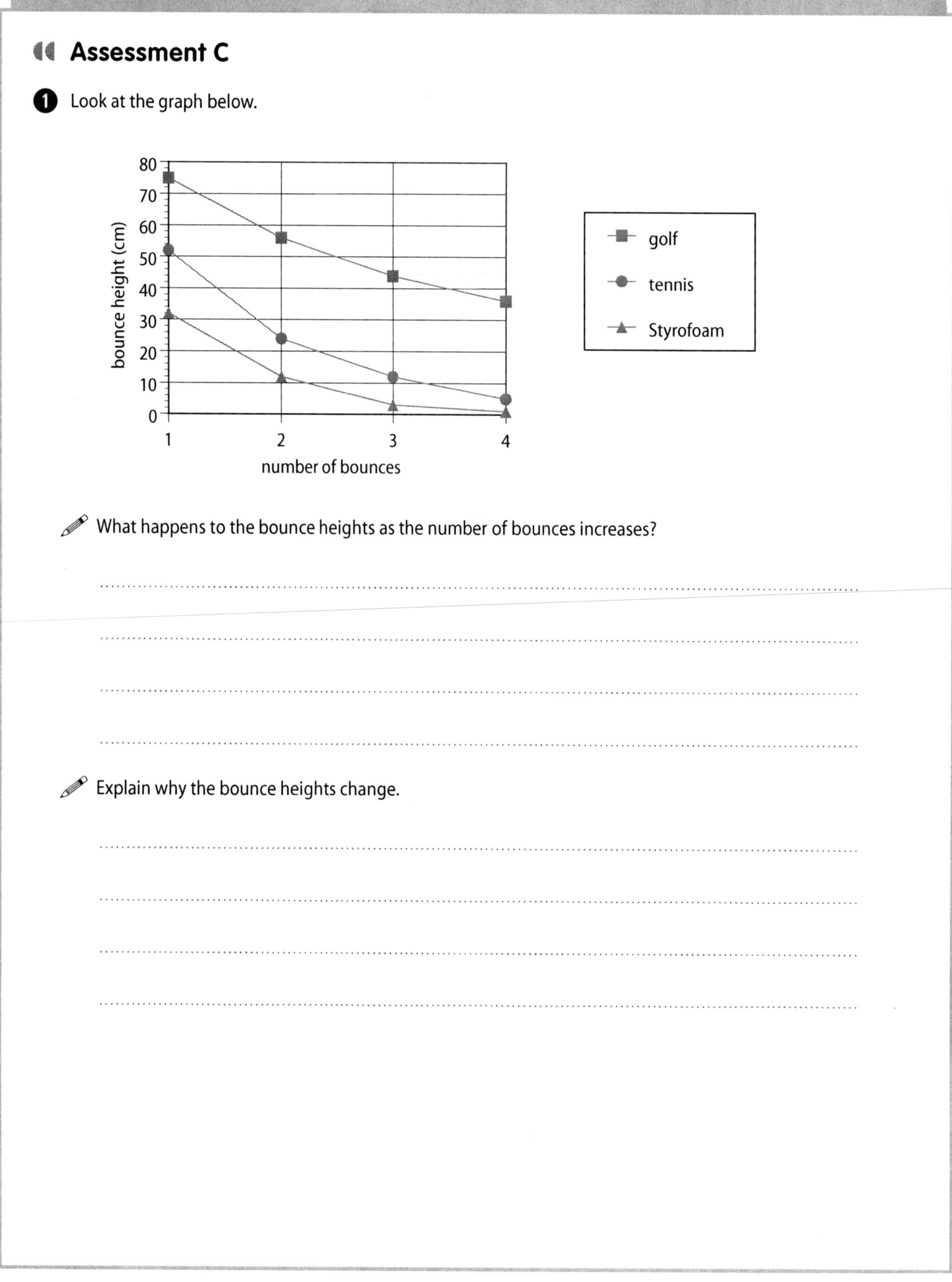

What happens to the bounce heights as the number of bounces increases?

....

....

....

....

Explain why the bounce heights change.

....

....

....

....

Sound Off

Let's rock 'n' roll! Students use rubber bands to make a toy instrument and explore sound.

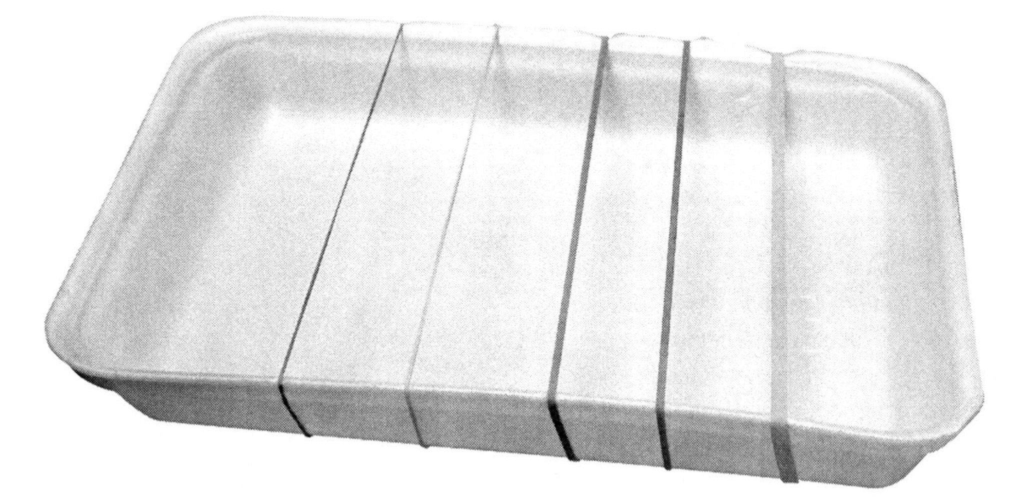

Grade Levels

Science activity appropriate for grades 3–5

Student Background

Students require no particular background preparation for this activity.

Time Required

Setup	10	minutes
Procedure	20	minutes
Cleanup	5	minutes

Assessment time is not included.

Key Science Topics

- frequency
- longitudinal wave
- pitch
- sound
- vibration

National Science Education Standards Overview

See *www.terrificscience.org/physicsez/* for details of how these standards relate to the activity.

Science as Inquiry

Abilities Necessary to Do Scientific Inquiry

K–4 *Plan and conduct a simple investigation.*
K–4 *Use data to construct a reasonable explanation.*
K–4 *Communicate investigations and explanations.*

5–8 *Design and conduct a scientific investigation.*
5–8 *Develop descriptions, explanations, predictions, and models using evidence.*
5–8 *Think critically and logically to make the relationships between evidence and explanations.*
5–8 *Communicate scientific procedures and explanations.*

Physical Science

K–4 *Properties of objects and materials*
K–4 *Position and motion of objects*

5–8 *Motions and forces*
5–8 *Transfer of energy*

Materials

For the Procedure

Part A, per student

- 12 inch (approximately 30 cm) ruler
- flexible meterstick or yardstick

Thick metersticks are not flexible enough to work.

Part B, per class

- Slinky®

Part B, per student

- 4–5 rubber bands with different lengths and widths
- clean Styrofoam® meat tray

Use any size tray that the rubber bands will easily fit around. If the rubber bands fit too tightly, the tray might bend or crack. Although other open containers (such as pencil boxes) will work, Styrofoam trays provide excellent sound amplification.

For the Assessment

Assessment A, per group

- rubber band
- scissors
- tape
- bowl that plastic food wrap will cling to
- plastic food wrap (such as Glad® Cling Wrap)

Assessment B, per class

- (optional) guitar

Safety and Disposal

No special safety or disposal procedures are required.

Procedure

This section provides teacher notes corresponding to each step of the student procedure. The procedure without teacher notes is included in the reproducible Student Notebook pages at the end of this activity and at www.terrificscience.org/physicsez/.

	Student Procedure	Teacher Notes
	Part A: Making Sounds	
①	Place a ruler flat on a table. Let about 4 inches (about 10 cm) of the ruler extend over the edge of the table. Hold the ruler securely to the table with one hand and pluck the end of the ruler that sticks out over the end of the table with the other hand. (Be careful not to pluck so hard that you risk breaking the ruler.) ✎ Record what you see and hear.	Students observe that the ruler vibrates up and down and makes a sound.
②	Try to cause the ruler to make different sounds. ✎ Record what you did and your observations. ✎ Do you notice a pattern in the sounds you made? ✎ On another piece of paper, draw a picture of what you think the ruler is doing when it makes the sound.	Students can make different sounds by allowing different lengths of ruler to extend over the edge of the table as they pluck. They should observe that the pitch of the sound becomes higher as the ruler gets shorter and lower as it gets longer. See Sample Answers for student drawing example.
③	Think about what will happen if you repeat step 1 with a meterstick and let more of it hang over the edge of the table. ✎ Predict what will happen. ✎ Try the experiment and record your observations. ✎ Compare your results to your prediction.	Based on what they observed in step 2, some students should be able to predict that a longer stick will make a lower-pitched sound.
	Part B: Making Music	
①	Stretch a rubber band between your fingers and pluck it. Pay attention to the sound it makes. ✎ Record your observations.	Students observe that the rubber band makes a soft sound. They should also be able to observe that the rubber band vibrates.

	Student Procedure	Teacher Notes
②	Wrap the same rubber band around a tray as shown. ✎ Predict what will happen when you pluck the rubber band. ✎ Pluck the rubber band and describe how this sound is different from the sound you made in step 1.	The rubber band wrapped around the tray makes a louder sound when plucked than the plain rubber band plucked in step 1. Students should also notice that the pitch is different than in step 1 since the rubber band is probably stretched farther.
③	Pluck the rubber band again, this time touching the back of the tray as you pluck. ✎ What do you feel? ✎ List some of the reasons the rubber band sounds different when on the tray.	Students might list several factors such as the hardness of the pluck or the amount the rubber band is stretched. Be sure they also realize that the tray acts as a sounding board to amplify the vibrations produced by the vibrating rubber band, resulting in a louder sound.
④	Pluck the rubber band as gently as possible and watch it carefully. Now pluck it much harder, and watch it again. ✎ What differences do you see and hear between the soft pluck and the harder pluck? ✎ On another piece of paper, draw pictures of the rubber band while it is making a soft sound and while it is making a louder sound.	Students should observe that the soft pluck produces a quieter sound and less vibration in the rubber band than the harder pluck does. See Sample Answers for student drawing example.
⑤	Pluck the rubber band again and observe it carefully. Now use one hand to pinch the rubber band about an inch from the edge of the tray. Continue to pinch the rubber band as you use the other hand to pluck again. ✎ What differences do you see and hear? ✎ On another piece of paper, draw what the rubber band looks like when you pinch it and then pluck.	Students should observe that the pitch of the sound is higher when the rubber band is pinched. They should also see that only part of the rubber band vibrates when it's pinched on one end. See Sample Answers for student drawing example.
⑥	What do you think would happen if you moved the place where you pinched the rubber band? ✎ Write your prediction. ✎ Try it by moving the pinched place to several locations. What do you observe? ✎ How is what you did here similar to what you did to cause the ruler to make different sounds?	Students should observe the pitch getting higher as they pinch off more and more of the rubber band. They should relate this to hanging less of the ruler over the edge of the table. In both cases, the pitch is higher when the vibrating part is shortened.

Student Procedure	Teacher Notes
⑦ Think about another way you could make the sound of the rubber band become higher or lower. TIP: Think about the way a musician tunes a guitar or violin. What's your idea? Describe what happens when you try your idea.	You may need to help students visualize that a musician tunes a stringed instrument using pegs that put more or less tension on the strings. They should be able to come up with ideas for tightening and loosening their rubber bands. This can work in several ways. • Stretch the rubber band more tightly across the top of the tray and pull the slack towards the back. Friction will keep the rubber band in place. You can do the opposite to loosen the rubber band on top. 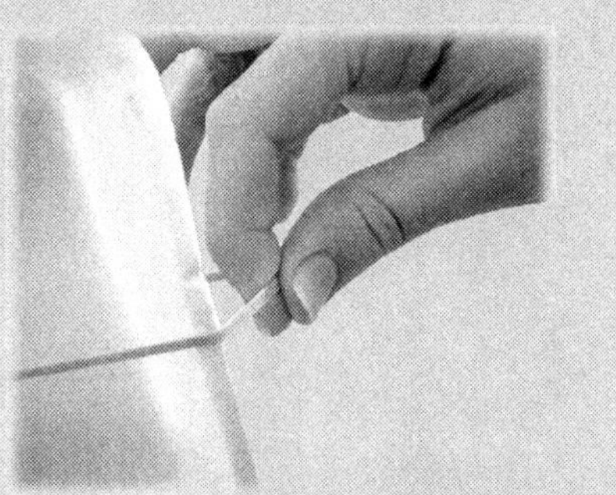 • Pull the sides of the rubber band away from the tray and put something between the rubber band and the tray to keep it tight. • Students may come up with other methods.
⑧ Select two, three, or more rubber bands having different thicknesses. Arrange them around your tray to make a simple guitar that produces the lowest sound at one end and the highest at the other. Pluck the rubber bands to test if your arrangement is correct and move them around if you need to. Play a simple song on your instrument.	Students should realize that rubber bands with different thicknesses, masses, and tensions vibrate differently and make different sounds.
Class Discussion	• Through discussion, relate pinching the rubber band to pressing a string against the neck of a guitar or violin to change the pitch. Explain that a shorter string vibrates more quickly than a longer one, producing a higher pitch. • Relate the tightening and loosening of the rubber bands to changing the tension on guitar strings with a tuning peg. Explain that a string that has more tension vibrates more quickly than a looser one of the same length, producing a higher pitch.

Student Procedure	Teacher Notes
Class Discussion (continued)	• Point out that the various rubber bands on the tray have different thicknesses and masses and are under different amounts of tension. These differences contribute to different frequencies of vibration and, therefore, different sounds. • The number of times something vibrates in a given period of time is its frequency. Scientists can measure the frequency of a sound with special equipment. When we hear a sound, we cannot tell what its frequency is, but we can tell how high or low it sounds to us. We use the word "pitch" to describe this highness or lowness. • Explain that the size of the vibration is called its amplitude. A soft pluck causes a small amplitude and a soft sound. A harder pluck causes a bigger amplitude and, therefore, a louder sound. • Now that students have explored sound, explain to them that the energy from a vibrating object (such as a plucked rubber band) transfers to gas particles. These gas particles pass their energy to adjacent particles, producing a longitudinal wave that we call a sound wave. • Illustrate a longitudinal wave by stretching a Slinky along the floor or a flat table. Have someone hold one end of the Slinky as you move the other end back and forth. A longitudinal wave will travel down the spring. Move one end of the Slinky back and forth. • Sound waves are variations in the density of the air just as the Slinky waves are variations in the distance between the coils.
Assessment A	
1 **Cut a rubber band and tape one end of it to a desk or table. Pull the rubber band until it is lightly stretched and pluck it. Pay attention to the sound it makes.** **Record your observations.**	

	Student Procedure	Teacher Notes
2	Seal the top of a bowl with plastic food wrap. Make sure the wrap is nice and tight across the top of the bowl. Tape one end of the rubber band to the center of the food wrap. ✎ Predict how the rubber band will sound if you pluck it now. Record your prediction and your reasons for it.	Students should predict that the rubber band will sound louder when attached to the food wrap because the food wrap will amplify the sound.
3	Pull the rubber band up until it is lightly stretched. Pluck the rubber band. ✎ What do you observe? ✎ How does your observation compare to your prediction? ✎ Explain what happens.	Students should recognize that the sound is amplified by the food wrap and bowl when the rubber band is plucked.
	Assessment B	
1	Look at the diagram of the guitar. If you have one available, compare the diagram to an actual guitar. ✎ On the diagram, mark which parts of the guitar create sound, amplify sound, and control pitch. Explain how these parts work.	See Sample Answers for an example of student labeling. Students should recognize that the strings create vibrations when plucked. These vibrations are transmitted to the body of the guitar, which amplifies the sound. The tuning pegs change pitch by increasing or decreasing string tension. Pitch is also changed by pressing on the frets. This changes the length of the vibrating portion of the string.

Sample Answers

Where students follow the same procedure using the same materials, these answers are close to answers you can expect. Where students design their own experiment or model, students' results will vary.

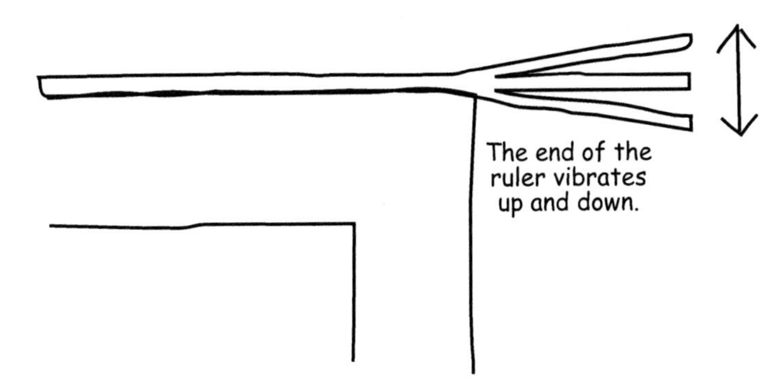

Example of student drawing for step 2 of Part A

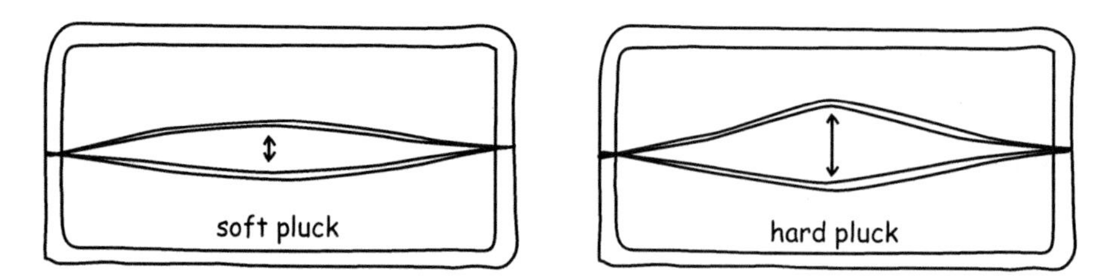

Example of student drawings for step 4 of Part B

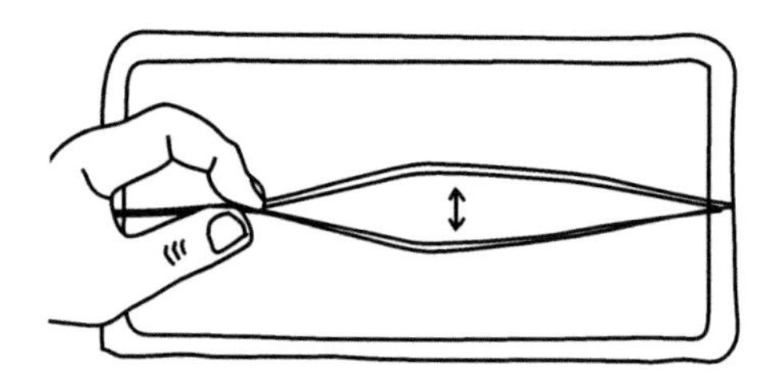

Example of student drawing for step 5 of Part B

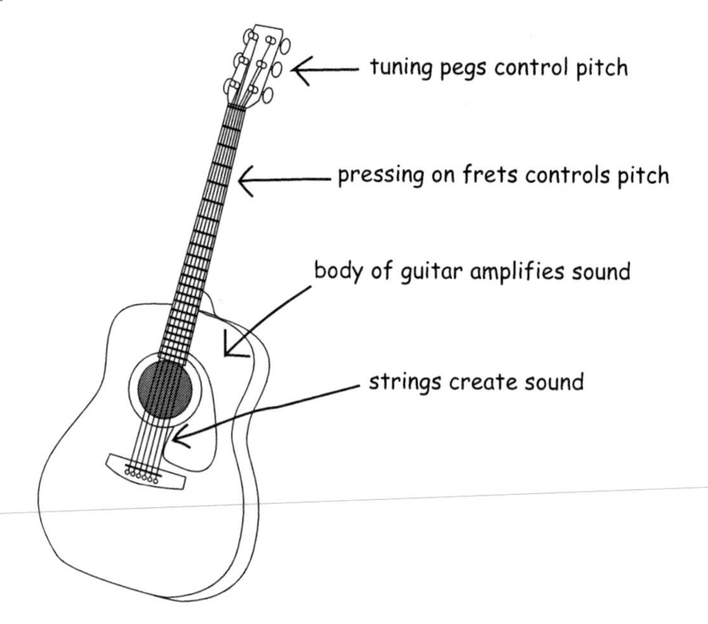

Example of student labeling for Assessment B

Explanation

This section is intended for teachers. Modify the explanation for students as needed.

When any object that can vibrate is disturbed, it will vibrate at its own special set of frequencies (rates of vibration), which together form its special sound. These frequencies are called the object's natural frequencies, and they depend on factors such as the material, elasticity, and shape of the object. Sound is produced when the energy from the vibrating object gets transferred to gas particles in the air that are moving freely and happen to collide with the vibrating object. These gas particles pass their energy to adjacent particles and eventually transmit the energy to the ear and cause the eardrum to vibrate, sending messages to the brain that let us know we have heard a sound. We give the name "music" to collections of sounds that are changed in a controlled manner and are considered pleasing by the listener. Of course, what is considered pleasing varies with culture, age, experience, and taste.

Musical instruments are built to produce pleasing sound vibrations through various methods. Stringed instruments, a large group of musical instruments that includes the violin family, guitars, banjos, harpsichords, and harps, produce sound when a string is caused to vibrate. Vibrations may be produced by drawing a bow across the strings (as with the violin family) or by plucking (as with guitars, banjos, harpsichords, and harps). Other parts of the instrument act as a soundboard to amplify the vibrations produced by the strings. (In the activity, students wrap rubber bands around a tray to amplify the sound.)

When playing a stringed instrument, different sounds (or notes) are produced by producing different rates of vibration in the strings. The rate of vibration is known as the frequency. Higher sounds are produced by faster rates of vibration (higher frequencies). Lower sounds are produced by slower rates of vibration (lower frequencies). The brain interprets frequency detected by the ear primarily in terms of a subjective quality called pitch. In a stringed instrument, many factors contribute to the creation of different pitches. The strings are under different amounts of tension and may have different masses. These differences contribute to different frequencies of vibration and therefore different notes. Changing the length of the string also changes the frequency of the vibration and, thus, the pitch. (See figure below.) Think about the pictures as representing the movement of the vibrating ruler.

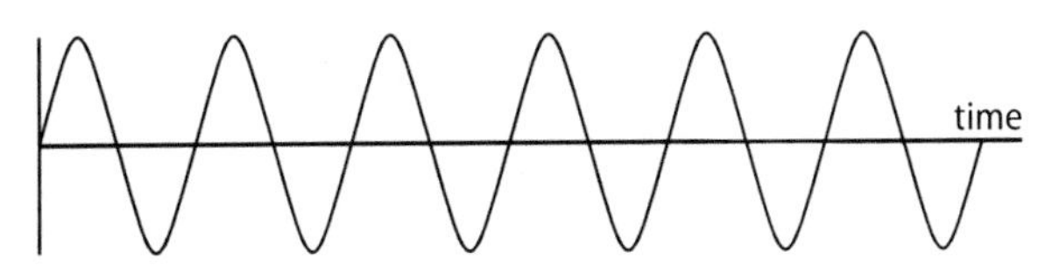

High frequency wave (higher pitch)

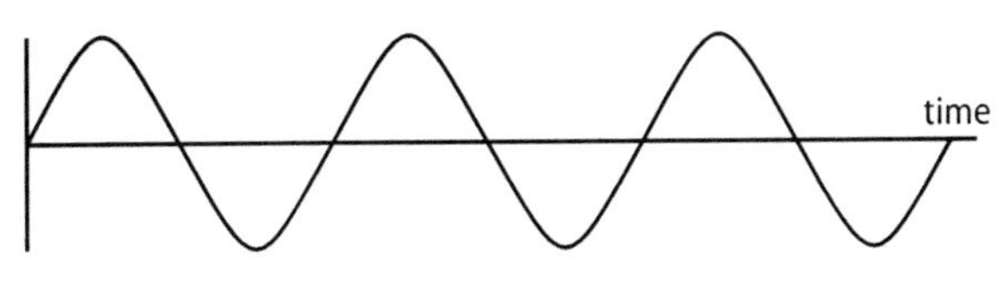

Low frequency wave (lower pitch)

The loudness or softness of a note can be changed by plucking or bowing the string more forcefully or more lightly. Plucking the string more forcefully makes it move a greater distance, increasing the amplitude (or height) of the vibration. When the string vibrates with greater amplitude, it passes more energy to surrounding air particles; thus, more energy reaches the ear. When the string is pulled only a small distance, it transfers less energy to the air particles and, consequently, to the ear.

Regardless of how forcefully strings are plucked or bowed, they are so small and thin that they don't impart their energy to very many air molecules. In order to produce a widely audible sound, they must be in firm contact with other parts of the instrument. The largest of these pieces is called the soundboard. Small vibrations of the string are transferred to the soundboard, where they are faithfully reproduced. The soundboard, being many times larger than the strings, imparts the motion to many times more air molecules. This makes a much louder sound than the strings alone can produce. When using a tuning fork, musicians can take advantage of this principle by striking the fork and holding the bottom end on a table or chair seat to amplify the sound.

Cross-Curricular Integration

Art:

- Have students design and make different kinds of musical instruments.

Language arts:

- Ask students to choose four different sounds: two that they like and two that they do not like. Have them use a web-type graphic organizer to record the following information: 1) name of the sound; 2) when they usually hear the sound; 3) kind of object, organism, or event producing the sound; 4) how they think the vibrations that create the sound are produced; and 5) why they like or dislike the sound. (See *www.terrificscience.org/physicsez/* for a reproducible master.) Have students use the completed graphic organizer as a guide to write a paragraph about each sound.
- Read aloud or suggest that students read the following books:
    - *Making Sounds,* by Julian Rowe and Molly Perham (grades 3–6)
      This book describes how different sounds are made and how they travel.
    - *Moses Goes to a Concert,* by Isaac Millman (grades K–3)
      Moses and his schoolmates, all deaf, attend a concert where the orchestra's percussionist is also deaf.

- *Rubber-Band Banjos and a Java Jive Bass: Projects and Activities on the Science of Music and Sound,* by Alex Sabbeth (grades 4–7)
  This book discusses scientific principles behind musical sounds, shows the construction of several musical instruments, and contains instructions for making some basic instruments.

Life science:

- Have students study how the ear works and what can cause hearing loss.

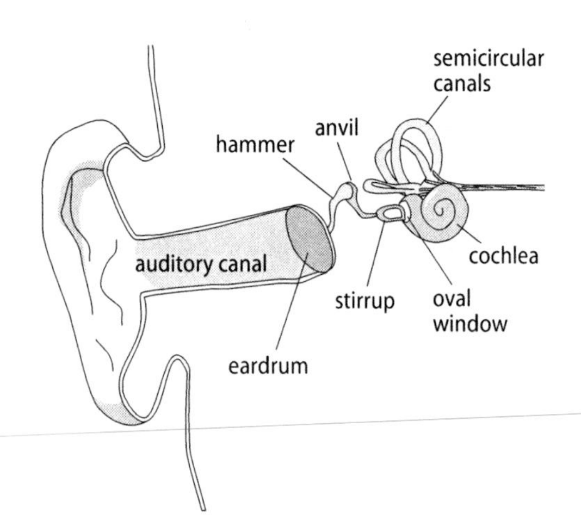

Components of the ear

Music:

- Have students study different types of instruments and discuss how they produce vibrations.
- Have students investigate the musical instruments created by different cultures in different historical periods.

References

Harlan, J. *Science Experiments for the Early Childhood Years,* 4th ed.; Merrill: Columbus, OH, 1988; pp 227–228.

Hewitt, P. *Conceptual Physics,* 9th ed.; Addison Wesley: San Francisco, 2002; pp 362–410.

Steinway & Sons Website. Technical Info. The Soundboard. http://www.steinway.com (accessed February 22, 2005).

Part A: Making Sounds

① Place a ruler flat on a table. Let about 4 inches (about 10 cm) of the ruler extend over the edge of the table. Hold the ruler securely to the table with one hand and pluck the end of the ruler that sticks out over the end of the table with the other hand. (Be careful not to pluck so hard that you risk breaking the ruler.)

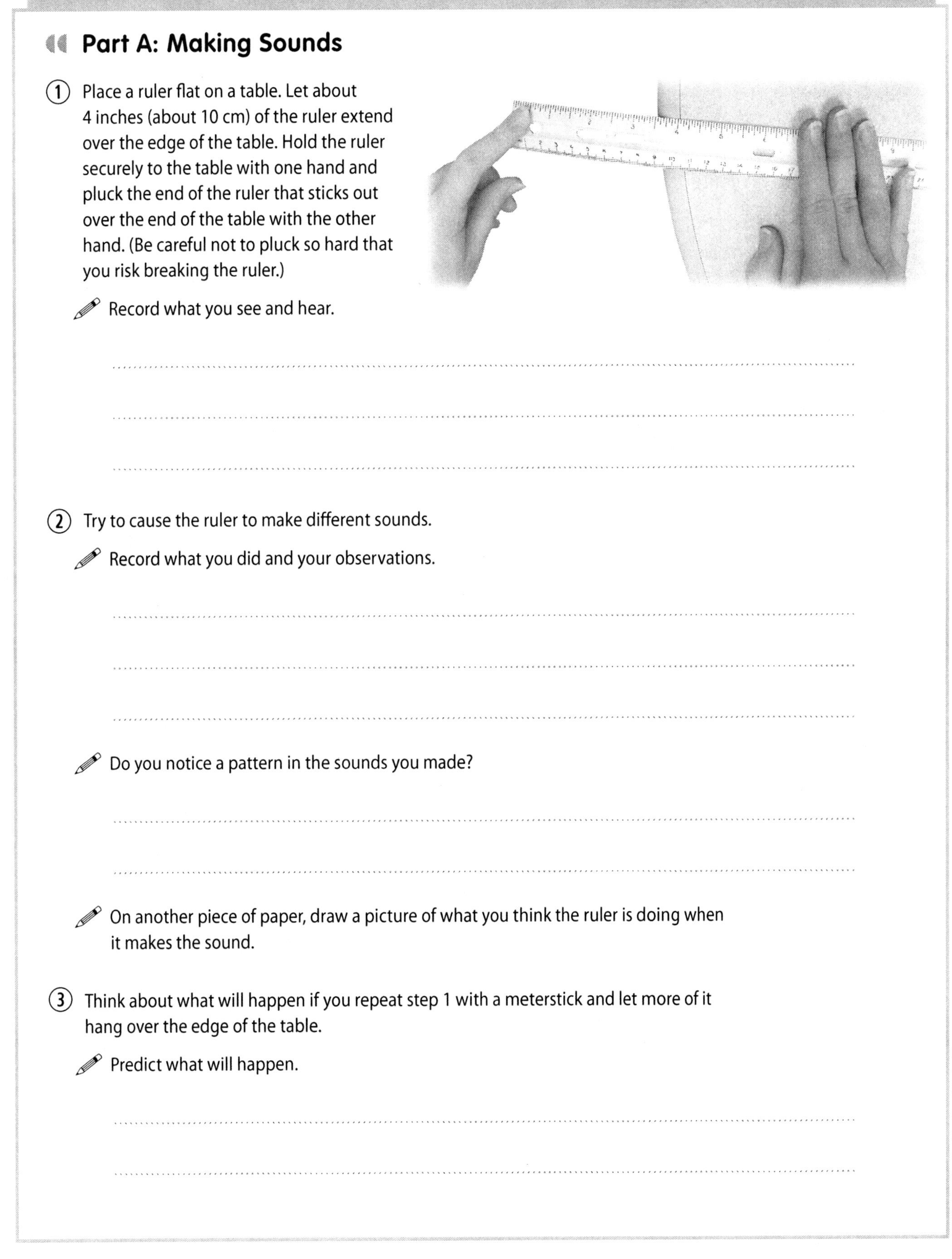

Record what you see and hear.

....

....

....

② Try to cause the ruler to make different sounds.

Record what you did and your observations.

....

....

....

Do you notice a pattern in the sounds you made?

....

....

On another piece of paper, draw a picture of what you think the ruler is doing when it makes the sound.

③ Think about what will happen if you repeat step 1 with a meterstick and let more of it hang over the edge of the table.

Predict what will happen.

....

....

Try the experiment and record your observations.

....

....

....

Compare your results to your prediction.

....

....

Part B: Making Music

1. Stretch a rubber band between your fingers and pluck it. Pay attention to the sound it makes.

   Record your observations.

   ....

   ....

   ....

2. Wrap the same rubber band around a tray as shown.

   Predict what will happen when you pluck the rubber band.

   ....

   ....

   ....

   Pluck the rubber band and describe how this sound is different from the sound you made in step 1.

   ....

   ....

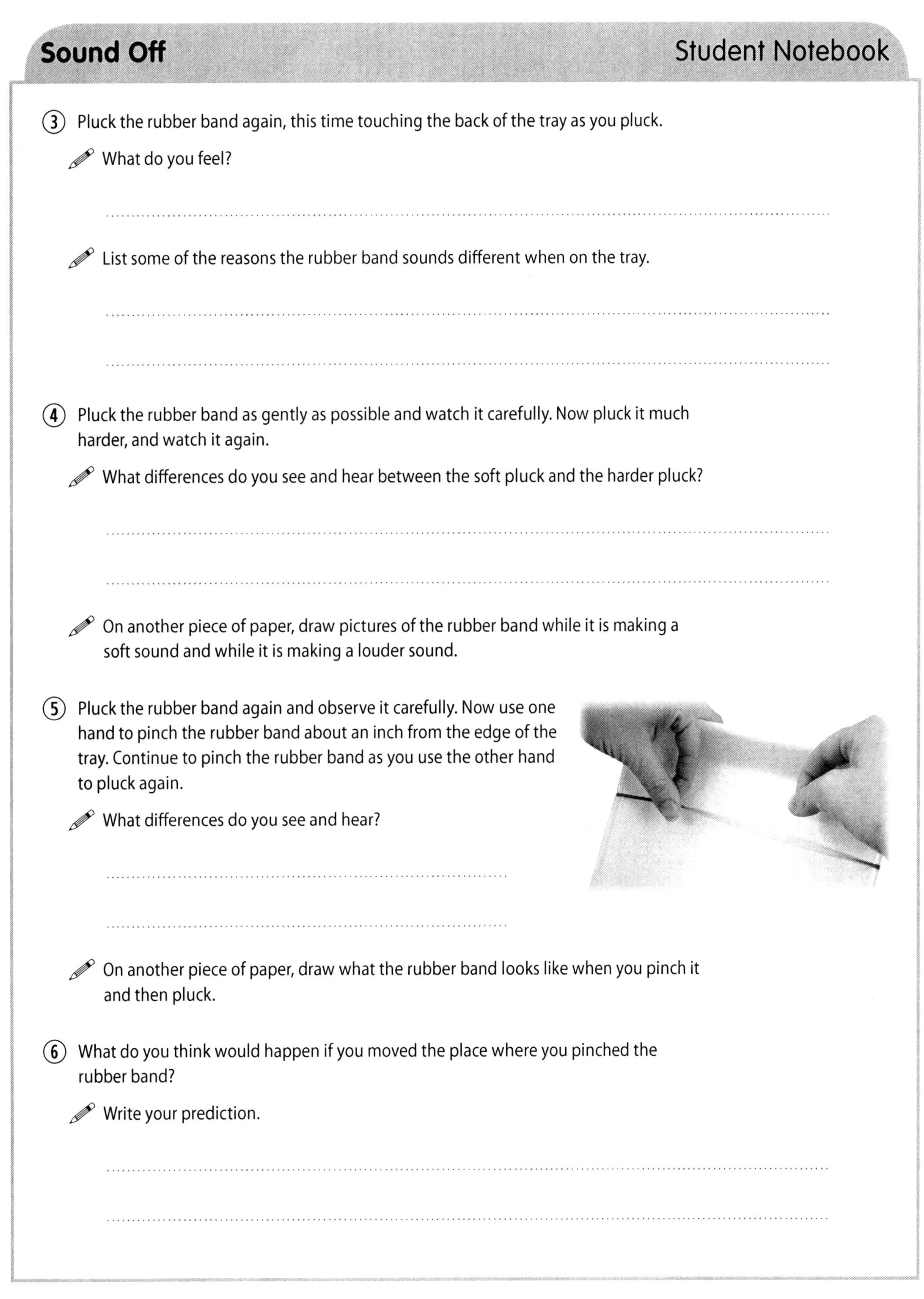

(3) Pluck the rubber band again, this time touching the back of the tray as you pluck.

✏ What do you feel?

....

✏ List some of the reasons the rubber band sounds different when on the tray.

....

....

(4) Pluck the rubber band as gently as possible and watch it carefully. Now pluck it much harder, and watch it again.

✏ What differences do you see and hear between the soft pluck and the harder pluck?

....

....

✏ On another piece of paper, draw pictures of the rubber band while it is making a soft sound and while it is making a louder sound.

(5) Pluck the rubber band again and observe it carefully. Now use one hand to pinch the rubber band about an inch from the edge of the tray. Continue to pinch the rubber band as you use the other hand to pluck again.

✏ What differences do you see and hear?

....

....

✏ On another piece of paper, draw what the rubber band looks like when you pinch it and then pluck.

(6) What do you think would happen if you moved the place where you pinched the rubber band?

✏ Write your prediction.

....

....

✎ Try it by moving the pinched place to several locations. What do you observe?

✎ How is what you did here similar to what you did to cause the ruler to make different sounds?

7. Think about another way you could make the sound of the rubber band become higher or lower.

TIP: Think about the way a musician tunes a guitar or violin.

✎ What's your idea?

✎ Describe what happens when you try your idea.

8. Select two, three, or more rubber bands having different thicknesses. Arrange them around your tray to make a simple guitar that produces the lowest sound at one end and the highest at the other. Pluck the rubber bands to test if your arrangement is correct and move them around if you need to. Play a simple song on your instrument.

Assessment A

1 Cut a rubber band and tape one end of it to a desk or table. Pull the rubber band until it is lightly stretched and pluck it. Pay attention to the sound it makes.

Record your observations.

....

....

2 Seal the top of a bowl with plastic food wrap. Make sure the wrap is nice and tight across the top of the bowl. Tape one end of the rubber band to the center of the food wrap.

Predict how the rubber band will sound if you pluck it now. Record your prediction and your reasons for it.

....

....

....

3 Pull the rubber band up until it is lightly stretched. Pluck the rubber band.

What do you observe?

....

....

How does your observation compare to your prediction?

....

....

Explain what happens.

....

....

....

....

Assessment B

1. Look at the diagram of the guitar. If you have one available, compare the diagram to an actual guitar.

   On the diagram, mark which parts of the guitar create sound, amplify sound, and control pitch. Explain how these parts work.

Balance This!

Students build and explore balance toys and discover how varying the amount and position of weight affects the toys' balance.

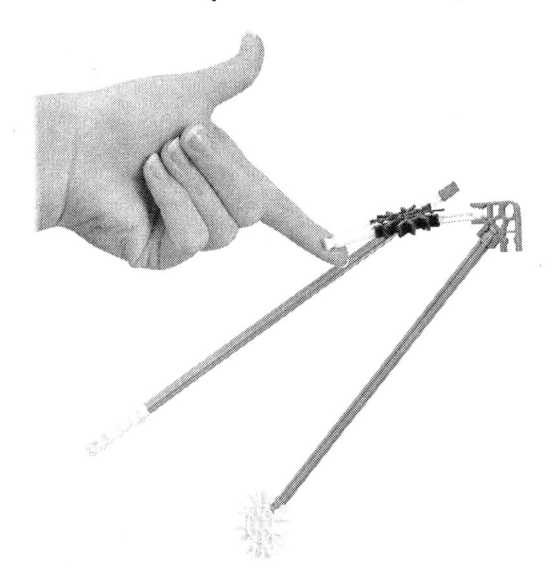

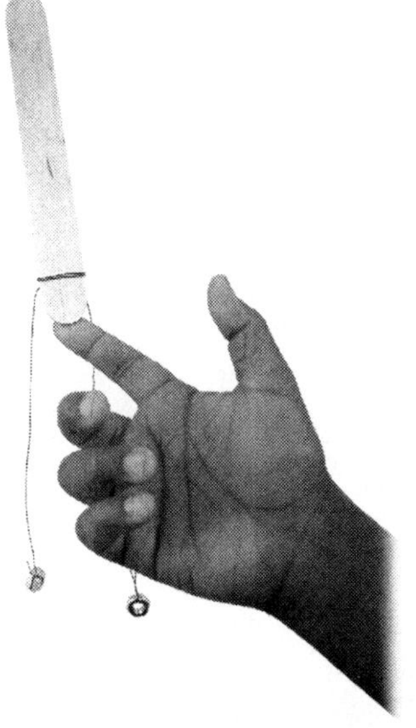

Grade Levels

Science activity appropriate for grades 3–6

Student Background

Students require no particular background preparation for this activity.

Time Required

Setup	20	minutes
Procedure	20	minutes
Cleanup	5	minutes

Assessment time is not included.

Key Science Topics

- balance
- center of gravity

National Science Education Standards Overview

See *www.terrificscience.org/physicsez/* for details of how these standards relate to the activity.

Science as Inquiry

Abilities Necessary to Do Scientific Inquiry

K–4 *Use data to construct a reasonable explanation.*
K–4 *Communicate investigations and explanations.*

5–8 *Develop descriptions, explanations, predictions, and models using evidence.*
5–8 *Think critically and logically to make the relationships between evidence and explanations.*
5–8 *Communicate scientific procedures and explanations.*

Physical Science

K–4 *Position and motion of objects*

5–8 *Motions and forces*

Science and Technology

Abilities of Technological Design

K–4 *Propose a solution.*
K–4 *Implement proposed solutions.*
K–4 *Communicate a problem, design, and solution.*

5–8 *Design a solution or product.*
5–8 *Implement a proposed design.*
5–8 *Communicate the process of technological design.*

Materials

For the Procedure

Activity Introduction, per class

- several commercial and homemade balance toys

Commercial toys come in many forms, including wire figures on bases, lightweight cardboard animals (such as butterflies) with weights on the wings, and molded plastic animals (such as birds with outstretched wings).

- chart or chalkboard

Per student

- materials to make one of two styles of balance toys (craft stick or K'NEX®):
  - craft stick version
    - craft stick
    - pipe cleaner or wire
    - 1 or more of the following to use as weights: 2 paper clips, 2 hex nuts, 2 flat washers
    - (optional) additional weights
  - K'NEX version
    - assorted K'NEX rods and connectors
- balance toys

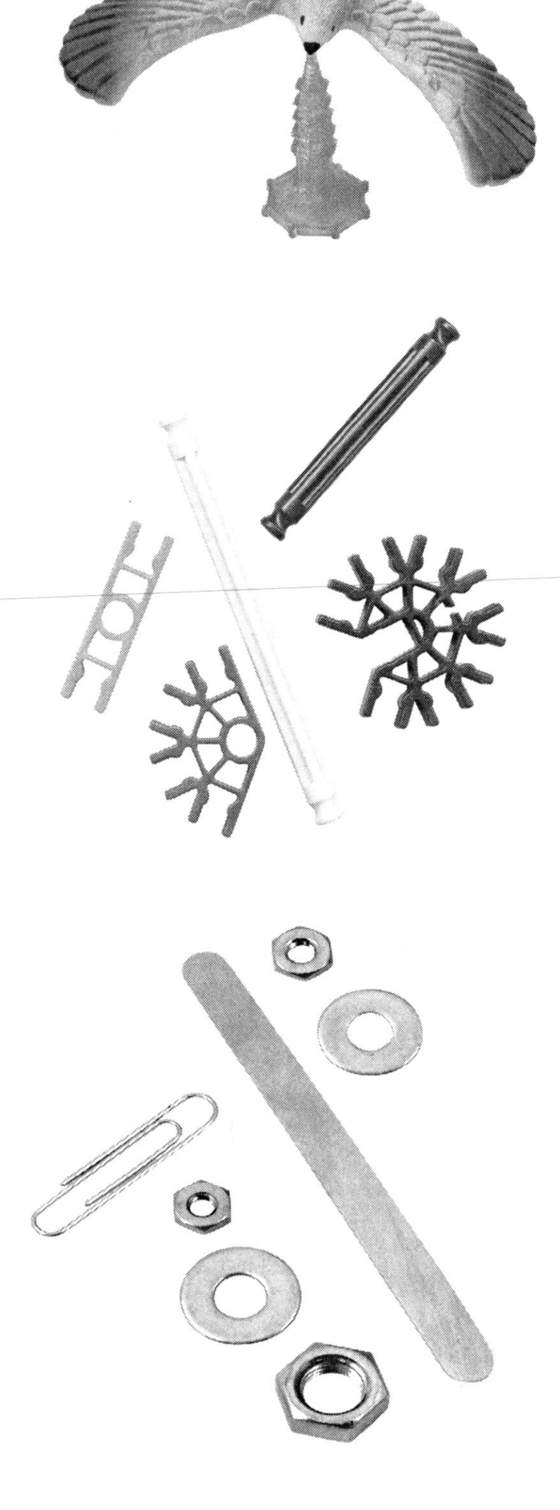

For the Assessment

Per student

- set of materials for each student to make a balance toy (such as straws, clay, Styrofoam® balls, wire, forks, cardboard, washers, pennies, corks, marshmallows, fruit, pencils, and toothpicks)

Safety and Disposal

Students should take care to work in their own space when bending pipe cleaner and wire, keeping the material well away from their eyes and the eyes of other students. No special disposal procedures are required.

Procedure

This section provides teacher notes corresponding to each step of the student procedure. The procedure without teacher notes is included in the reproducible Student Notebook pages at the end of this activity and at www.terrificscience.org/physicsez/.

	Student Procedure	Teacher Notes
	Activity Introduction	• Give students opportunities to play with an assortment of commercial and homemade balance toys prior to the activity. • As a class, have students discuss the balance toys, sharing their ideas about how the toys balance. Record their ideas on a chart or chalkboard. Point out each toy's support point.
	Make a Balance Toy	
①	**Try to balance a craft stick or K'NEX rod (we'll just say "stick" for both) so that it stands up on your finger.** **What happens?**	If using K'NEX, avoid distributing the longer rods (such as red and gray) in this step. Students will need to use the longer rods later. Students should discover that balancing a stick upright on the finger is very difficult.
②	**Try to balance a balance toy on your finger. Does it stay on your finger if you move your hand around? Try some other balance toys.** **How do these balance toys behave? How is this different than what the stick did in step 1?**	Balance toys should easily balance on the finger, even when the toys are tipped.
③	**Compare the balance toys and the stick to find out why one may be harder to balance than the other.** **TIP: On the items you were able to balance, look below your finger.** **What do the balance toys all have in common that is different than the stick?**	Students will see that balance toys placed on the finger have some parts that hang or extend below the finger. Although students may not realize it yet, these parts lower the center of gravity so that the toys balance.

Student Procedure	Teacher Notes
(4) What can you add to your stick to try to make it balance more like the balance toys? Test your ideas. **TIP: Be creative! Share ideas! If the stick balances on your finger...you've done it!** **On another piece of paper, draw your balancing stick design. Show and label the support point and your finger.**	Do not demonstrate a solution. Some teachers are initially concerned about giving students such an open-ended task and are pleasantly surprised at their students' creativity. See Sample Answers for examples of student solutions. Through their exploration, students should see that objects balance more easily when parts hang or extend below the support point. For craft stick and pipe cleaner models, variables include position of pipe cleaner on stick; length of pipe cleaner (students can combine pipe cleaners to make longer wires); and number, type, and arrangement of weights. For K'NEX models, variables include length, number, and arrangement of rods; number and weight of connectors (using the purple or blue interlocking connectors makes it especially easy to add weight); and angles of connections.
Class Discussion	• Introduce the idea that every object or person has something called a center of gravity that is located in one particular spot on the object or person. In rough terms, the center of gravity can be thought of as the average location of the mass of the object. • An object balances when its center of gravity is below or directly above the point of support. • When the center of gravity for an object or person is above a narrow support point, balancing may be difficult. Ask students to give examples of cases where people balance above a support point (such as a gymnast on a balance beam or a circus tightrope walker on a high wire without a balancing pole). • When the center of gravity is below the support point, balancing is much easier. • For older students, you may want to include additional information from the Explanation.

Student Procedure	Teacher Notes
⑤ Look at the three pictures. In each picture, something is balancing on something else. ✎ Circle the thing or person that is balancing on something else in each picture. ✎ For each picture, draw an arrow to the support point for the balancing object. ✎ For each picture, record whether the center of gravity of the balancing object is above or below the support point. ✎ Which do you think is least likely to fall off its support point: the woman, the bananas, or the toy horse? Why do you think so?	Students should circle the woman, the plate of bananas, and the horse. The support points are located where the woman's foot touches the log (the center of gravity is above the support point), where the plate of bananas touch the girl's head (the center of gravity is above the support point), and where the horse touches the finger (the center of gravity is below the support point). Students may have a variety of answers to the last question. They may realize that the horse is the least likely to fall because its center of gravity is below the support point. They may also notice that the bananas have a relatively wide support on the girl's head. Although the center of gravity is above the support point for both the bananas and the woman, students may decide that the bananas are less likely to fall than the woman balancing on a narrow log.
⑥ Look at the drawing you made of your balancing stick in step 4. ✎ Is the balancing stick's center of gravity above or below the support point?	The balancing stick's center of gravity should be below the support point.
⑦ Now try changing your stick's center of gravity by changing the position or amount of added weight. ✎ On another piece of paper, draw a picture showing how your stick balances after this change.	See Sample Answers for some design solutions. (See Explanation for a discussion.)
Assessment	
❶ Build an object that can balance on the tip of a pencil eraser using the materials your teacher gives you. The object must stay balanced even when you move the pencil around. ✎ Draw your balance toy invention below. Show and label the support point and all of your toy's parts. ✎ Is the center of gravity above or below the support point?	Decide on a set of materials to give each student. See Sample Answers for two examples. One balancing object is made from a Styrofoam ball, toothpick, straws, and clay. Another is made from a cherry, toothpick, and plastic forks. If the object balances easily, its center of gravity is below the support point.

Sample Answers

Where students follow the same procedure using the same materials, these answers are close to answers you can expect. Where students design their own experiment or model, students' results will vary.

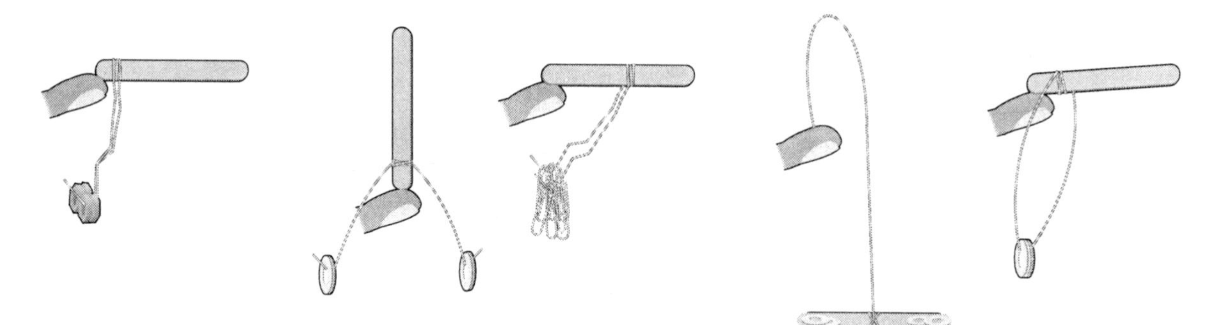

Examples of student solutions for the craft stick version

Examples of student solutions for the K'NEX version

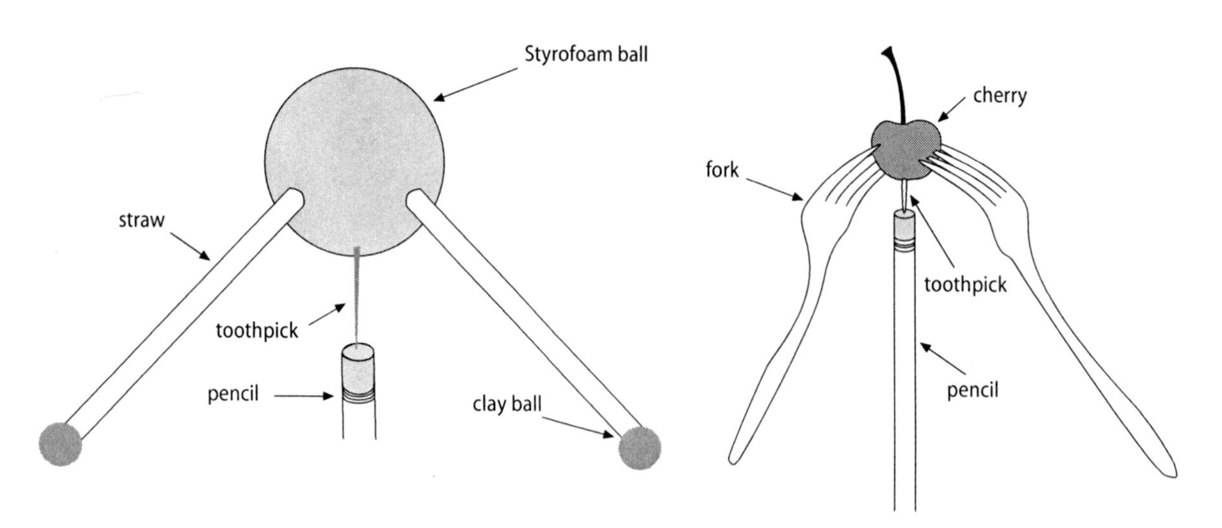

Examples of student balance toy inventions for the Assessment

Explanation

This section is intended for teachers. Modify the explanation for students as needed.

Center of Gravity

The center of gravity of an object or connected group of objects is the point at which the force of gravity acts on the object as if all the weight were concentrated at that point. The center of gravity may be located where no actual material exists.

An object balances when its center of gravity is below or directly above the point of support. An object with its center of gravity below its point of support is more stable. Balancing objects that have their centers of gravity below the point of support act like simple pendulums. When the object is tipped, the center of gravity swings to one side like the ball of a simple pendulum. Like a pendulum, the forces acting on the object bring it back to equilibrium, and the center of gravity is once again directly below the support point.

An object can balance with its center of gravity above the point of support, but the object is unstable unless the area of support is large. If the object is tipped even a little, gravity pulls it away from the balanced position. For example, tightrope walkers and gymnasts on a balance beam are unstable because their center of gravity is above their point of support (the rope or balance beam). They need to be very skilled at keeping their center of gravity directly over the rope or beam. The bananas on the girl's head (shown in step 5) are more stable than the woman on the log because the support area is larger compared to the size of the object.

Making Balance Toys

In step 1, students try to balance the stick on end with no other weight added. Under these conditions, the stick's center of gravity is well above the point of support (the finger), and the stick falls. The stick balances when additional weight (in the form of weights or K'NEX pieces) is used to move the center of gravity to a point below the support point.

In step 7, changing the position of the added weight changes the behavior of the balancing stick. If part of the additional weight is moved upward and outward, the center of gravity of the whole system is redistributed and moves upward and outward. In response, the whole system tilts or rotates to move the center of gravity back below the stick (again, like the swing of a pendulum). If the stick balanced upright on the support point before moving the weight, it may balance at an angle after moving the weight. If the center of gravity is moved upward to a point above the point of support, the whole system may rotate off its support point.

Cross-Curricular Integration

Art:

- Have students turn their balancing sticks into balancing people or animals by attaching tagboard or card stock shapes to the front of the stick. For an example, see the clown templates at the end of this activity and at *www.terrificscience.org/physicsez/.*

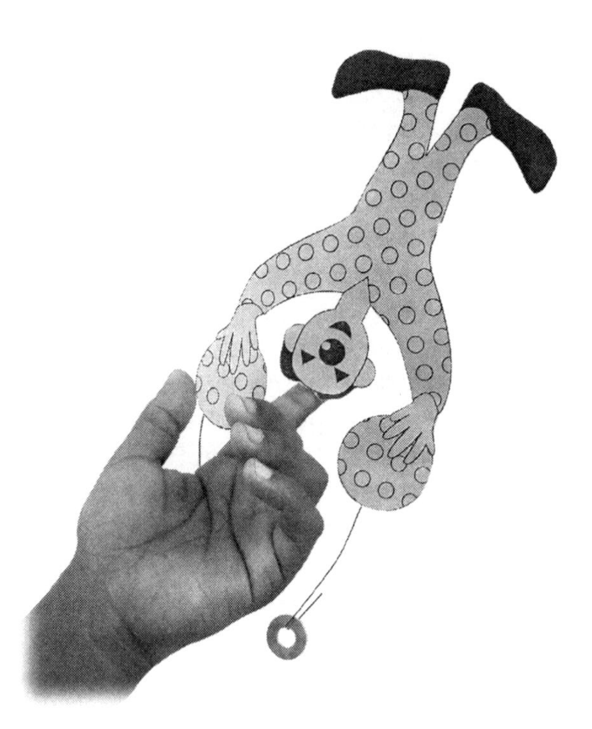

- Introduce the sculptor Alexander Calder. Emphasize that Calder is famous for art that depends on balancing shapes, and show them pictures of both hanging and standing pieces. (See examples on the Calder Foundation website; in the book *Calder, 1898–1976,* by Jacob Baal-Teshuva; or a similar resource.) Have students use K'NEX or other objects to make a hanging or standing piece of balance art.

Language arts:

- Tell students that they will be packaging a balance toy, including all necessary parts and directions for assembly. Have students use their completed Student Notebook pages to decide what to place in the bag and to help write the directions. Students should write the directions in paragraph form, remembering to use the language of direction (first, next, then, and last). Ask students to place all of their balance toy supplies and the assembly directions into a zipper-type plastic bag.
- Read *Mirette on the High Wire,* by Emily Arnold McCully (grades K–3), and/or *High-Wire Henry,* by Mary Calhoun (grades K–3). Discuss how the balancing events in the stories relate to the activity.
- Have students write a cinquain poem about their balancing sticks or rods. The standard form for a cinquain is as follows:

  Noun
  Adjective, adjective
  Three-word sentence
  Four participles
  Noun

  A sample poem for this activity would be as follows:

  Mass
  Upside-down, silly
  Will it balance?
  Moving, falling, wobbling, standing
  Center of gravity

- Read aloud or suggest that students read the following book:
  - *Mobiles: Building and Experimenting with Balancing Toys,* by Bernie Zubrowski (grades 4–7)
    Students are introduced to basic principles of balancing by constructing and experimenting with toys and mobiles.

Life science:

- Have students study how we can change the center of gravity of our bodies to remain balanced while walking, running, standing, or bending over.
- Discuss the use of crutches or prosthetic limbs to maintain balance and mobility.

Physical education:

- Have students take turns walking on a balance beam. Then have them try walking on the beam while carrying a heavy weight in one hand.

Social studies:

- Have students study the history of the circus, especially the history of balancing acts.

Reference

Hewitt, P. *Conceptual Physics,* 9th ed.; Addison Wesley: San Francisco, 2002; pp 133–137.

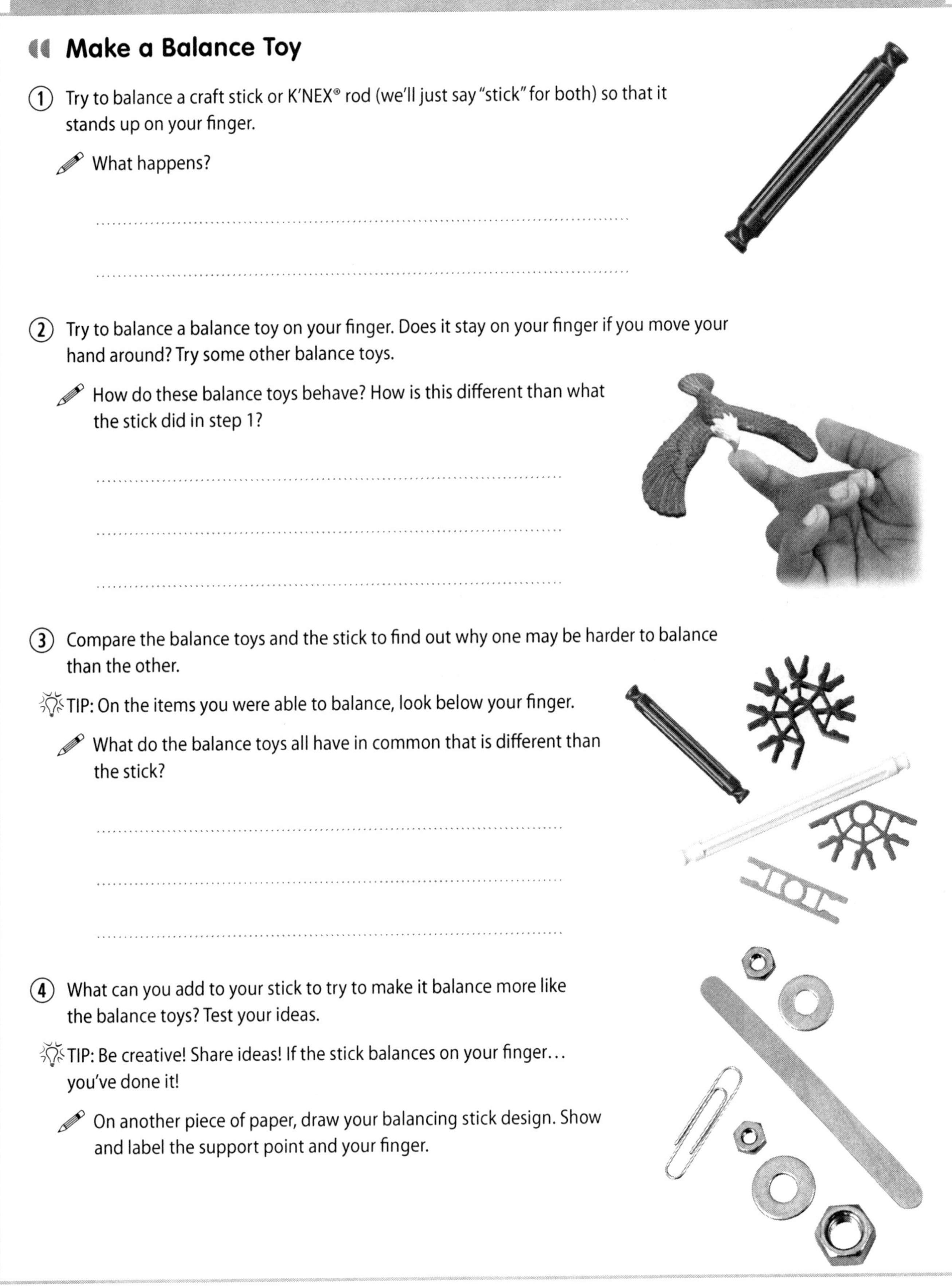

Make a Balance Toy

① Try to balance a craft stick or K'NEX® rod (we'll just say "stick" for both) so that it stands up on your finger.

What happens?

....

....

② Try to balance a balance toy on your finger. Does it stay on your finger if you move your hand around? Try some other balance toys.

How do these balance toys behave? How is this different than what the stick did in step 1?

....

....

....

③ Compare the balance toys and the stick to find out why one may be harder to balance than the other.

TIP: On the items you were able to balance, look below your finger.

What do the balance toys all have in common that is different than the stick?

....

....

....

④ What can you add to your stick to try to make it balance more like the balance toys? Test your ideas.

TIP: Be creative! Share ideas! If the stick balances on your finger… you've done it!

On another piece of paper, draw your balancing stick design. Show and label the support point and your finger.

⑤ Look at the three pictures. In each picture, something is balancing on something else.

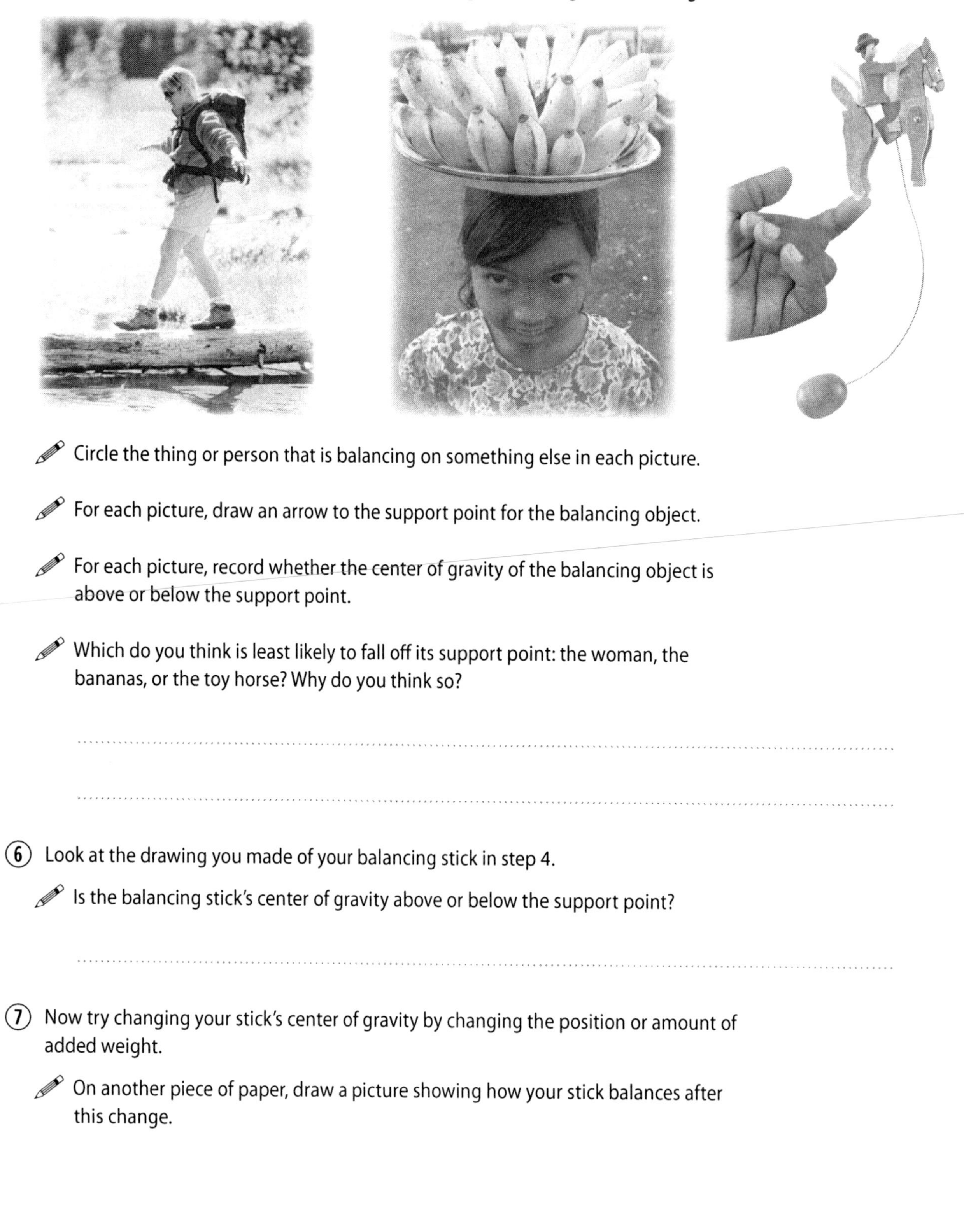

✎ Circle the thing or person that is balancing on something else in each picture.

✎ For each picture, draw an arrow to the support point for the balancing object.

✎ For each picture, record whether the center of gravity of the balancing object is above or below the support point.

✎ Which do you think is least likely to fall off its support point: the woman, the bananas, or the toy horse? Why do you think so?

⑥ Look at the drawing you made of your balancing stick in step 4.

✎ Is the balancing stick's center of gravity above or below the support point?

⑦ Now try changing your stick's center of gravity by changing the position or amount of added weight.

✎ On another piece of paper, draw a picture showing how your stick balances after this change.

Assessment

1. Build an object that can balance on the tip of a pencil eraser using the materials your teacher gives you. The object must stay balanced even when you move the pencil around.

   Draw your balance toy invention below. Show and label the support point and all of your toy's parts.

   Is the center of gravity above or below the support point?

   ....

Clown Templates

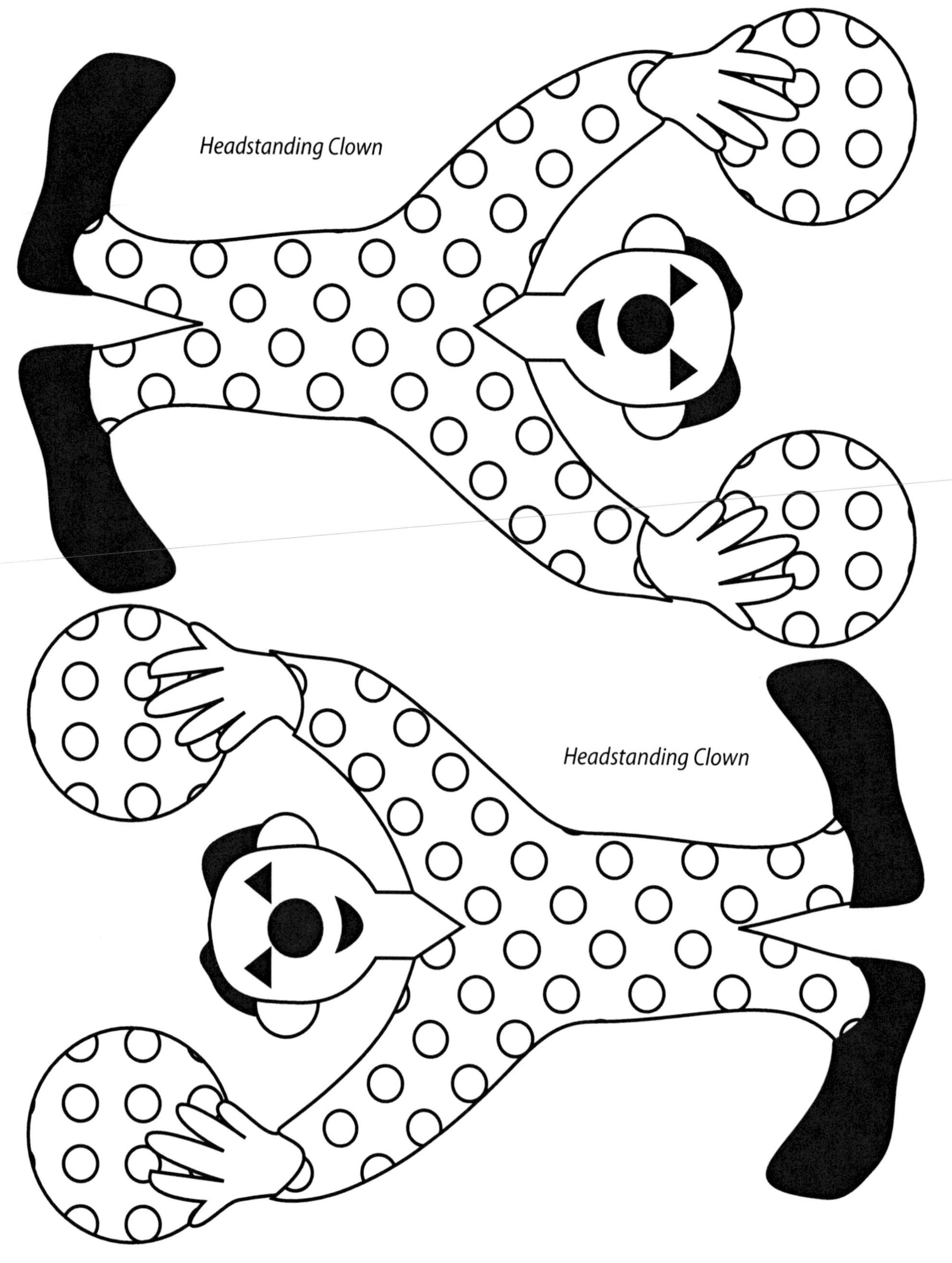

Skyhook

Students explore gravity and balance while playing with the skyhook, a popular pioneer toy.

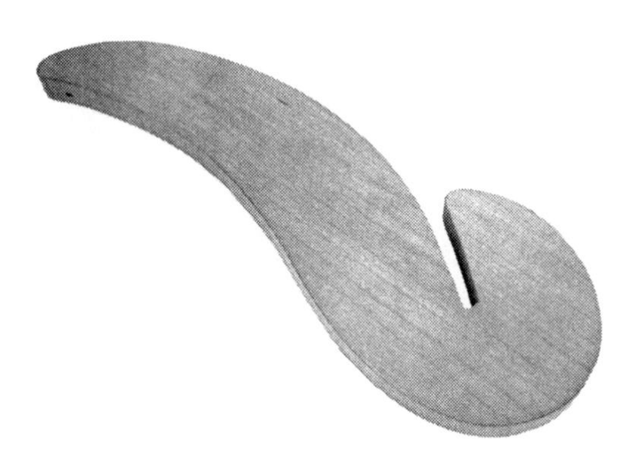

Grade Levels

Science activity appropriate for grades 3–8

Student Background

Students should have experience with balance toys and center of gravity, topics covered in "Balance This!" in this book.

Time Required

Setup	5	minutes
Procedure	30	minutes
Cleanup	5	minutes

Assessment time is not included.

Key Science Topics

- balance
- center of gravity
- gravity

National Science Education Standards Overview

See *www.terrificscience.org/physicsez/* for details of how these standards relate to the activity.

Science as Inquiry

Abilities Necessary to Do Scientific Inquiry

K–4 Communicate investigations and explanations.

5–8 Develop descriptions, explanations, predictions, and models using evidence.

5–8 Think critically and logically to make the relationships between evidence and explanations.

5–8 Communicate scientific procedures and explanations.

Physical Science

K–4 Position and motion of objects

5–8 Motions and forces

Science and Technology

Abilities of Technological Design

K–4 Propose a solution.

K–4 Implement proposed solutions.

K–4 Evaluate a product or design.

K–4 Communicate a problem, design, and solution.

5–8 Design a solution or product.

5–8 Implement a proposed design.

5–8 Evaluate completed technological designs or products.

5–8 Communicate the process of technological design.

Materials

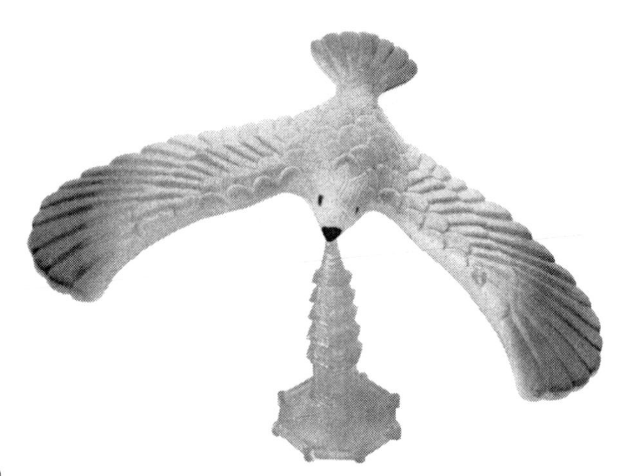

For the Procedure

Per class

- commercial or homemade balance toy
- (optional) wooden skyhook

A pattern to make a traditional wooden skyhook is provided at www.terrificscience.org/physicsez/. Commercially made wooden skyhooks are also available. For supply source suggestions, see www.terrificscience.org/supplies/.

Per group

- piece of corrugated cardboard about 16 cm × 16 cm
- scissors
- glue stick
- assorted objects (such as string, cloth, oaktag, poster board, newspaper, pipe cleaners, cereal boxes, cloth belts, leather belts, paper clips, binder clips, rubber bands, pennies, nuts, and bolts)

For the Assessment

Per group

- wooden clothespin (the kind without a spring)
- assorted objects (such as string, cloth, oaktag, poster board, newspaper, pipe cleaners, cereal boxes, cloth belts, leather belts, paper clips, binder clips, rubber bands, pennies, nuts, and bolts)

Safety and Disposal

Students should take care to work in their own space when bending pipe cleaner and wire, keeping the material well away from their eyes and the eyes of other students. No special disposal procedures are required.

Getting Ready

- Practice using the balance toy and skyhook prior to class.
- Cut some of the cloth, oaktag, poster board, and newspaper into strips that fit into the skyhook's slot.

Procedure

This section provides teacher notes corresponding to each step of the student procedure. The procedure without teacher notes is included in the reproducible Student Notebook pages at the end of this activity and at www.terrificscience.org/physicsez/.

	Student Procedure	Teacher Notes
	Balancing Act	
①	**Observe the toy balanced on your teacher's finger. Answer the following questions.** ✎ **Where and what is the support point?** ✎ **Discuss the relationship between the toy's support point and center of gravity.** ✎ **How is the toy's weight distributed?**	Balance a balance toy on your finger. Depending on student experience, have students work in groups to answer the questions and share with the class or just discuss the topics with the class. Some students will recognize that your finger is the support point (the point where the object is being held up). An object with its center of gravity below its support point is stable because, when the object is tipped, gravity pulls it back into position. The toy's weight is probably distributed so that some is below the support point (depending on the toy you demonstrate), making the toy very stable.
②	**Follow the "How to Make a Skyhook" instruction sheet to make a special balance toy.**	
③	**Using assorted materials provided by your teacher, try to make the skyhook balance on a finger.** ✎ **On another piece of paper, show how you made the skyhook balance. Label the support point and all of the objects you used.** ✎ **Discuss the relationship between your design's support point, center of gravity, and weight distribution.**	Allow students to experiment with the materials on their own. (See Sample Answers for design examples.) Some will discover a solution without further suggestions. If students seem stuck, suggest that they try hanging something from the skyhook's slot. The skyhook balances when two conditions are met: • Long, stiff objects hanging from the skyhook's slot (such as leather belts and strips of poster board) twist out and back to move some weight behind the support point. Now, like a seesaw, there is weight on either side of the support point. (See Explanation.) • The system's center of gravity (the skyhook and its attached objects) is well below the support point. (See Explanation.) After all or most of the groups balance their skyhooks, you may want to demonstrate a wooden skyhook. Explain that the skyhook is an early American toy that pioneer boys and girls played with because it was easy to make and seemed almost magical to use.

Student Procedure	Teacher Notes
Assessment	
❶ Use what you learned about the skyhook to make a wooden clothespin balance on your finger. ✎ In the space below, show how you made the clothespin balance. Label the support point and all of the objects you used. ✎ Discuss the relationship between your design's support point, center of gravity, and weight distribution.	Encourage students to come up with a different design than what they used for their skyhook design. Just like with the skyhook, the clothespin will balance when the center of gravity of the entire system (the clothespin and its attached objects) is below the support point. Stiff objects placed in the clothespin's slot (such as leather belts, strips of oaktag, and poster board) work best. (See Sample Answers for design examples.)

Sample Answers

Where students follow the same procedure using the same materials, these answers are close to answers you can expect. Where students design their own experiment or model, students' results will vary.

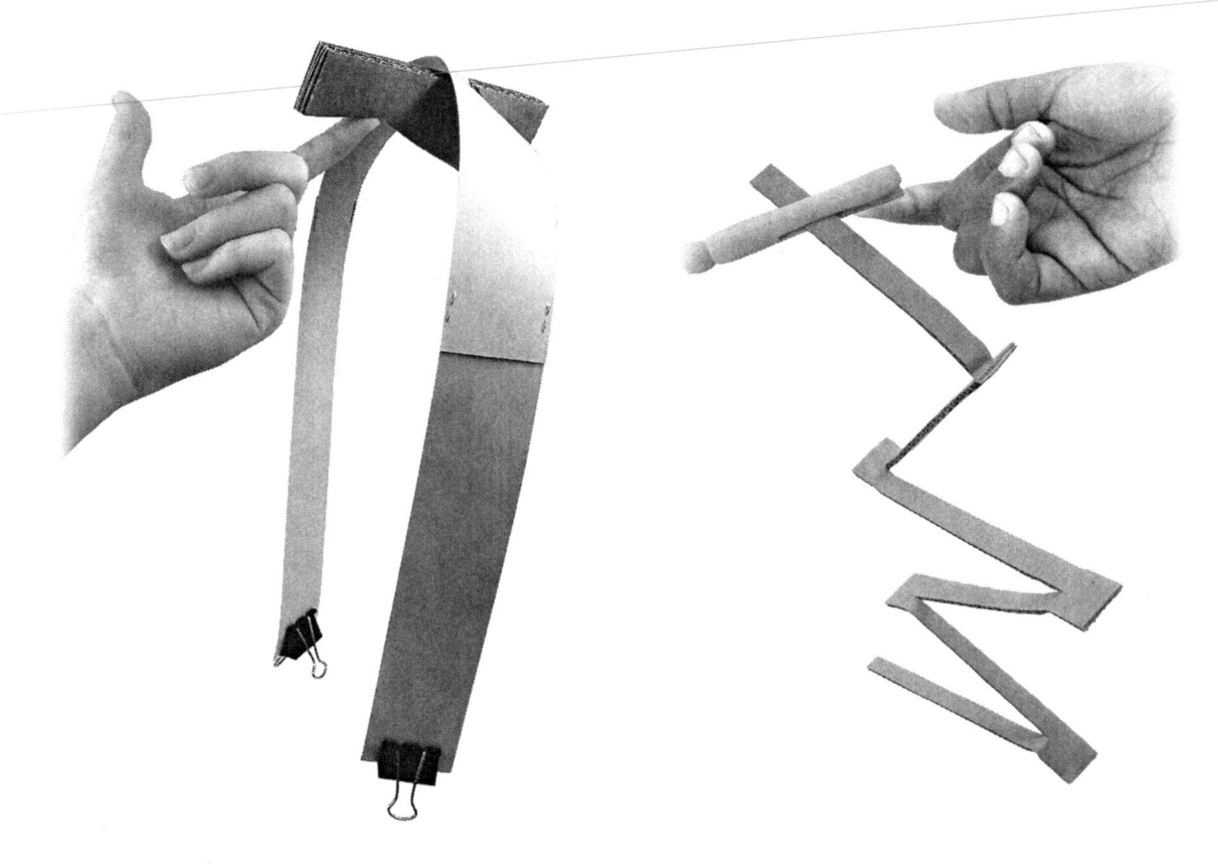

Example of student designs for step 3 and Assessment.
Weight added to the ends of the strips makes the system more stable.

Explanation

This section is intended for teachers. Modify the explanation for students as needed.

The skyhook seems to defy gravity. How does it work? To answer, we can consider the skyhook and a leather belt. Look at the skyhook and belt as a system and consider the center of gravity of the system. When we talk about the weight of an object, we mean the sum of all the individual gravitational forces on all the individual molecules. Luckily, we really don't have to deal with that complexity most of the time. The center of gravity is a single point at which the entire weight of an object appears to act. (See Explanation in "Balance This!" in this book for more on this topic.)

When the skyhook without a belt is placed on the finger, the center of gravity is beyond the support point on the fingertip since most of the skyhook's weight is in front of the support point. Thus, the skyhook falls. If you carefully observe the skyhook system after a belt is added, you can see that much of the belt is directly under the support point and some of the belt is slightly behind the support point. (See figure at top right.) This moves the center of gravity to a position directly below the support point (directly below your finger), a condition necessary for the skyhook and belt to balance. So even though our eyes tell us that the loaded skyhook should be tipping forward and falling, the whole skyhook system is in balance.

The skyhook-belt system balances much like two people sitting on a seesaw. The finger (point of support) is the fulcrum. The weight of the skyhook and part of the belt sits on one side of the fulcrum. The rest of the weight of the belt sits on the other side of the fulcrum. (See Explanation in "Seesaw Forces" in this book for more on this topic.)

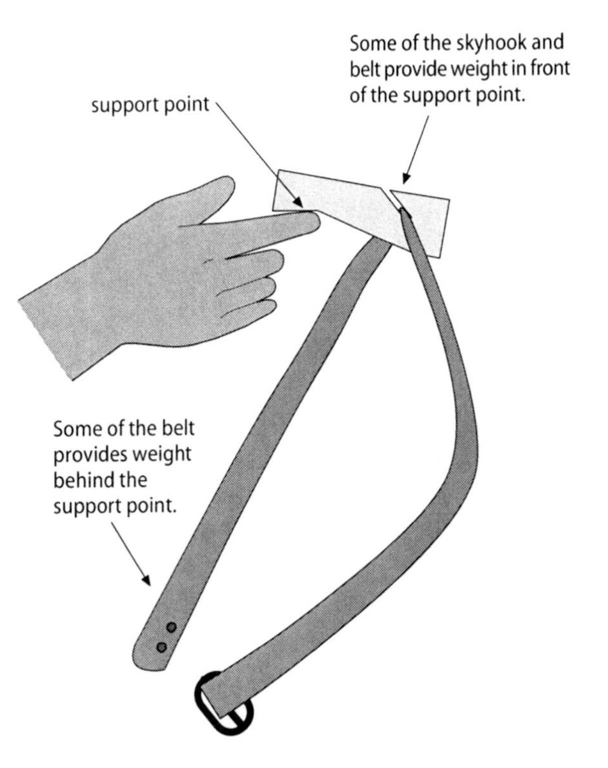

Balance the skyhook and belt on your finger.

Students attempting to balance the skyhook should realize that hanging something long and stiff from the skyhook's slot works best. The angle of the slot forces the stiff object to twist so that the ends of the object are below and behind the support point. Adding weight to the ends of the stiff object makes the system more stable. By using the long, stiff object with weight attached to its ends, the system's center of gravity is well below the support point and the skyhook balances.

Once students create a stable system, have them gently shake their fingers to move the system. The movement causes the system's center of gravity to shift away from under the finger. However, the system will act like a pendulum by swinging the center of gravity back to a point directly under the finger, thus allowing the system to stay balanced.

Cross-Curricular Integration

Language arts:

- Have students read about toy inventors such as Ole Kirk Christiansen (LEGO® products), Richard and Betty James (Slinky®), Jack Odell (Matchbox® cars), and Ruth Handler (Barbie®). Then, pose this question to students: If you wrote a letter to the person who invented the skyhook, what would you say? Think about why you like or dislike the toy, questions you would ask, your suggestions of ways to improve it, and your ideas for a new toy based on balance and center of gravity. Have students complete a web-type graphic organizer before beginning the letter writing process. (See *www.terrificscience.org/physicsez/*.)
- Read aloud or suggest that students read the following books:
  - *Toys!: Amazing Stories Behind Some Great Inventions,* by Don Wulffson (grades 4–8)
    This book offers history and interesting trivia about the creation of a wide selection of classic and commercial toy inventions.
  - *Steven Caney's Toy Book,* by Steven Caney (grades 4–7)
    This book contains instructions for making 51 fun and creative toys and games out of common, easily available materials.

Life science:

- Discuss different animals' ability to balance themselves (squirrels and cats) or to balance other objects (dolphins and sea lions).

Social studies:

- Use the skyhook to introduce a unit on American pioneers.

Reference

Hewitt, P.G. *Conceptual Physics,* 9th ed.; Addison Wesley: Los Angeles, 2002; pp 133–137.

Balancing Act

① Observe the toy balanced on your teacher's finger. Answer the following questions.

Where and what is the support point?

Discuss the relationship between the toy's support point and center of gravity.

How is the toy's weight distributed?

② Follow the "How to Make a Skyhook" instruction sheet to make a special balance toy.

③ Using assorted materials provided by your teacher, try to make the skyhook balance on a finger.

On another piece of paper, show how you made the skyhook balance. Label the support point and all of the objects you used.

Discuss the relationship between your design's support point, center of gravity, and weight distribution.

Assessment

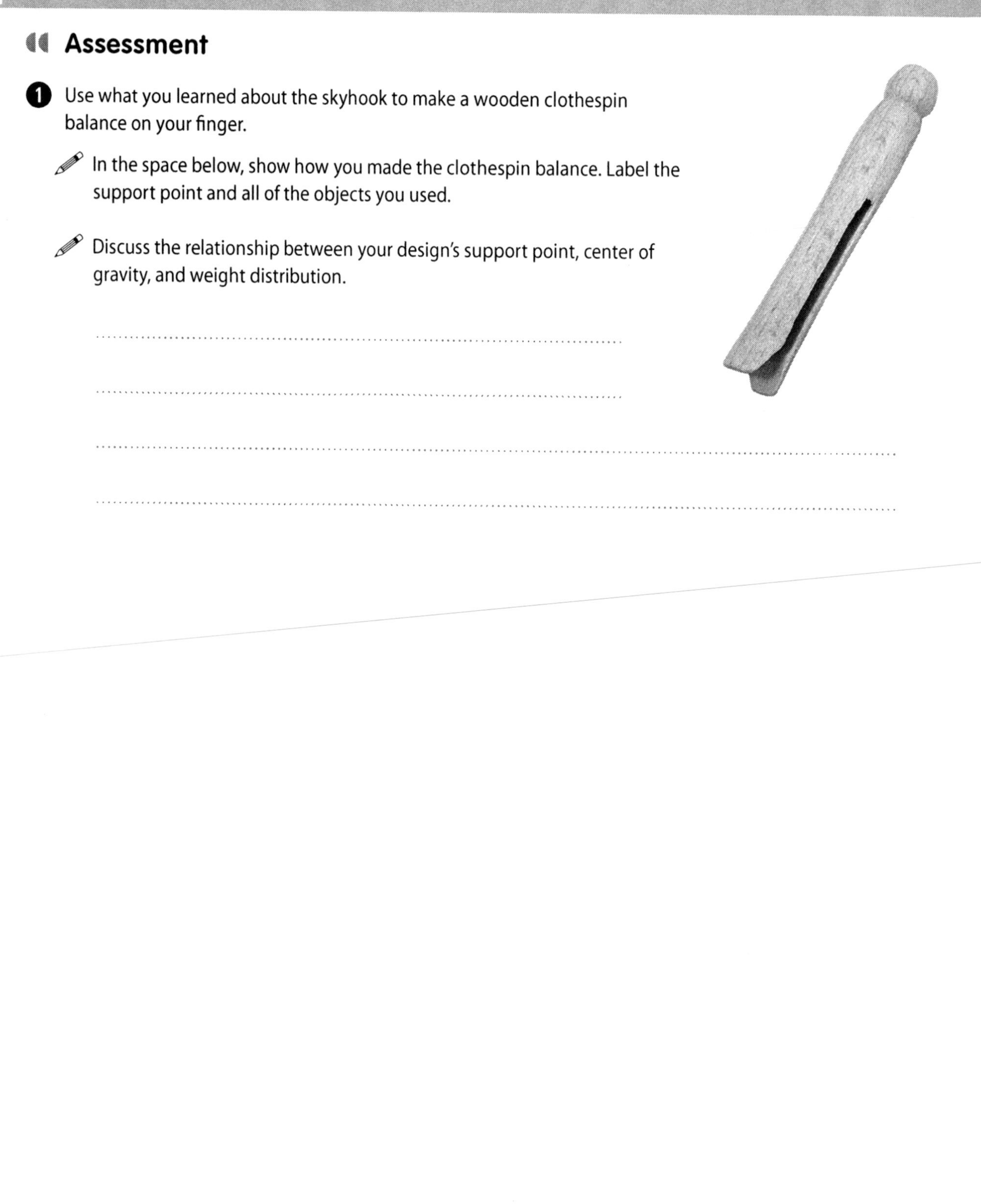

1. Use what you learned about the skyhook to make a wooden clothespin balance on your finger.

   In the space below, show how you made the clothespin balance. Label the support point and all of the objects you used.

   Discuss the relationship between your design's support point, center of gravity, and weight distribution.

How to Make a Skyhook

- Trace the following pattern on corrugated cardboard three times.
- Carefully cut out the three pieces.
- Stack and glue the three pieces together.

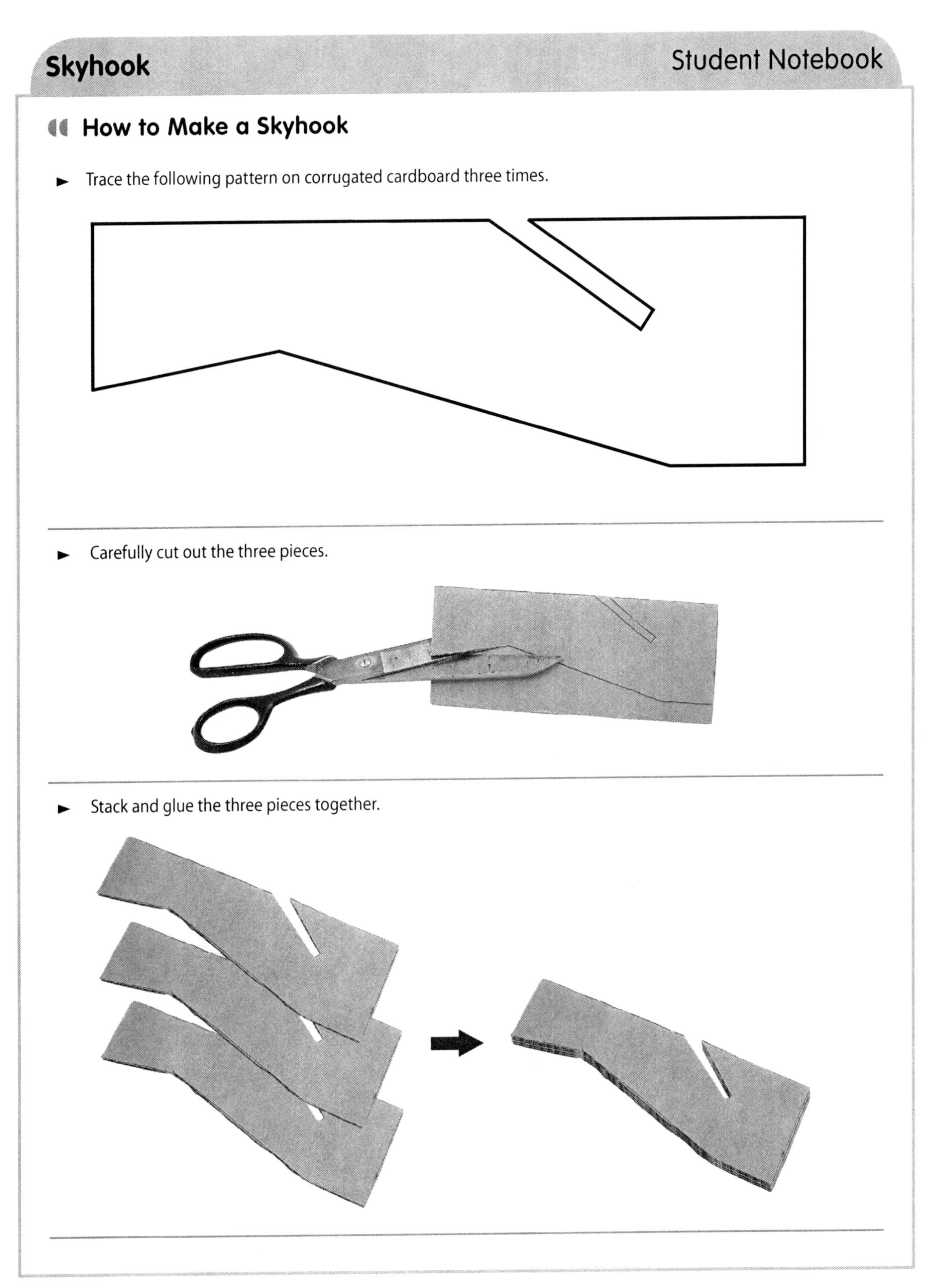

Gear Up, Gear Down

Students explore the operation of gears.

Grade Levels

Science activity appropriate for grades 4–8

Student Background

Students require no particular background preparation for this activity.

Time Required

Setup	10	minutes
Build K'NEX® model	15	minutes
Part A	25	minutes
Part B	25	minutes
Cleanup	10	minutes

Assessment time is not included.

Key Science Topics

- energy
- simple machines

National Science Education Standards Overview

See *www.terrificscience.org/physicsez/* for details of how these standards relate to the activity.

Science as Inquiry

Abilities Necessary to Do Scientific Inquiry

K–4 *Employ simple equipment and tools to gather data and extend the senses.*
K–4 *Use data to construct a reasonable explanation.*
K–4 *Communicate investigations and explanations.*

5–8 *Use appropriate tools and techniques to gather, analyze, and interpret data.*
5–8 *Think critically and logically to make the relationships between evidence and explanations.*
5–8 *Communicate scientific procedures and explanations.*
5–8 *Use mathematics in all aspects of scientific inquiry.*

Physical Science

K–4 *Position and motion of objects*

5–8 *Motions and forces*

Science and Technology

Abilities of Technological Design

K–4 *Evaluate a product or design.*
K–4 *Communicate a problem, design, and solution.*

5–8 *Evaluate completed technological designs or products.*
5–8 *Communicate the process of technological design.*

Materials

For Getting Ready

- black permanent marker

For the Procedure

Activity Introduction, per group

- assorted gear-based construction toys

For supply source suggestions, see www.terrificscience.org/supplies/.

Part A, per group

- set of gears, such as
    - assorted K'NEX gears and rods

      For specifics, see K'NEX Parts Inventory at www.terrificscience.org/physicsez/ or www.terrificscience.org/supplies/.
    - commercial gears (1 small, 2 medium-sized, and 1 large) that can be secured in place
- meterstick or measuring tape
- 2 colors of ¼-inch self-adhesive dots or dots made from adhesive paper using a hole punch
- large rubber band
- (optional) string or ribbon

Part B, per group

- assorted K'NEX rods, connectors, and gears

For specifics, see K'NEX Parts Inventory at www.terrificscience.org/physicsez/ or see www.terrificscience.org/supplies/.

- K'NEX assembly diagrams

For the Assessment

Per class

- (optional) object that operates with gears (such as oscillating lawn sprinkler, electric screwdriver, and rotary eggbeater)

Safety and Disposal

No special safety or disposal procedures are required.

Getting Ready

- For Part A, mark one tooth on each gear with a black permanent marker.
- For Part B, print a set of K'NEX assembly diagrams for each group. (See *www.terrificscience.org/physicsez/*.) Since K'NEX pieces are color coded, the assembly diagrams need to be printed in color.

Depending on your schedule, you may want to have students build models in advance.

Procedure

This section provides teacher notes corresponding to each step of the student procedure. The procedure without teacher notes is included in the reproducible Student Notebook pages at the end of this activity and at www.terrificscience.org/physicsez/.

Student Procedure	Teacher Notes
Activity Introduction	Give students an opportunity for free exploration with gear-based toys. Discuss their observations as a class. Observations may include the following: • Gears can have different diameters. • Gears can have different numbers of teeth. • Gears that are connected to each other can turn in different directions. • One gear can turn many other gears. • Some gears turn faster than others.
Part A: How Do Gears Operate?	
① **Gears are a type of wheel, and wheels are simple machines. Count the number of teeth on each of the three different-sized gears.** **Record the number of teeth for each gear in the data table.**	Younger students may need to be reminded to begin counting from the marked tooth in order to keep track of where they started counting.
② **Lay one gear flat. Turn the gear with your hand so the gear goes around one time.** **How can you measure the distance that one tooth of the gear travels when the gear goes around one time?** **Try your idea and record the travel distance measurement for each gear in the data table.**	Give students a chance to figure out the problem. (See Sample Answers for example data.) If they are stuck, help them reach the idea that the distance the tooth travels is the circumference of the gear. Students can measure the circumference of each gear in several ways. For example: • They can draw a straight line with a meterstick, roll the gear along the line so that the gear rotates one time (mark where the marked tooth starts and where it ends), and measure the distance the gear traveled. • They can begin at the marked tooth and place a string (or ribbon) along the edge of the gear until the string goes around once. The string can then be measured. • If students understand how to calculate the circumference of a circle, they can measure the gear's diameter and multiply by π.

Student Procedure	Teacher Notes
③ Lay the two medium-sized gears side by side. Mesh the gear teeth together so that the marked tooth from each gear touches the other. Use a sticker dot to label one gear as the driven gear. Use a sticker dot of another color to label the other gear as the driver gear. Watching both marked teeth, turn the driver gear with your hand. TIP: As you turn, gently press the driver gear towards the driven gear to keep both gears meshed together. If you are using K'NEX gears, keep the driven gear in place by putting a red K'NEX rod upright in the center of that gear. ✎ Record the dot colors for the driven and driver gears. ✎ Compare the directions that the two gears turn.	Students observe that the gears do not turn the same way. One gear turns the other gear in the opposite direction.
④ Begin again with the marked teeth touching. Watching the marked teeth, turn the driver gear with your hand so that its mark makes one complete revolution. ✎ How many turns does the driven gear make?	Students observe that the driven gear makes one turn.
⑤ Repeat step 3, this time using one medium-sized gear and one large gear. Use colored dots as in step 3 to label the large gear as the driven gear and the medium-sized gear as the driver gear. Turn the medium-sized, driver gear with your hand so that its mark makes one complete revolution. ✎ How many turns does the large gear make? ✎ Compare the speed of the large gear to the speed of the driver gear.	Since the large gear does not make a complete revolution, the large gear moves more slowly than the medium-sized gear. Introduce the idea that using gears of different sizes to decrease the speed of the driven gear's rotation is called gearing down.
⑥ Repeat step 3 again, but switch the colored dots so the medium-sized gear is the driven gear and the large gear is the driver gear. Turn the large, driver gear with your hand so that its mark makes one complete revolution. ✎ How many turns does the medium-sized gear make? ✎ Compare the speed of the medium-sized gear to the speed of the driver gear.	The medium-sized gear makes more than one revolution and moves faster than the large gear. Introduce the idea that using gears of different sizes to increase the speed of the driven gear's rotation is called gearing up.

Student Procedure	Teacher Notes
⑦ Make a simple gear train as shown by placing a small gear between a large gear and a medium-sized gear. Mesh the gear teeth together. As in step 3, use colored dots to label the large gear on the left as the driven gear and the medium-sized gear on the right as the driver gear. ✏ Predict what would happen to the two other gears when you rotate the driver gear one turn clockwise. On the image above, use arrows to show your predictions about the directions all of the gears will turn and the number of turns for each gear. ✏ Keeping the gears' teeth meshed together and the driven gear and small gear in place, test your prediction. Record your results and describe how your results compare with your predictions.	Students discover that rotating the driver gear one turn clockwise causes the small gear in the middle to turn more than one revolution counter-clockwise and the large, driven gear to turn less than one revolution clockwise. The small middle gear has no effect on the speed of the outer two gears. It is called an idler gear and serves the purpose of making the driven gear turn in the same direction as the driver gear.
⑧ Can one gear drive another without touching it? Using two gears and a rubber band, work with a partner to get a driver gear to move a driven gear. Don't forget—this time the two gears cannot touch each other! ✏ Draw and label your design on another piece of paper. ✏ Compare the directions that the two gears turn. ✏ What objects have you seen or used with gears that work in a similar way? TIP: Think of a chain instead of a rubber band.	Give students ample opportunity to solve the problem. (See Sample Answers for student drawing example.) Encourage groups to observe each other and share ideas. If students get really stuck, suggest the following: • Position the large rubber band so that it surrounds the outside of both gears. • With a partner, pull the gears away from each other until the rubber band clings to the edges of the gears and the gears' teeth are not meshed together. Hold the gears in this position. (If students are using K'NEX, they can hold the gears in place using a K'NEX rod in the middle of each gear. If they are using other gears, they can use pencils or similar items that fit in the holes at the center of the gears.) • Turn the driver gear by hand. Both gears in this configuration turn the same way. The pedals and rear wheel of a bicycle are similarly connected using an arrangement of gears and a chain. This gear configuration is another way to transmit force from one location to another.

	Student Procedure	Teacher Notes
	Class Discussion	• Discuss with the class that larger gears turn fewer revolutions per second than smaller gears because larger gears travel a longer distance during one complete turn. The data table in step 1 shows that the large gear has a longer travel distance (circumference) than the medium-sized and small gears. • Explain to students that a gear train is useful for transmitting force from one place to another. As you add more gears to the train, an object turned by the driven gear can be a greater distance from the driver gear. The gear train also allows control of the direction that the driver and driven gears turn. If the number of gears is odd, the driven gear will turn in the same direction as the driver gear.
	Part B: Gearing Up to Crank a Fan	
1	**Build the crank fan model with two medium-sized gears according to the K'NEX assembly diagrams.**	Students will need color K'NEX assembly diagrams to build the model. (See *www.terrificscience.org/physicsez/*.)
2	**Turn the crank handle. Observe the motion of the two gears and the fan blades.** ✎ **Compare the directions that the two gears turn.** ✎ **Do the fan blades turn in the same direction as the crank?** ✎ **Does the speed of the fan blades seem to be slower than, the same as, or faster than the turning speed of the crank handle? Record your answer in the data table.**	The gears do not move in the same direction. The fan blades do not move in the same direction as the crank handle. The fan blades seem to move at the same speed as the crank handle.
3	**Stop the motion and attach a sticker dot to one of the fan blades. Using the sticker dot as a guide, select a starting position for the crank handle and the fan blade. Turn the crank handle slowly through six turns and count the number of times the fan blades turn.** ✎ **Record this information in the data table.** ✎ **In the data table, record the number of teeth on the driver gear (the gear attached to the crank handle) and the number of teeth on the driven gear (the gear attached to the fan blades).**	See Sample Answers.

Student Procedure	Teacher Notes
④ Replace the two medium-sized gears with one large gear and one small gear as shown in the K'NEX assembly diagrams. Put the crank handle on the shaft with the large gear and the fan blades on the shaft with the small gear. Put the crank fan on a book or at the edge of a table so the fan has room to rotate.	When replacing the gears, have students refer to the K'NEX assembly diagrams.
⑤ Turn the crank handle. ✎ In the data table, record whether the turning speed of the fan blades is slower than, the same as, or faster than the turning speed of the crank handle.	See Sample Answers.
⑥ Repeat step 3 with this new setup. ✎ Record your information in the data table.	See Sample Answers.
⑦ Switch the fan blades and crank so the crank is on the shaft with the small gear and the blades are on the shaft with the large gear. Repeat steps 5 and 6. ✎ Record your information in the data table.	See Sample Answers.
⑧ Calculate the gear ratio of each crank fan design by dividing the number of teeth on the driver gear by the number of teeth on the driven gear. ✎ Record your answers in the data table. ✎ Which of the various arrangements do you think is best and why?	See Sample Answers.
Assessment A	
❶ Review your observations and data for Parts A and B of this activity. ✎ Make a list of reasons why gears are useful; then, discuss your reasons with the class.	After they work individually or in small groups, give students a chance to share and discuss ideas as a class. Through discussion, relate what students did in Part A to the following reasons why gears are useful. • reversing the direction of motion (step 3 of Part A) • gearing down—decreasing the speed of rotation (step 5 of Part A) • gearing up—increasing the speed of rotation (step 6 of Part A) • keeping the direction of motion (and speed of rotation, if desired) synchronized (step 8 of Part A)

Student Procedure	Teacher Notes
Optional ☞	If possible, give students a chance to explore examples of a few common objects that use gears to operate. Ideally, allow students to disassemble the objects to see the gears. Examples include an oscillating lawn sprinkler (shows gearing down), electric screwdriver (shows gearing down), and rotary eggbeater (shows gearing up).
Assessment B	
1 On another piece of paper, write a short essay on the history of bicycles. Include answers to the following questions: • In early bicycles, where were the pedals attached? • How many times did the front wheel turn with each rotation of the pedals? • What controlled how far each turn of the pedals could move the bicycle? • Why were the front wheels of bicycles made larger and larger? • What innovation allowed bicycles to have two wheels of equal size but move just as fast as ones with the big front wheel? • How do "gearing up" and "gearing down" apply to modern bicycles?	In early bicycles, the pedals were attached to the front wheel. The front wheel turned once with each rotation of the pedals. The diameter of the front wheel controlled how far each turn of the pedals could move the bicycle. The front wheels were made larger and larger so that each turn of the pedals would move the bicycle farther and riders could travel faster. Pedals that turn a chain attached to the rear wheels allow use of two equal-sized wheels. For bicycles with gear-shifting capabilities, the gears that turn with the pedals and the gears attached to the rear wheel come in different sizes. The gears attached to the pedals drive the gears attached to the rear wheel. The rider can adjust which size gears are being used by shifting. If the front gear is smaller than the rear gear, the bicycle moves less than a full turn of the rear wheel with each turn of the pedals. This is gearing down. If the front gear is larger than the rear gear, the bicycle moves more than a full turn of the rear wheel with each turn of the pedals. This is gearing up.

Sample Answers

Where students follow the same procedure using the same materials, these answers are close to answers you can expect. Where students design their own experiment or model, students' results will vary.

	Number of Teeth	Travel Distance (Circumference)
small gear	14	7.9 cm
medium-sized gear	34	17.6 cm
large gear	82	40.8 cm

Example of data for Part A

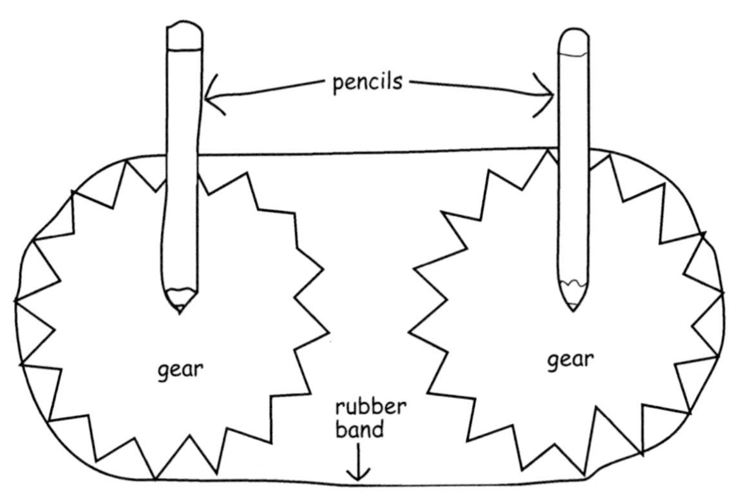

Example of student drawing for Part A

	Crank Fan Design		
	Two Medium-Sized Gears	Large Gear on the Crank Handle	Small Gear on the Crank Handle
turning speed of the fan blades compared to the crank handle	same	faster	slower
turns of the fan blades when crank handle is turned six times	6	36	1
number of teeth on the driver gear (gear on the crank handle)	34	82	14
number of teeth on the driven gear (gear on the fan blades)	34	14	82
approximate gear ratio	34:34 or 1:1	82:14 or about 6:1	14:82 or about 1:6

Example of data for Part B

Explanation

This section is intended for teachers. Modify the explanation for students as needed.

During this activity, students observe some properties of gears. A gear is a simple machine. It is a toothed wheel that precisely meshes with another toothed wheel. Two or more gears are used to change the turning speed of the gear or to change direction of motion.

Gears can be connected to one another in two ways. They can either mesh with one another or be connected by a chain. (See figure below.) Two gears that mesh will turn in opposite directions. Two gears that are connected by a chain turn in the same direction.

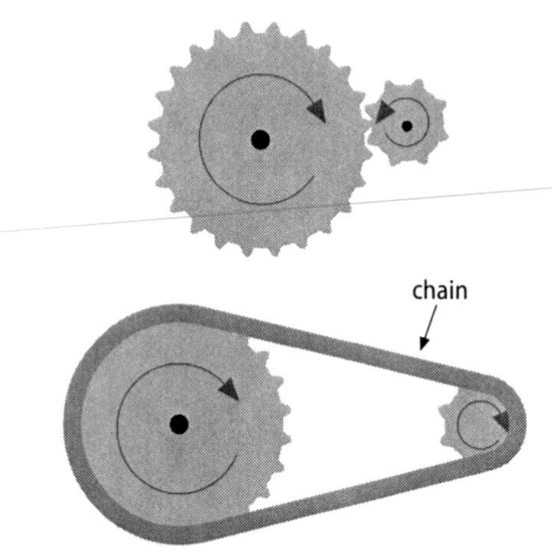

Gears can mesh with one another (top) or be connected by a chain (bottom).

Gears of different sizes can be used to change turning speed. A small gear connected to a large gear will turn faster than the large gear. This is because the total distance traveled by a tooth on the small gear must be the same as the distance traveled by a tooth on the large one. If the circumference of one circle is three times bigger than the other, then the smaller gear must turn three times around for every complete turn of the larger one. One turn of a 24-toothed gear will turn an 8-toothed gear three times. (See figure at top right.) Thus, the 8-toothed gear's speed is three times faster than the 24-toothed gear.

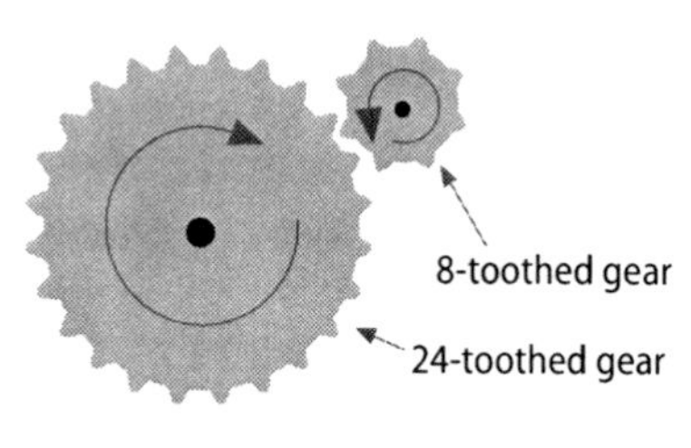

Gears can be used to change speed and direction.

Depending on whether the driver gear is larger or smaller than the driven gear, we call these changes in turning speed "gearing up" or "gearing down." Gearing up happens when a larger driver gear turns a smaller driven gear. One turn of the driver gear turns the driven gear more than once; therefore, the driven gear turns faster than the driver gear. You can use the ratio of the number of teeth in the driver gear to the number of teeth in the driven gear to figure out the relative speeds of the gears. For example, when the large gear drives the small gear in the crank fan model, this gear ratio is 82:14 or about 6:1. (The fan blades make six turns for every one turn of the crank handle so the fan blades turn six times faster than you turn the crank.)

Gearing down happens when a smaller driver gear turns a larger driven gear. One turn of the driver gear turns the driven gear less than one rotation; therefore, the driven gear turns slower than the driver gear. For example, when the small gear drives the large gear in the crank fan model, this gear ratio is 14:82 or about 1:6. (The fan blades make ⅙ of a turn for every one turn of the crank handle so the fan blades turn six times slower than you turn the crank.)

In an electric screwdriver or oscillating lawn sprinkler, the power source (an electric motor or water, respectively) turns a small driver gear that drives a set of larger gears. These larger gears in turn drive other, still larger gears. The effect of these gear trains is to slow (gear down) the small gear's initial rotation and to turn the screw or move the sprinkler head at a practical speed.

Cross-Curricular Integration

Social studies:

- Study the invention of labor-saving machines. Students may enjoy reading *Steven Caney's Invention Book* (grades 4–6).
  This book lists ideas for inventions, gives projects, and tells the stories behind some well-known inventions.

References

HowStuffWorks Website. How Bicycles Work. http://www.howstuffworks.com (accessed February 19, 2005).

HowStuffWorks Website. How Gear Ratios Work. http://www.howstuffworks.com (accessed February 19, 2005).

HowStuffWorks Website. How Oscillating Sprinklers Work. http://www.howstuffworks.com (accessed February 19, 2005).

HowStuffWorks Website. Inside a Wind-Up Alarm Clock. http://www.howstuffworks.com (accessed February 19, 2005).

HowStuffWorks Website. Inside an Electric Screwdriver. http://www.howstuffworks.com (accessed February 19, 2005).

Part A: How Do Gears Operate?

① Gears are a type of wheel, and wheels are simple machines. Count the number of teeth on each of the three different-sized gears.

Record the number of teeth for each gear in the data table.

	Number of Teeth	Travel Distance (Circumference)
small gear		
medium-sized gear		
large gear		

② Lay one gear flat. Turn the gear with your hand so the gear goes around one time.

How can you measure the distance that one tooth of the gear travels when the gear goes around one time?

....

....

Try your idea and record the travel distance measurement for each gear in the data table.

③ Lay the two medium-sized gears side by side. Mesh the gear teeth together so that the marked tooth from each gear touches the other. Use a sticker dot to label one gear as the driven gear. Use a sticker dot of another color to label the other gear as the driver gear. Watching both marked teeth, turn the driver gear with your hand.

TIP: As you turn, gently press the driver gear towards the driven gear to keep both gears meshed together. If you are using K'NEX® gears, keep the driven gear in place by putting a red K'NEX rod upright in the center of that gear.

Record the dot colors for the driven and driver gears.

....

Compare the directions that the two gears turn.

....

4. Begin again with the marked teeth touching. Watching the marked teeth, turn the driver gear with your hand so that its mark makes one complete revolution.

   ✎ How many turns does the driven gear make?

   ....

5. Repeat step 3, this time using one medium-sized gear and one large gear. Use colored dots as in step 3 to label the large gear as the driven gear and the medium-sized gear as the driver gear. Turn the medium-sized, driver gear with your hand so that its mark makes one complete revolution.

   ✎ How many turns does the large gear make?

   ....

   ✎ Compare the speed of the large gear to the speed of the driver gear.

   ....

   ....

   ....

6. Repeat step 3 again, but switch the colored dots so the medium-sized gear is the driven gear and the large gear is the driver gear. Turn the large, driver gear with your hand so that its mark makes one complete revolution.

   ✎ How many turns does the medium-sized gear make?

   ....

   ✎ Compare the speed of the medium-sized gear to the speed of the driver gear.

   ....

   ....

   ....

⑦ Make a simple gear train as shown by placing a small gear between a large gear and a medium-sized gear. Mesh the gear teeth together. As in step 3, use colored dots to label the large gear on the left as the driven gear and the medium-sized gear on the right as the driver gear.

Predict what would happen to the two other gears when you rotate the driver gear one turn clockwise. On the image above, use arrows to show your predictions about the directions all of the gears will turn and the number of turns for each gear.

Keeping the gears' teeth meshed together and the driven gear and small gear in place, test your prediction. Record your results and describe how your results compare with your predictions.

....

....

....

⑧ Can one gear drive another without touching it? Using two gears and a rubber band, work with a partner to get a driver gear to move a driven gear. Don't forget—this time the two gears cannot touch each other!

Draw and label your design on another piece of paper.

Compare the directions that the two gears turn.

....

What objects have you seen or used with gears that work in a similar way?

TIP: Think of a chain instead of a rubber band.

Part B: Gearing Up to Crank a Fan

1. Build the crank fan model with two medium-sized gears according to the K'NEX assembly diagrams.

2. Turn the crank handle. Observe the motion of the two gears and the fan blades.

   Compare the directions that the two gears turn.

   Do the fan blades turn in the same direction as the crank?

   Does the speed of the fan blades seem to be slower than, the same as, or faster than the turning speed of the crank handle? Record your answer in the data table.

	Crank Fan Design		
	Two Medium-Sized Gears	Large Gear on the Crank Handle	Small Gear on the Crank Handle
turning speed of the fan blades compared to the crank handle			
turns of the fan blades when crank handle is turned six times			
number of teeth on the driver gear (gear on the crank handle)			
number of teeth on the driven gear (gear on the fan blades)			
approximate gear ratio			

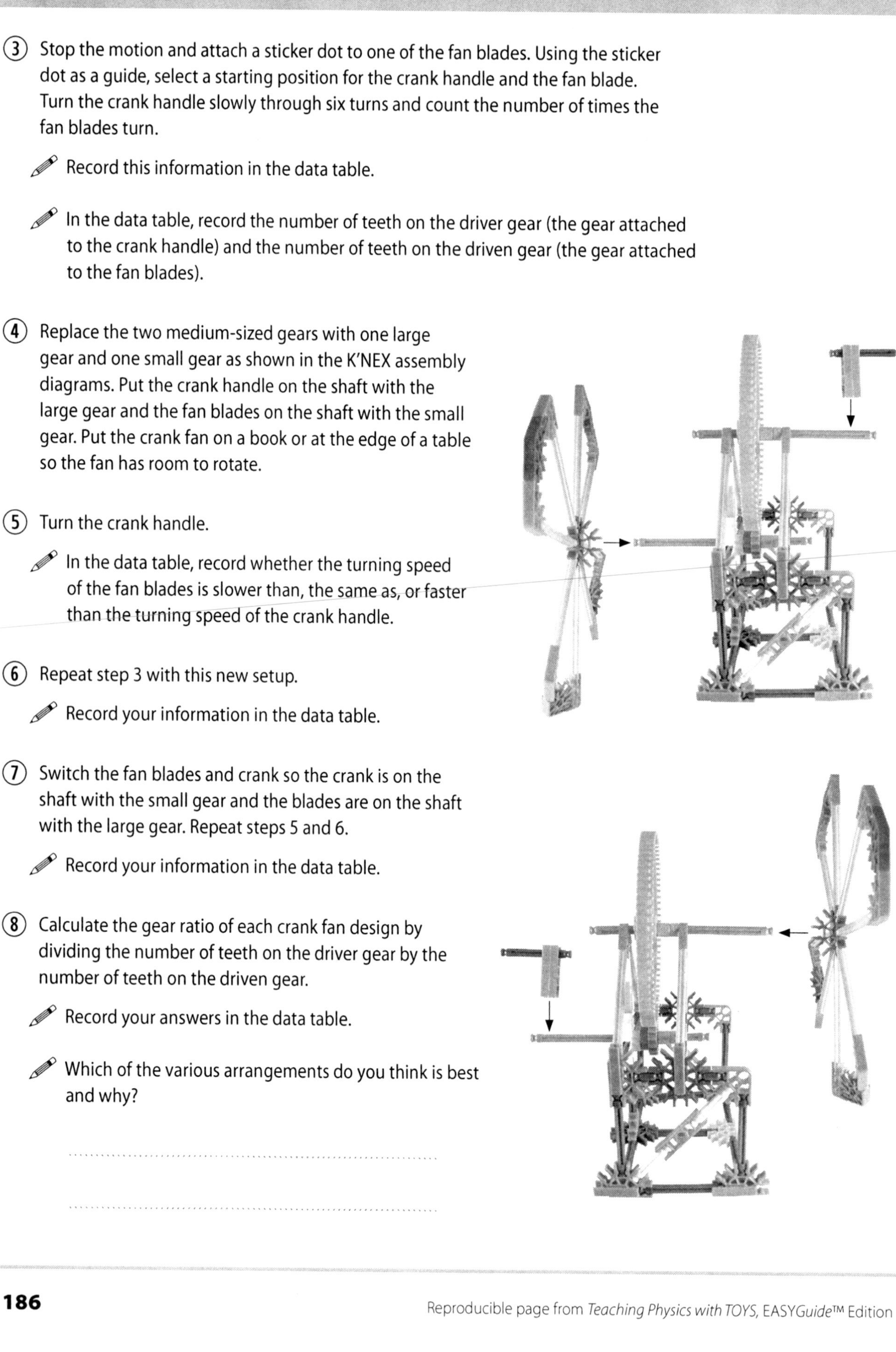

③ Stop the motion and attach a sticker dot to one of the fan blades. Using the sticker dot as a guide, select a starting position for the crank handle and the fan blade. Turn the crank handle slowly through six turns and count the number of times the fan blades turn.

✎ Record this information in the data table.

✎ In the data table, record the number of teeth on the driver gear (the gear attached to the crank handle) and the number of teeth on the driven gear (the gear attached to the fan blades).

④ Replace the two medium-sized gears with one large gear and one small gear as shown in the K'NEX assembly diagrams. Put the crank handle on the shaft with the large gear and the fan blades on the shaft with the small gear. Put the crank fan on a book or at the edge of a table so the fan has room to rotate.

⑤ Turn the crank handle.

✎ In the data table, record whether the turning speed of the fan blades is slower than, the same as, or faster than the turning speed of the crank handle.

⑥ Repeat step 3 with this new setup.

✎ Record your information in the data table.

⑦ Switch the fan blades and crank so the crank is on the shaft with the small gear and the blades are on the shaft with the large gear. Repeat steps 5 and 6.

✎ Record your information in the data table.

⑧ Calculate the gear ratio of each crank fan design by dividing the number of teeth on the driver gear by the number of teeth on the driven gear.

✎ Record your answers in the data table.

✎ Which of the various arrangements do you think is best and why?

....

....

Assessment A

1. Review your observations and data for Parts A and B of this activity.

   Make a list of reasons why gears are useful; then, discuss your reasons with the class.

Assessment B

1. On another piece of paper, write a short essay on the history of bicycles. Include answers to the following questions:

   - In early bicycles, where were the pedals attached?
   - How many times did the front wheel turn with each rotation of the pedals?
   - What controlled how far each turn of the pedals could move the bicycle?
   - Why were the front wheels of bicycles made larger and larger?
   - What innovation allowed bicycles to have two wheels of equal size but move just as fast as ones with the big front wheel?
   - How do "gearing up" and "gearing down" apply to modern bicycles?

Measuring Mass

Students use nonstandard and standard units of measure to determine and compare the masses of different materials.

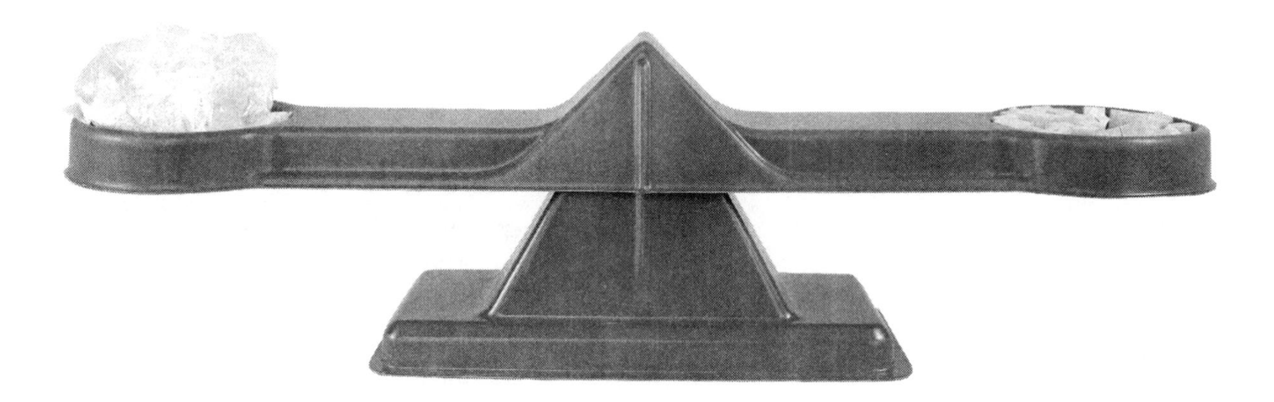

Grade Levels

Science activity appropriate for grades 3–4

Student Background

It is helpful if students have had some experience with pan balances.

Time Required

Setup	30	minutes
Part A	20	minutes
Part B	20	minutes
Cleanup	10	minutes

Assessment time is not included.

Key Science Topics

- mass
- standard units of measure

National Science Education Standards Overview

See *www.terrificscience.org/physicsez/* for details of how these standards relate to the activity.

Science as Inquiry

Abilities Necessary to Do Scientific Inquiry

K–4 *Employ simple equipment and tools to gather data and extend the senses.*

K–4 *Use data to construct a reasonable explanation.*

K–4 *Communicate investigations and explanations.*

Physical Science

K–4 *Properties of objects and materials*

Materials

For Getting Ready
Per class
- modeling dough (such as Play-Doh®)
- measuring cup

Per group
- 6 small zipper-type plastic bags
- about ¼ cup uncooked rice
- 20 g or less of 5 different materials such as:
  - sand
  - puffed rice
  - unpopped popcorn
  - popped popcorn
  - small buttons

For the Procedure
Activity Introduction, per class
- pan balance
- 4 objects (of equal mass) large enough to be seen from the back of the classroom when in the pan balance (such as Ping-Pong balls)
- 2 balls of modeling dough of equal mass prepared in Getting Ready

Part A, per group
- 6 bags of different materials prepared in Getting Ready
- pan balance
- assorted objects that fit on the balance pan (such as pennies, paper clips, and marbles)
- gram unit cubes
- graph paper

Graph paper masters for copying are available at www.terrificscience.org/physicsez/.

Part B, per group
- 3 different objects (of different mass)
- bag of rice from Part A
- gram unit cubes
- pan balance

For the Assessment
Per group
- 4 different objects (of different mass) not used in Part B
- gram unit cubes
- pan balance

Safety and Disposal

No special safety or disposal procedures are required.

Getting Ready

- Prepare two balls of modeling dough of equal mass.
- Prepare six zipper-type plastic bags for each group.
  - Put approximately ¼ cup uncooked rice in one bag.
  - In each of the five remaining bags, put 20 g or less of one of the five different materials.
  - Determine the mass of each bag ahead of time using gram unit cubes as units of mass. Adjust the amount of material in each bag so it balances with a whole number of cubes. Make sure that all the bags of rice have the same mass. (Bags of other material do not need to be equal in mass.)

Procedure

This section provides teacher notes corresponding to each step of the student procedure. The procedure without teacher notes is included in the reproducible Student Notebook pages at the end of this activity and at www.terrificscience.org/physicsez/.

Student Procedure	Teacher Notes
Activity Introduction	If your students do not have experience with a pan balance, conduct a class demonstration as follows: • Introduce the pan balance and its use (to measure the mass of objects). Look at its parts. • Demonstrate that the pan balance tips down on the side holding the greater mass. Also show students that the pans balance when the masses on both sides of the balance are equal. Use your four objects of equal mass in the following ratios to illustrate these concepts: ∘ two objects in one pan and no objects in the other pan, ∘ two objects in one pan and one object in the other pan, and ∘ two objects in one pan and two objects in the other pan. • Demonstrate that mass does not change when something changes shape or is broken into pieces. Begin with two balls of modeling dough that have the same mass. Use a pan balance to show the class that the two balls have the same mass, and then make one of the balls into a snake. After asking the class to estimate whether the mass of the ball and the mass of the snake are different, use the pan balance to reveal the result. Break the snake into pieces and ask the class to estimate whether the masses of the ball and the dough pieces are different. Use the pan balance to reveal the result.
Part A: Setting the Standard	
① **Place a bag of rice on one side of a pan balance. Add any assortment of objects to the other side until the pans balance.** **Record the type and number of objects that equal the mass of the bag of rice.**	Provide each group with a bag of rice (all bags of equal mass) and assorted objects such as pennies, paper clips, and marbles. Don't provide the gram unit cubes until step 3. Each group will probably use a different combination of objects to balance the bag of rice. Differences in student results are expected and will lead to the discussion after step 2.

	Student Procedure	Teacher Notes
2	Share your results with the class. ✎ Do all of the bags of rice have the same mass? Explain your answer.	Students should realize that, since each group used a different combination of objects to balance the mass of the rice, deciding whether all of the bags of rice have the same mass is very difficult.
	Class Discussion	• Tell the class that, when doing experiments, scientists have to be able to compare results. Ask the class what they could have done differently to be able to compare their results. • Through discussion, help students see that everyone would have to balance the mass of the rice using the same kind of object. Explain that this object becomes a standard unit of measure. • Tell the class that they will use the cubes as a standard unit of measure in the rest of the activity.
3	This time, determine the mass of the same bag of rice using cubes. ✎ Record the number of cubes that equals the mass of the bag of rice.	Students should all get the same results based on the mass determined in Getting Ready.
4	Share your results with the class. ✎ Do all of the bags of rice have the same mass? Explain your answer.	By using the same standard unit of measure, students can determine that all bags of rice have the same mass.
5	Determine the mass of different bags of materials using the pan balance and gram unit cubes. TIP: One gram unit cube equals one gram (g). ✎ On another piece of paper, create a data table and record your data. On graph paper, make a bar graph showing your results. ✎ Can you compare the masses of the different bags? Explain your answer. ✎ Look at your bar graph. List the bags from least massive to most massive.	Students should realize that, since the same unit of measure is used to measure all of the bags, they can compare the masses of the bags. (See Sample Answers for data and graph examples.)
	Class Discussion	• Discuss the idea that, although the cubes are fine as a unit of measure in the classroom where we are sharing information only among ourselves, scientists need a standard unit of mass that everyone in the world agrees on.

Student Procedure		Teacher Notes
	Class Discussion (continued)	• Introduce the gram as a standard unit of mass. Explain that each of the cubes they have been using has a mass of 1 gram. • For the rest of the activity, students should report their results in grams.
	Part B: Educated Estimates	
①	Use one hand to pick up the object provided by your teacher. Hold the bag of rice from Part A in your other hand and compare the masses. Based on what you know about the mass of the bag of rice, estimate the mass of the object and record your estimate in the data table.	Students should make their best estimate of the object's mass based on the known mass of the bag of rice (from step 5 in Part A).
②	Use one hand to pick up the object again. In your other hand, pick up the number of gram unit cubes that equal your mass estimate in step 1. If you are not satisfied with your first estimate, record a second estimate in the data table.	Some students may change their estimates after comparing the object's mass to the number of gram unit cubes equaling their first estimate.
③	Use a pan balance to determine the actual mass of the object. Record the actual mass in the data table.	
④	Repeat steps 1–3 with two more objects. How do your estimates compare to the actual masses?	Some students may notice that they got better at estimating the mass as they got to the second and third objects.
	Assessment	
❶	Select a set of four different objects. Without using a balance, estimate the mass of each individual object. Record the names of your objects in the data table. Record your estimates in the data table column labeled "Step ❶ First Estimate." Add the four estimated masses together to calculate the estimated mass for the total set of objects. Record this number.	Students should hold the four individual objects in their hands to estimate the individual masses. Students should then add the estimated masses of all four objects together to calculate the estimated mass of the total set of objects.

	Student Procedure	Teacher Notes
2	Determine the mass of the first object using the pan balance. Think about how this information might change your previous estimates. ✎ Record the actual mass of the first object in the last column of the data table. ✎ Now that you know the mass of the first object, you may not be satisfied with your first estimates of the other three objects. If needed, make and record second mass estimates for the remaining three objects. Use the mass of the first object plus the estimated masses of the other three objects to recalculate the estimated mass for the total set of objects. ✎ Explain the reasons for your adjustments.	If students have trouble understanding how knowing the mass of one object might affect their estimates, pose these questions. • What if one object has a mass greater than your estimate for the entire set? • What if one object has a mass of about half of your estimate for the set, but the other three objects feel heavier than the one for which you already determined the mass?
3	Use the pan balance and gram unit cubes to measure the mass of the total set of objects. Think about how this information might change your previous estimates. ✎ Record the actual mass of the total set of objects in the last column of the data table. ✎ Now that you know the mass of the set, you may not be satisfied with your second estimates. If needed, make and record third mass estimates for the remaining three objects. ✎ Explain the reasons for your adjustments.	
4	Using the pan balance, measure the mass of each remaining object. ✎ Record the actual masses in the last column of the data table. ✎ Compare your estimates to the actual masses. How did your accuracy change over the course of this assessment?	

Sample Answers

Where students follow the same procedure using the same materials, these answers are close to answers you can expect. Where students design their own experiment or model, students' results will vary.

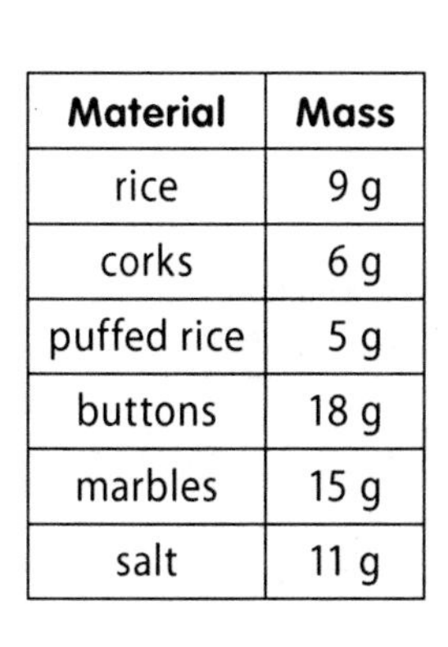

Material	Mass
rice	9 g
corks	6 g
puffed rice	5 g
buttons	18 g
marbles	15 g
salt	11 g

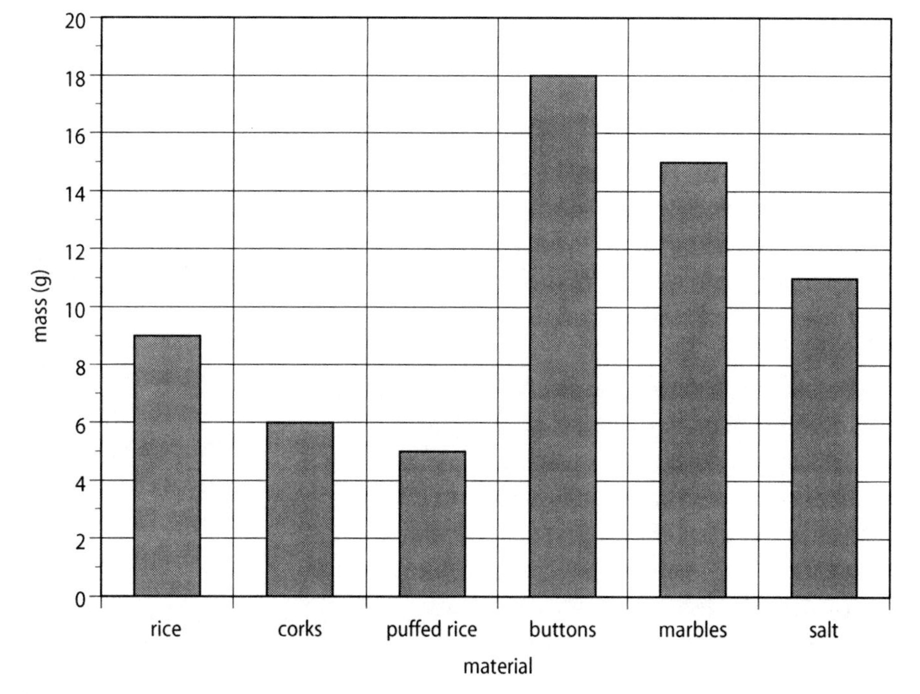

Example of data and graph for Part A

Explanation

This section is intended for teachers. Modify the explanation for students as needed.

Mass and Weight

Mass is a measure of the quantity of matter in an object. Even though we sometimes use the terms heavy and light when referring to mass, mass is not the same thing as weight. Mass is an intrinsic property of the object, but weight is not. For simplicity, we have used grams as a standard unit of mass in the activity. However, the kilogram (equal to 1,000 grams) is actually the scientific standard for mass.

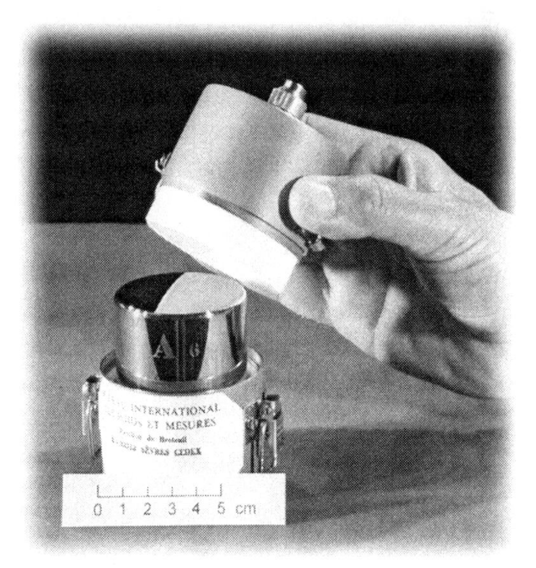

The standard kilogram is kept at the International Bureau of Weights and Measures near Paris, France.

The weight of an object is the gravitational force that the earth exerts on an object. An object's weight depends on the mass of the object and its location. For instance, the weight of an object will vary slightly if it is moved from sea level to the top of a high mountain, where it is further from the center of the earth. The object would also have a different weight on other planets or moons, because the gravitational force there is different than on Earth. Weight is measured with a spring scale and has units of newtons in *Le Système International d'Unités,* usually abbreviated as the SI (also called metric) system, and pounds in the United States Customary System (USCS).

Teaching the Concept of Mass

Younger students should not be expected to fully grasp the distinction between mass and weight, but teachers should be careful to use the terms correctly themselves. Early experiences help students develop an awareness of mass as a property of objects.

The first step in developing an understanding of mass is directly comparing the mass of two objects using a balance. Students require practical experience to understand conservation of mass (such as changing the shape of a ball of modeling dough and seeing that the mass remains equal). They also need practical experience to disassociate mass from size and accept that a small object can be more massive than a large one.

Once students have developed the understandings of mass just discussed, they are ready to explore nonstandard units as measuring devices. Nonstandard units can reinforce basic concepts about mass. But once they become proficient at measuring mass with nonstandard units, they are ready to see that such units are limited as a means of communication. Students are now ready to practice measuring mass using standard units.

Students should be encouraged to estimate the masses of objects in order to develop a sense of the size of the unit. Estimating is also good practice for situations in daily life where exact measurements are not always available.

Cross-Curricular Integration

Language arts:

- Read aloud or suggest that students read the following book:
  - *Who Sank the Boat?* by Pamela Allen (grades K–3) *Good animal friends go for a boat ride, but the boat sinks. Someone sank the boat!*

Math:

- Have students practice measuring and estimating.

References

David, A. *Understanding Mathematics Five Teachers' Resource Book;* Shortland: New Zealand, 1984; pp 107–109.

Ferruggia, A. et al. *Silver Burdett Science;* Teacher Ed.; Silver Burdett: Morristown, NJ, 1987; pp 66–67.

Hewitt, P.G. *Conceptual Physics,* 9th ed.; Addison Wesley: San Francisco, 2002, pp 58–59, 151–161, 739.

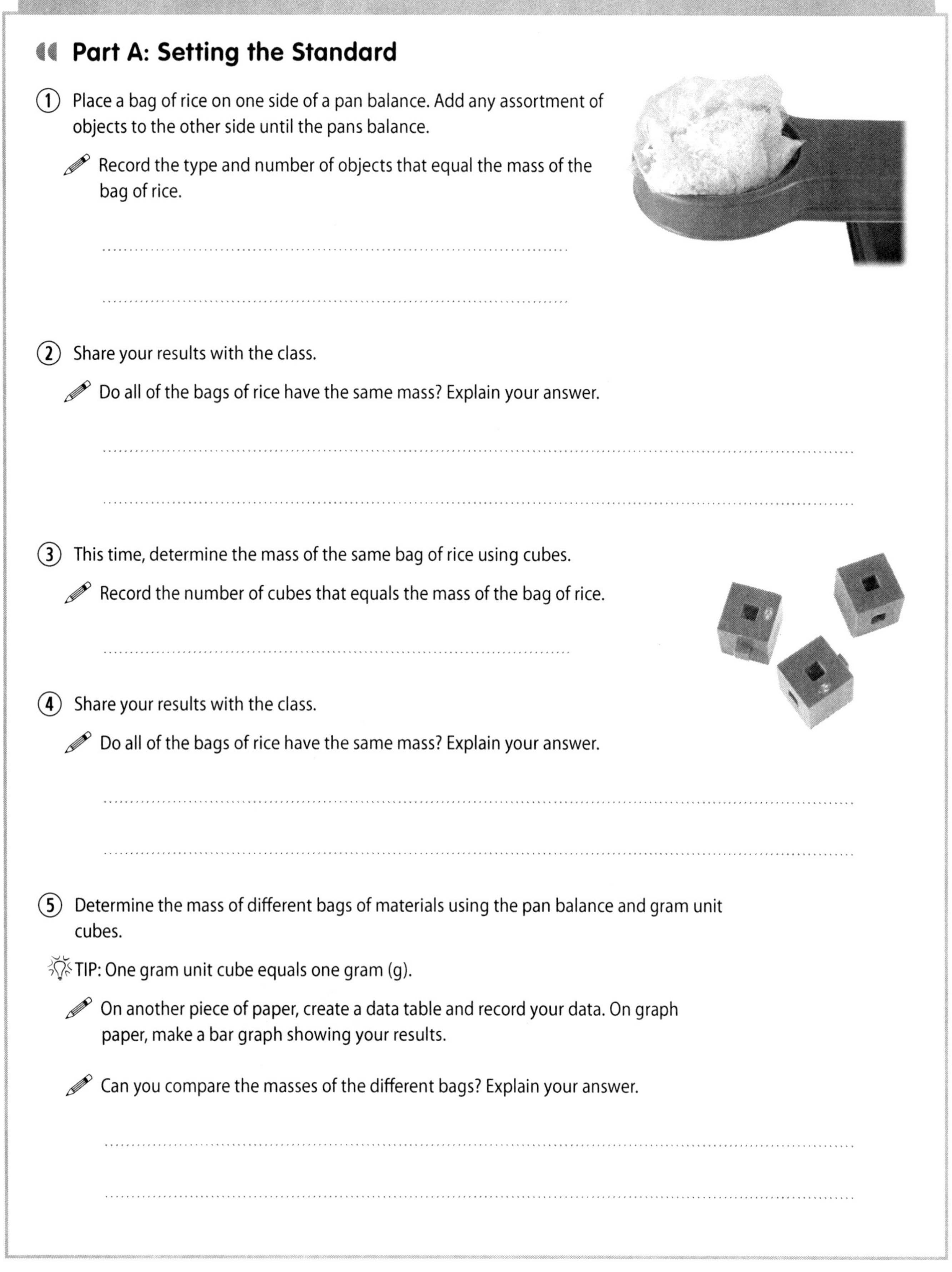

Part A: Setting the Standard

① Place a bag of rice on one side of a pan balance. Add any assortment of objects to the other side until the pans balance.

Record the type and number of objects that equal the mass of the bag of rice.

② Share your results with the class.

Do all of the bags of rice have the same mass? Explain your answer.

③ This time, determine the mass of the same bag of rice using cubes.

Record the number of cubes that equals the mass of the bag of rice.

④ Share your results with the class.

Do all of the bags of rice have the same mass? Explain your answer.

⑤ Determine the mass of different bags of materials using the pan balance and gram unit cubes.

TIP: One gram unit cube equals one gram (g).

On another piece of paper, create a data table and record your data. On graph paper, make a bar graph showing your results.

Can you compare the masses of the different bags? Explain your answer.

✎ Look at your bar graph. List the bags from least massive to most massive.

Part B: Educated Estimates

① Use one hand to pick up the object provided by your teacher. Hold the bag of rice from Part A in your other hand and compare the masses.

✎ Based on what you know about the mass of the bag of rice, estimate the mass of the object and record your estimate in the data table.

Object	Step ① First Estimate	(Optional) Step ② Second Estimate	Actual Mass

② Use one hand to pick up the object again. In your other hand, pick up the number of gram unit cubes that equal your mass estimate in step 1.

✎ If you are not satisfied with your first estimate, record a second estimate in the data table.

③ Use a pan balance to determine the actual mass of the object.

✎ Record the actual mass in the data table.

④ Repeat steps 1–3 with two more objects.

✎ How do your estimates compare to the actual masses?

Assessment

1. Select a set of four different objects. Without using a balance, estimate the mass of each individual object.

   Record the names of your objects in the data table. Record your estimates in the data table column labeled "Step 1 First Estimate." Add the four estimated masses together to calculate the estimated mass for the total set of objects. Record this number.

Object	Step 1 First Estimate	(Optional) Step 2 Second Estimate	(Optional) Step 3 Third Estimate	Actual Mass
Total Set of Objects				

2. Determine the mass of the first object using the pan balance. Think about how this information might change your previous estimates.

   Record the actual mass of the first object in the last column of the data table.

   Now that you know the mass of the first object, you may not be satisfied with your first estimates of the other three objects. If needed, make and record second mass estimates for the remaining three objects. Use the mass of the first object plus the estimated masses of the other three objects to recalculate the estimated mass for the total set of objects.

   Explain the reasons for your adjustments.

3 Use the pan balance and gram unit cubes to measure the mass of the total set of objects. Think about how this information might change your previous estimates.

- Record the actual mass of the total set of objects in the last column of the data table.

- Now that you know the mass of the set, you may not be satisfied with your second estimates. If needed, make and record third mass estimates for the remaining three objects.

- Explain the reasons for your adjustments.

4 Using the pan balance, measure the mass of each remaining object.

- Record the actual masses in the last column of the data table.

- Compare your estimates to the actual masses. How did your accuracy change over the course of this assessment?

Seesaw Forces

Students make toy seesaws that demonstrate how levers work and how balance is achieved.

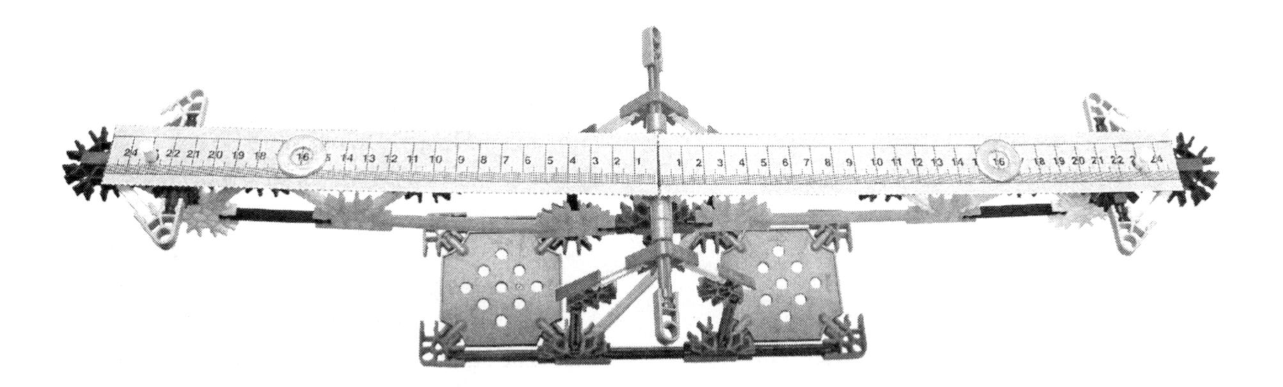

Grade Levels

Science activity appropriate for grades 4–7

Student Background

Students should have a basic understanding of gravity and how a balance works. The activity "Measuring Mass" in this book teaches these concepts.

Time Required

Setup	15	minutes
Build K'NEX® model	15	minutes
Part A	15	minutes
Part B	15	minutes
Cleanup	10	minutes

Assessment time is not included.

Key Science Topics

- balance
- first-class lever (simple machine)

National Science Education Standards Overview

See *www.terrificscience.org/physicsez/* for details of how these standards relate to the activity.

Science as Inquiry

Abilities Necessary to Do Scientific Inquiry

K–4 *Employ simple equipment and tools to gather data and extend the senses.*

K–4 *Use data to construct a reasonable explanation.*

5–8 *Use appropriate tools and techniques to gather, analyze, and interpret data.*

5–8 *Develop descriptions, explanations, predictions, and models using evidence.*

Physical Science

5–8 *Motions and forces*

Science and Technology

Abilities of Technological Design

K–4 *Evaluate a product or design.*

5–8 *Evaluate completed technological designs or products.*

Materials

For Getting Ready
Per group
- card stock
- scissors
- rubber cement (for teacher use only)

For the Procedure
Per class
- scissors
- tape
- hole punch

Part A, per group
- assorted K'NEX rods, connectors, and panels

For specifics, see K'NEX Parts Inventory at www.terrificscience.org/physicsez/ or www.terrificscience.org/supplies/.
- K'NEX assembly diagrams
- 2 card stock rulers prepared in Getting Ready
- index card
- 3 identical metal washers

Washers with an outside diameter of about 2.5 cm work well.

Part B, per group
- K'NEX seesaw
- 3, 5-ounce paper cups
- small objects that fit into a 5-ounce cup (such as paper clips and buttons)
- 2 green K'NEX rods

For the Assessment
Per group
- K'NEX seesaw with equal-sized arms and card stock rulers attached
- 2 different small toys that fit on the seesaw

Make sure each group uses two toys that have different masses.

Safety and Disposal

Rubber cement has strong fumes. Use it in a well-ventilated room, preferably before the students enter. No special disposal procedures are required.

Getting Ready

- Print a set of K'NEX assembly diagrams for each group. (See *www.terrificscience.org/physicsez/*.) Since K'NEX pieces are color coded, the assembly diagrams need to be printed in color.

Depending on your schedule, you may want to have students build models in advance.
- Photocopy the ruler template onto card stock. (The ruler template is provided on the last page of this activity and at *www.terrificscience.org/physicsez/*.) Each group needs a left- and right-side ruler. Carefully cut out the rulers along the dashed lines.
- Paint a thin coating of rubber cement down the middle of each ruler. You only need to cover the area where the numbers are printed. This coating keeps the washers from slipping when the seesaws tip. Let dry.
- Test to make sure that the small objects you select to use in Part B are heavy enough to make a substantial difference in steps 2–5.

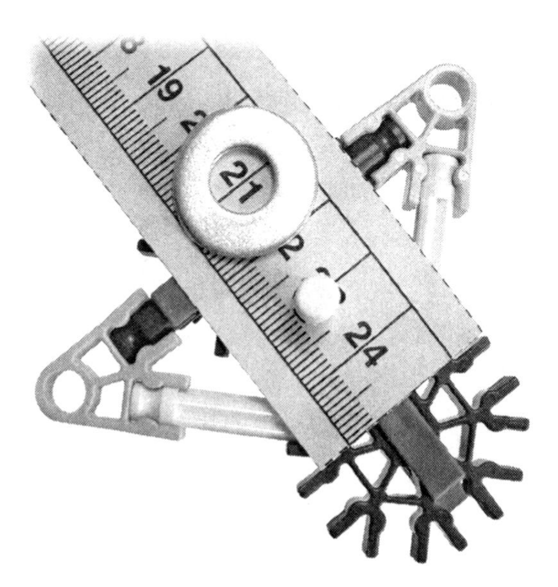

Procedure

This section provides teacher notes corresponding to each step of the student procedure. The procedure without teacher notes is included in the reproducible Student Notebook pages at the end of this activity and at www.terrificscience.org/physicsez/.

	Student Procedure	Teacher Notes
	Part A: Balance the Seesaw	
1	Build and prepare the seesaw model according to the "How to Make a Seesaw" instruction sheet.	Students will need color K'NEX assembly diagrams to build the model. (See *www.terrificscience.org/physicsez/*.) The "How to Make a Seesaw" instruction sheet provides further instructions on preparing the seesaw.
2	Carefully put one washer on each end of the seesaw so that the seesaw remains balanced. ✎ Record the distance each washer is from the center of the seesaw. ✎ Why do you think the seesaw remains balanced?	Students will probably say that the seesaw balances because the washers are the same distance from the seesaw's fulcrum (pivot point). Be sure students also realize that the washers must have the same mass to balance the seesaw in this position.
3	Predict what would happen if you stacked two washers at one end of the seesaw and kept only one washer at the other end. ✎ Write your prediction.	Students may realize that the end of the seesaw with two washers will be lower than the end with one washer.
4	Check your prediction by putting another washer on top of a washer you already have on the seesaw. ✎ What happens?	The seesaw will go down on the side with the two washers because two washers have more mass than one washer.
5	Balance the seesaw without adding or taking away any washers. The only rule is that you must keep the two stacked washers together. ✎ On the image below, draw where you put the washers to balance the seesaw. Show which side has two washers.	Students must move the two washers as a unit. The stacked washers and single washer can be put anywhere on the seesaw. Through trial and error, students will find out that the seesaw balances when the single washer is twice as far from the fulcrum as the two washers are.
6	Stack two washers together on the seesaw's left arm at the 12-cm location. Place a single washer on the right arm so that the seesaw balances. ✎ In the data table, record the location of the single washer. Continue the procedure until you finish the data table.	See Sample Answers for expected results.

	Student Procedure	Teacher Notes
7	Look at the numbers in the step 6 data table. ✎ What relationship do you see between each pair of distances?	Students should state that the seesaw balances when the single washer is twice as far from the fulcrum as the two washers are.
8	Place two washers on the seesaw's left arm at the 6-cm location. Predict where a washer needs to go on the right arm so that the seesaw balances. ✎ Record your prediction in the data table. ✎ Test your prediction and record the result in the data table. Repeat predicting and testing with the 4-cm location to finish the data table.	See Sample Answers for expected results.
9	Complete the columns in the following table. ✎ Compare the results of the seesaw's left and right arms. What patterns do you see?	See Sample Answers for expected results. Students should observe that when the seesaw is balanced, the products of Columns A and B equal the products of Columns C and D. Also, the products decrease as the washers are moved towards the fulcrum. (See Explanation for age-appropriate discussions.)
	Part B: More About Levers	**Part B briefly introduces the relationship between force, distance, and work. This concept is more appropriate for older students (grades 7–9). After Part B, older students can continue with a more quantitative exploration in the "Levers at Work" activity in this book. For younger students, Part B can be done simply as an observation on how adjusting the seesaw arms affects effort force.**
1	Poke a hole in the bottom of each of two 5-ounce paper cups. Remove the ruler from the seesaw and push one cup over the white upright rod at each end of the seesaw as shown.	
2	Fill a third 5-ounce cup with objects to act as a load. Place the load (cup of objects) in one of the cups on the seesaw. The weight of this load is known as the load force. Push all the way down on the opposite end of the seesaw to feel the resistance. This push is known as the effort force. ✎ Does the effort arm (the one you pushed all the way down on) move down more than, the same as, or less than the arm that contains the load moves up?	Load force is sometimes referred to as resistance force. Mention this term if it is used in your classroom textbook or if students are familiar with the term from previous experience. Since the arms are of equal length, they move up and down the same distance (d). Students should try to remember how much effort force they need to lift the load, since they will be comparing step 2 and step 3 effort forces.

	Student Procedure	Teacher Notes
3	Take the load cup and place it on the other end of the seesaw. Push all the way down on the end of the seesaw without the load. ✏ Is there any difference in the amount of effort force in step 2 and the amount of effort force in step 3? ✏ Is there any difference in the distance the arms move when you push one side all the way down?	There is no difference in the amount of effort force in steps 2 and 3. The effort force used to lift the loads in steps 2 and 3 should be equal, since the loads in steps 2 and 3 have the same load force and are the same distance from the fulcrum. The arms still move up and down the same distance.
4	On one side of the seesaw's arms, replace the two long yellow K'NEX rods nearest the fulcrum with two short green ones. Place the load on the long arm. Push on the short arm and feel the resistance. ✏ Compared to the equal arm seesaw, is it now easier or harder to lift the load? ✏ Does the short arm move up and down more or less than the long arm?	Students will feel that it is harder to lift the load. (More force is needed to lift the load.) Explain that, since the distance from the fulcrum is now shorter on the effort side of the seesaw, more effort force is needed to lift the load. The short arm moves up and down a short distance. The long arm moves up and down a long distance.
5	Place the load on the short arm. Push on the long arm and feel the resistance. ✏ Compared to step 4, is it now easier or harder to lift the load? ✏ Does the long arm move up and down more or less than the short arm?	Students will feel that it is easier to lift the load than in step 4. Explain that, because the distance from the fulcrum is now longer on the effort side of the seesaw than on the load side, less effort force is needed to lift the load. As in step 4, the short arm moves up and down a short distance. The long arm moves up and down a long distance.
	Class Discussion	Younger students should realize that the seesaw allows them to lift a load using less force. Through discussion, help them conclude that the load is easier to move when the arm they push on moves down a greater distance than the arm with the load moves. Help older students see that the amount of work done by a smaller force over a long distance (applied to the long arm) is the same amount of work as a large force over a short distance (applied to the short arm.) Students should realize that using the seesaw as in step 5 does not reduce the amount of work needed to lift the load. The seesaw just makes lifting the load easier.

	Student Procedure	Teacher Notes
	Assessment A	
1	Get two small toys from your teacher. Using a seesaw with equal-sized arms, put one toy on one end of the seesaw and the other toy on the other end. ✎ What happens?	The object with more mass will cause its side of the seesaw to go down.
2	Based on your experience with the seesaw, predict which toy you will need to move closer to the fulcrum for the seesaw to balance. ✎ Write down your prediction and explain your reasons for it.	Based on what they've learned, some students will know to move the toy with the most mass towards the fulcrum.
3	Test your prediction. ✎ What happens?	The toy having more mass should be moved closer to the fulcrum to make the seesaw balance.
	Assessment B	
1	Look at the picture. ✎ What can you tell about Person A compared to Person B? ✎ Draw an arrow on the picture showing in what direction one person needs to move in order to balance the seesaw.	See Sample Answers for example answer. Based on the photo, Person B has more mass than Person A so moving Person B closer to the center of the seesaw (closer to the fulcrum) should make the seesaw balance.
2	Look at the picture. ✎ On another piece of paper, draw a diagram showing how you could lift the rock using the ruler. Label the fulcrum, load, effort force, and load force.	See Sample Answers for example student diagram. Students should draw the ruler as a lever with the spool of thread underneath. The rock should be on the end of the ruler that touches the ground. The point where the spool touches the ruler is the fulcrum. The rock (the load) provides the load force on one end of the ruler. The person pushing the other end of the ruler down to lift the rock provides the effort force.

Sample Answers

Where students follow the same procedure using the same materials, these answers are close to answers you can expect. Where students design their own experiment or model, students' results will vary.

Distance from Fulcrum (Left Arm)—Two Stacked Washers	Distance from Fulcrum (Right Arm)—Two Stacked Washers
12 cm	24 cm
10 cm	20 cm
8 cm	16 cm

Example of data for step 6 of Part A

Distance from Fulcrum (Left Arm)—Two Stacked Washers	Distance from Fulcrum (Right Arm)—Two Stacked Washers	
	Prediction	Actual
6 cm	varies	12 cm
4 cm	varies	8 cm

Example of data for step 8 of Part A

Seesaw's Left Arm			Seesaw's Right Arm		
Ⓐ	Ⓑ	Ⓐ × Ⓑ	Ⓒ	Ⓓ	Ⓒ × Ⓓ
Number of Washers	Distance from Fulcrum	Product of Column A and B	Number of Washers	Distance from Fulcrum	Product of Column C and D
2	12 cm	24	1	24 cm	24
2	10 cm	20	1	20 cm	20
2	8 cm	16	1	16 cm	16
2	6 cm	12	1	12 cm	12
2	4 cm	8	1	8 cm	8

Example of data for step 9 of Part A

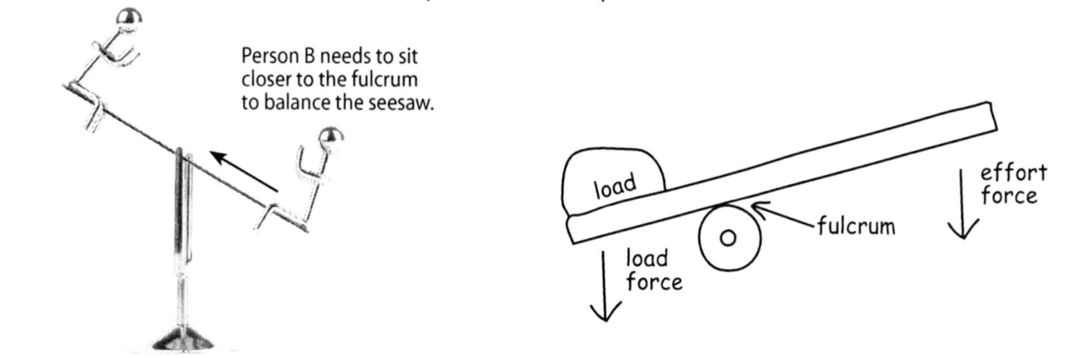

Example of student answers for Assessment B

Explanation

This section is intended for teachers. Modify the explanation for students as needed.

Seesaws and Balance

In Part A of this activity, students use a familiar type of first-class lever—a seesaw—to explore how masses balance on either side of the fulcrum. A first-class lever has the fulcrum (or pivot point) located on the lever arm between the load force (F_L) and the effort force (F_E). The first-class lever configuration is:

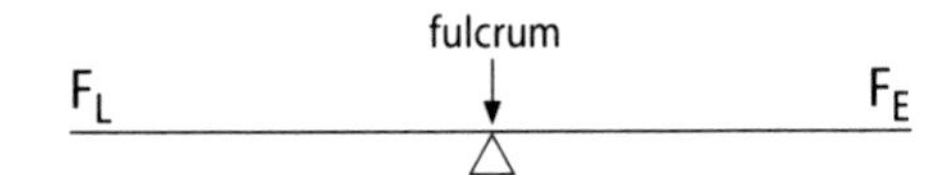

At the simplest level, students can conclude that something heavier can balance with something lighter when the heavier object is closer to the fulcrum. More specifically, they observe that if the load on one side is twice as heavy (two stacked washers) as the load on the other side (one washer), the lighter load needs to be twice as far from the fulcrum as the heavier one. If students have played on a seesaw, they can relate these observations to trying to seesaw with a bigger or smaller person. Through further data collection and calculations, students see a more general pattern emerge: the number of washers multiplied by the distance from the fulcrum on one side equals the number of washers multiplied by the distance from the fulcrum on the other side.

Torque

These observations provide a basis for introducing the concept of torque in high school physics. Torques cause objects to rotate, just as forces cause objects to move. In order for the seesaw to remain level, the clockwise torque applied by the washers on one side of the fulcrum must equal the counterclockwise torque applied by the washers on the other side. In other words, the total torque must be zero. If the torque on one side is greater than the torque on the other side, the seesaw will rotate so that the side with the larger torque is down.

The size of the torque produced by a force (such as the weight of the washers) depends on how far from the fulcrum the force is applied. If the force is applied close to the fulcrum, the torque produced is small. The farther the force is from the fulcrum, the larger the torque. Students should be able to relate this to opening a door. If one pushes on the door near the hinges, it takes a very large force to rotate the door. If one pushes at the doorknob (farther away from the hinges), then a much smaller force will rotate the door into its open position.

The distance used to find the torque is called the moment arm of the force. When the seesaw is horizontal, the moment arm is simply the distance from the force to the fulcrum. (See figure below.) The torque is the force times the moment arm. The fact that the torques on the two sides of the seesaw must be equal in order for the seesaw to balance explains why the moment arm for two stacked washers must be half the moment arm for a single washer.

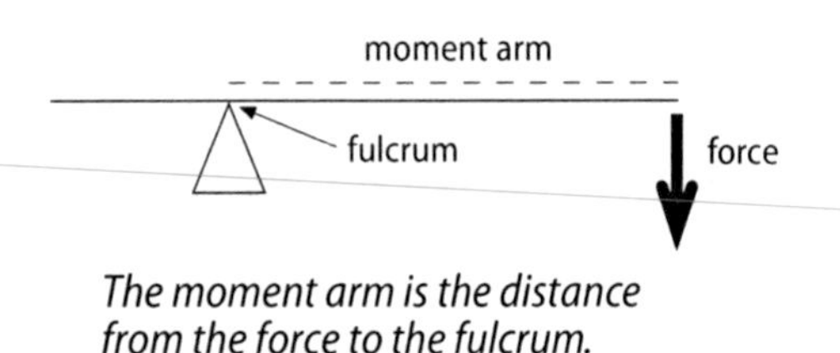

The moment arm is the distance from the force to the fulcrum.

Work

In Part B of this activity, students observe that a load can feel easier or harder to lift depending on how far the load force and the effort force are from the fulcrum. They also see a relationship between the length of the arm and the distance through which the effort force is applied. Younger students can simply conclude that, when used as a lever, a seesaw makes a load easier to lift if the load is closer to the fulcrum than the effort force.

For older students, these observations provide a basis for introducing the concept of work. While we use the term work to mean many things in everyday conversation, the scientific definition of work states that, for constant forces, work is the force applied to an object times the distance the object moves in the same direction as the applied force.

$$\text{work} = \text{force} \times \text{distance}$$

Two kinds of forces act on a lever to result in work. The push on one side of the lever provides the effort force (F_E) and the load on the other side of the lever provides the load force (F_L). The distance that the effort force moves down is d_E, and the distance that the load force moves up is d_L. Note that the distance used to calculate work is different than the distance used to calculate torque.

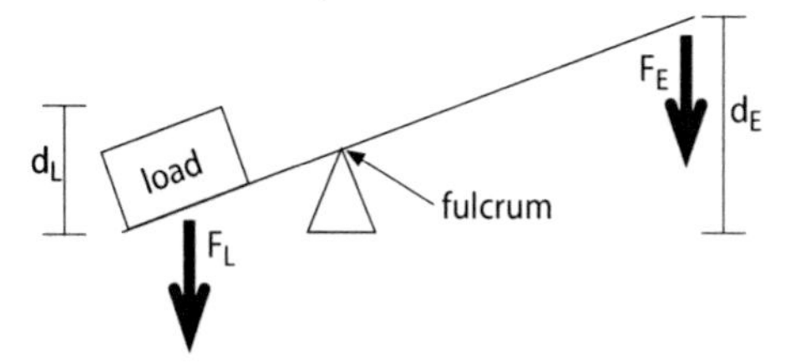

Distances on a seesaw used to calculate work

Since "work in" always equals "work out," we can also say

$$F_L \times d_L = F_E \times d_E$$

or

$$(\text{large force}) \times (\text{small distance moved}) = (\text{small force}) \times (\text{large distance moved})$$

For example, if $F_L = 2$ newtons, $d_L = 0.1$ m and $d_E = 0.2$ m, then F_E must equal 1 newton:

$$(2 \text{ newtons}) \times (0.1 \text{ m}) = (1 \text{ newton}) \times (0.2 \text{ m})$$
$$0.2 \text{ newton m} = 0.2 \text{ newton m}$$

We can now see that using a first-class lever does not reduce the amount of work needed to get a job done, but it can make the work feel easier by changing the amount of effort force required. For example, it is nearly impossible to pull a nail out of a piece of wood just using your hand. However, if you use a claw hammer (a first-class lever), you can apply a small force to the handle of the hammer, moving the handle through a large distance. The claw applies a large force to the nail and moves it through a small distance. The same amount of work is done on either side of the fulcrum.

For older students, the activity "Levers at Work" in this book provides more quantitative activities involving the concept of work.

Cross-Curricular Integration

Physical education:

- Take the students to a playground that has a seesaw. Experiment with balancing.

Part A: Balance the Seesaw

① Build and prepare the seesaw model according to the "How to Make a Seesaw" instruction sheet.

Helpful Tip
For easier reading, center the hole of the washer on the desired number.

② Carefully put one washer on each end of the seesaw so that the seesaw remains balanced.

Record the distance each washer is from the center of the seesaw.

....

Why do you think the seesaw remains balanced?

....

....

③ Predict what would happen if you stacked two washers at one end of the seesaw and kept only one washer at the other end.

Write your prediction.

....

....

④ Check your prediction by putting another washer on top of a washer you already have on the seesaw.

What happens?

....

....

Helpful Tip
Sometimes the seesaw sticks, so make sure to tap the seesaw each time you add or take objects away to help the seesaw rebalance itself.

⑤ Balance the seesaw without adding or taking away any washers. The only rule is that you must keep the two stacked washers together.

On the image below, draw where you put the washers to balance the seesaw. Show which side has two washers.

24 22 21 20 19 18 17 16 15 14 13 12 11 10 9 8 7 6 5 4 3 2 1 1 2 3 4 5 6 7 8 9 10 11 12 13 14 15 16 17 18 19 20 21 22 24

⑥ Stack two washers together on the seesaw's left arm at the 12-cm location. Place a single washer on the right arm so that the seesaw balances.

✏ In the data table, record the location of the single washer. Continue the procedure until you finish the data table.

Distance from Fulcrum (Left Arm)—Two Stacked Washers	**Distance from Fulcrum (Right Arm)—Two Stacked Washers**
12 cm	
10 cm	
8 cm	

⑦ Look at the numbers in the step 6 data table.

✏ What relationship do you see between each pair of distances?

....

....

....

⑧ Place two washers on the seesaw's left arm at the 6-cm location. Predict where a washer needs to go on the right arm so that the seesaw balances.

✏ Record your prediction in the data table.

Distance from Fulcrum (Left Arm)—Two Stacked Washers	**Distance from Fulcrum (Right Arm)—Two Stacked Washers**	
	Prediction	**Actual**
6 cm		
4 cm		

✏ Test your prediction and record the result in the data table. Repeat predicting and testing with the 4-cm location to finish the data table.

⑨ Complete the columns in the following table.

Seesaw's Left Arm			Seesaw's Right Arm		
Ⓐ	Ⓑ	Ⓐ × Ⓑ	Ⓒ	Ⓓ	Ⓒ × Ⓓ
Number of Washers	**Distance from Fulcrum**	**Product of Column A and B**	**Number of Washers**	**Distance from Fulcrum**	**Product of Column C and D**
2	12 cm		1		
2	10 cm		1		
2	8 cm		1		
2	6 cm		1		
2	4 cm		1		

✎ Compare the results of the seesaw's left and right arms. What patterns do you see?

....

....

Part B: More About Levers

① Poke a hole in the bottom of each of two 5-ounce paper cups. Remove the ruler from the seesaw and push one cup over the white upright rod at each end of the seesaw as shown.

② Fill a third 5-ounce cup with objects to act as a load. Place the load (cup of objects) in one of the cups on the seesaw. The weight of this load is known as the load force. Push all the way down on the opposite end of the seesaw to feel the resistance. This push is known as the effort force.

✎ Does the effort arm (the one you pushed all the way down on) move down more than, the same as, or less than the arm that contains the load moves up?

....

....

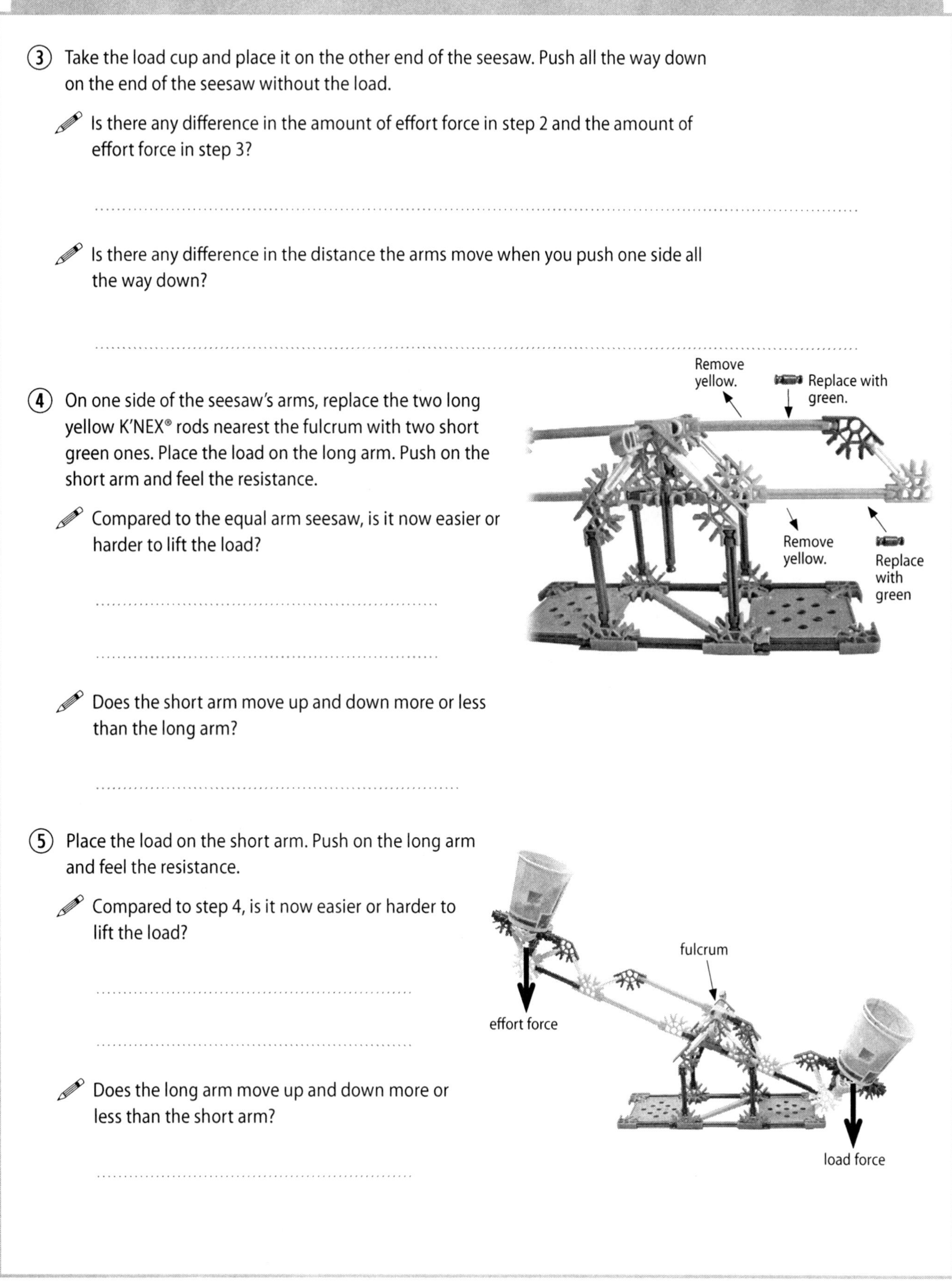

③ Take the load cup and place it on the other end of the seesaw. Push all the way down on the end of the seesaw without the load.

✎ Is there any difference in the amount of effort force in step 2 and the amount of effort force in step 3?

....

✎ Is there any difference in the distance the arms move when you push one side all the way down?

....

④ On one side of the seesaw's arms, replace the two long yellow K'NEX® rods nearest the fulcrum with two short green ones. Place the load on the long arm. Push on the short arm and feel the resistance.

✎ Compared to the equal arm seesaw, is it now easier or harder to lift the load?

....

....

✎ Does the short arm move up and down more or less than the long arm?

....

⑤ Place the load on the short arm. Push on the long arm and feel the resistance.

✎ Compared to step 4, is it now easier or harder to lift the load?

....

....

✎ Does the long arm move up and down more or less than the short arm?

....

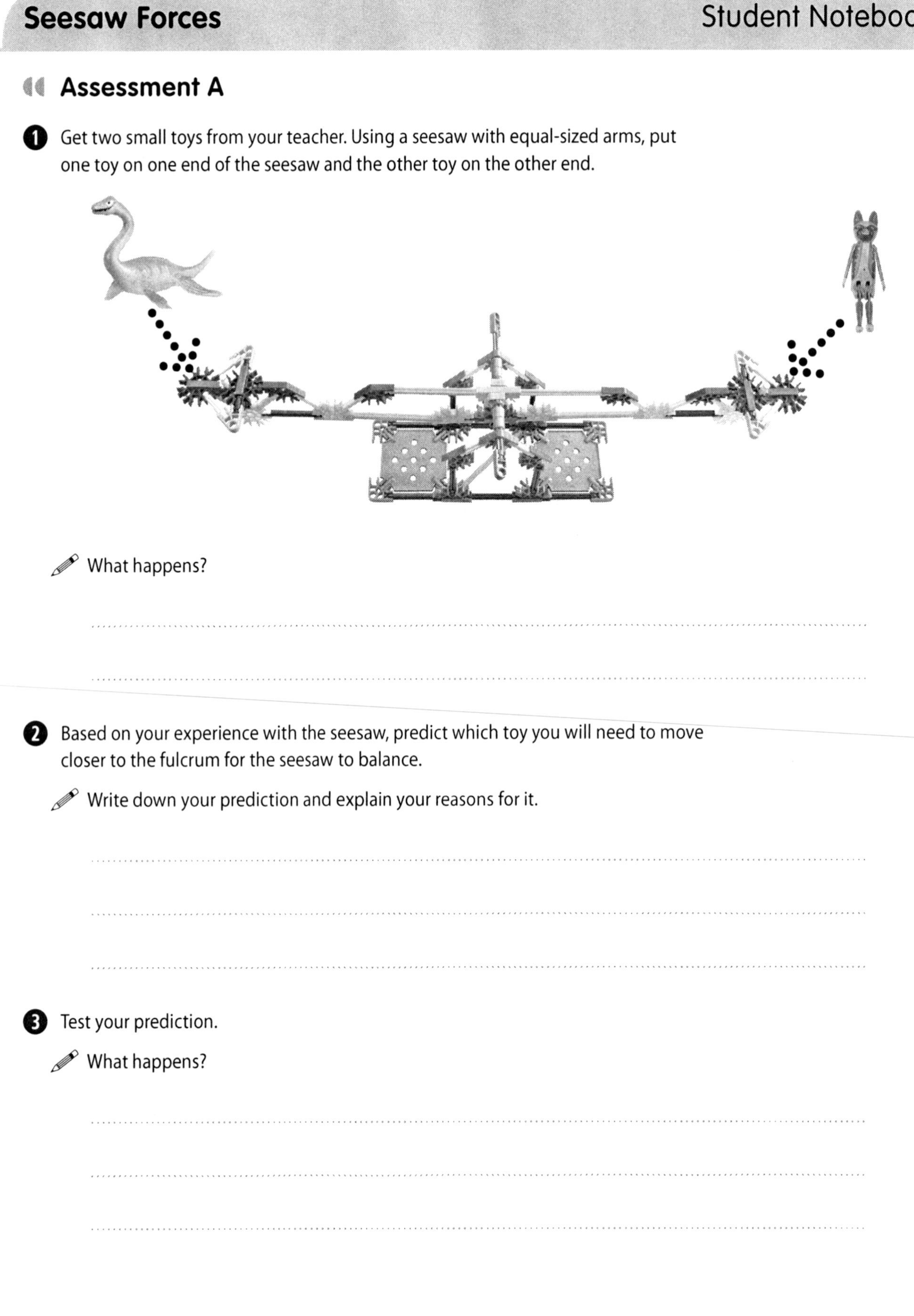

Assessment A

1 Get two small toys from your teacher. Using a seesaw with equal-sized arms, put one toy on one end of the seesaw and the other toy on the other end.

What happens?

....

....

2 Based on your experience with the seesaw, predict which toy you will need to move closer to the fulcrum for the seesaw to balance.

Write down your prediction and explain your reasons for it.

....

....

....

3 Test your prediction.

What happens?

....

....

....

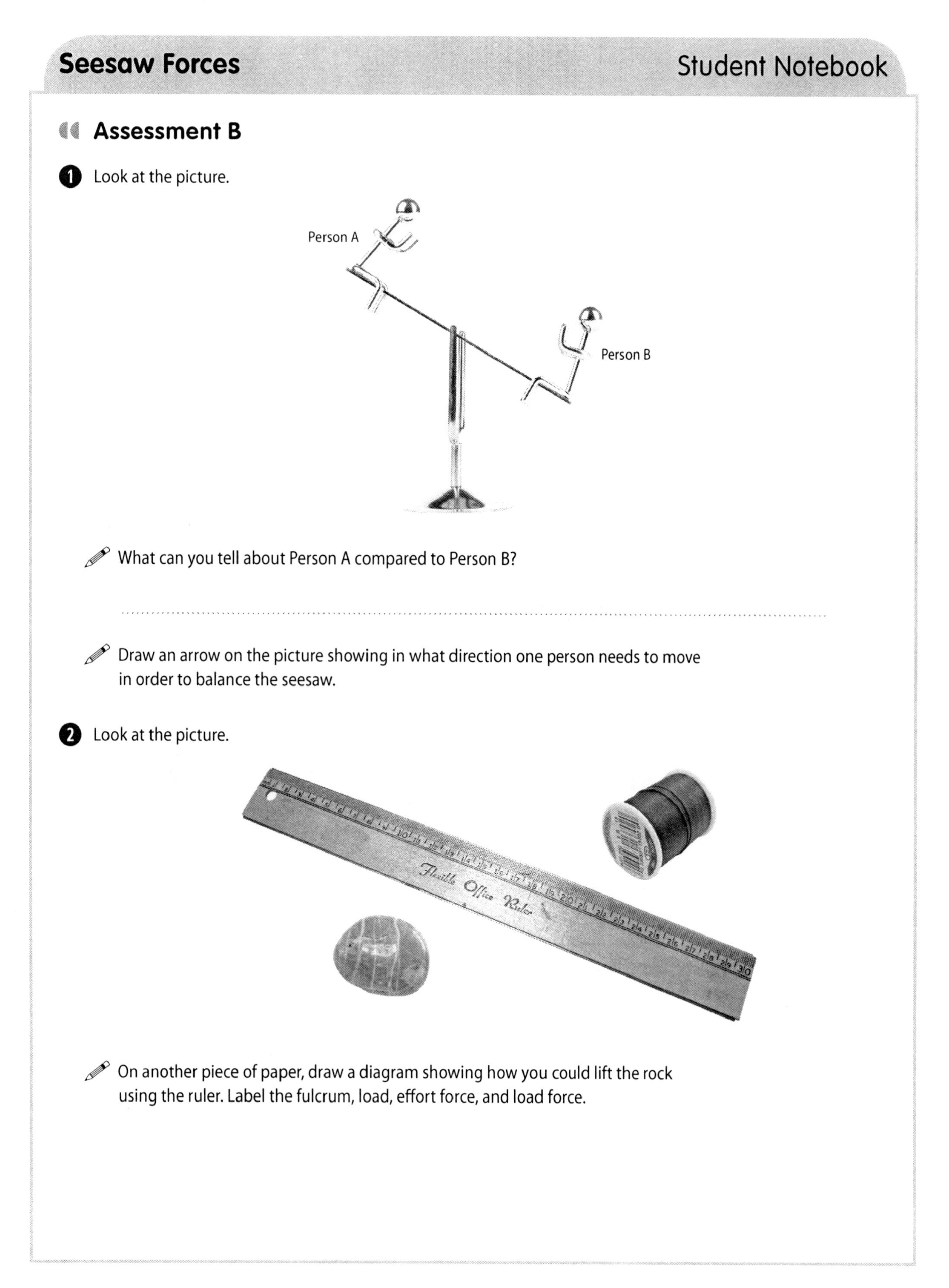

Assessment B

1 Look at the picture.

What can you tell about Person A compared to Person B?

....

Draw an arrow on the picture showing in what direction one person needs to move in order to balance the seesaw.

2 Look at the picture.

On another piece of paper, draw a diagram showing how you could lift the rock using the ruler. Label the fulcrum, load, effort force, and load force.

How to Make a Seesaw

A Assemble the seesaw.

- Build the seesaw model according to the K'NEX assembly diagrams.
- Place the card stock rulers on top of one another and check for identical length and width. If they are not identical, trim them.
- Place the two rulers as shown and tape them together to form one long ruler. Notice that the numbers increase as they move out from the center. Punch a hole where shown at each end of the ruler.

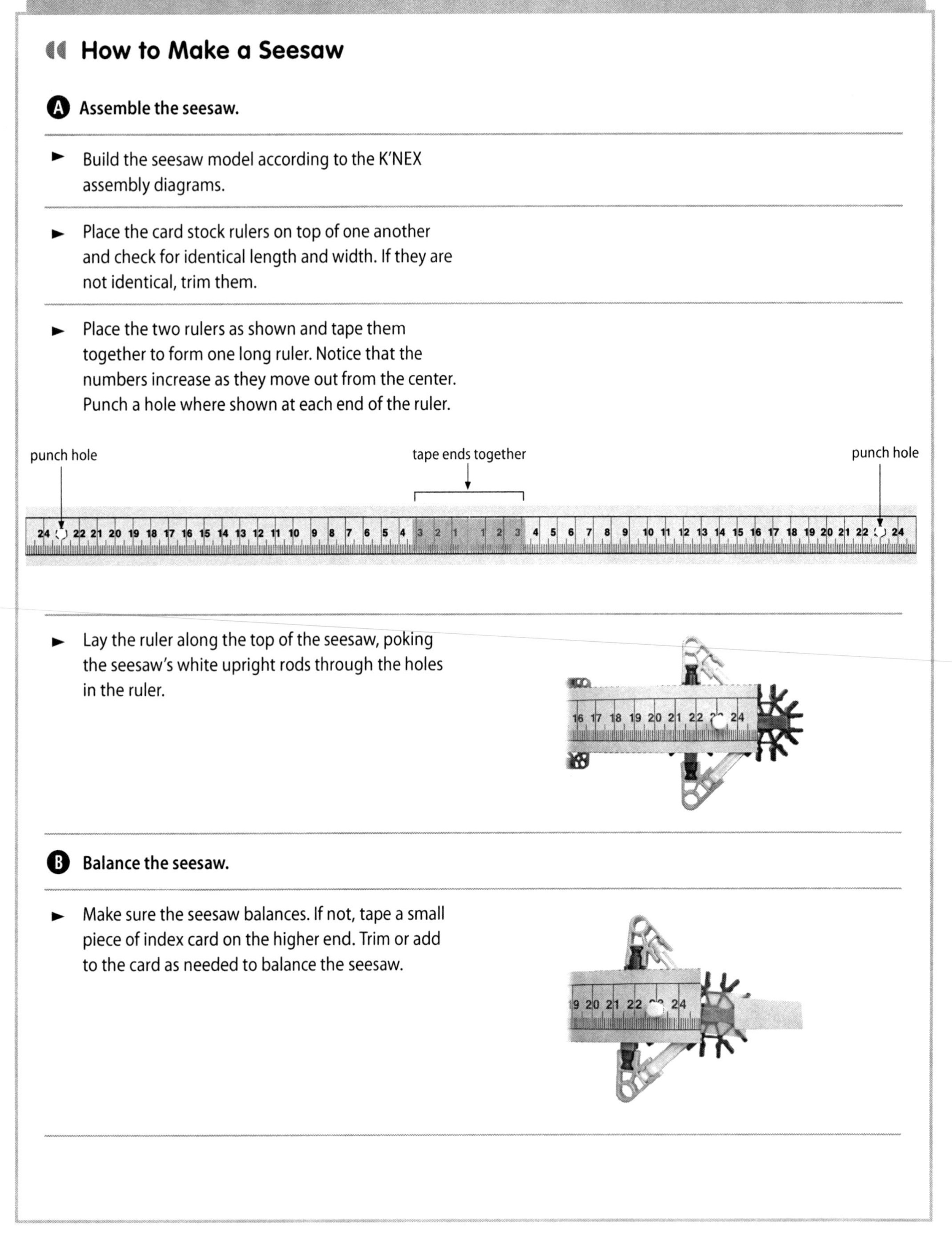

- Lay the ruler along the top of the seesaw, poking the seesaw's white upright rods through the holes in the ruler.

B Balance the seesaw.

- Make sure the seesaw balances. If not, tape a small piece of index card on the higher end. Trim or add to the card as needed to balance the seesaw.

Ruler Templates

Rulers are at actual scale. Be sure to reproduce at 100%.

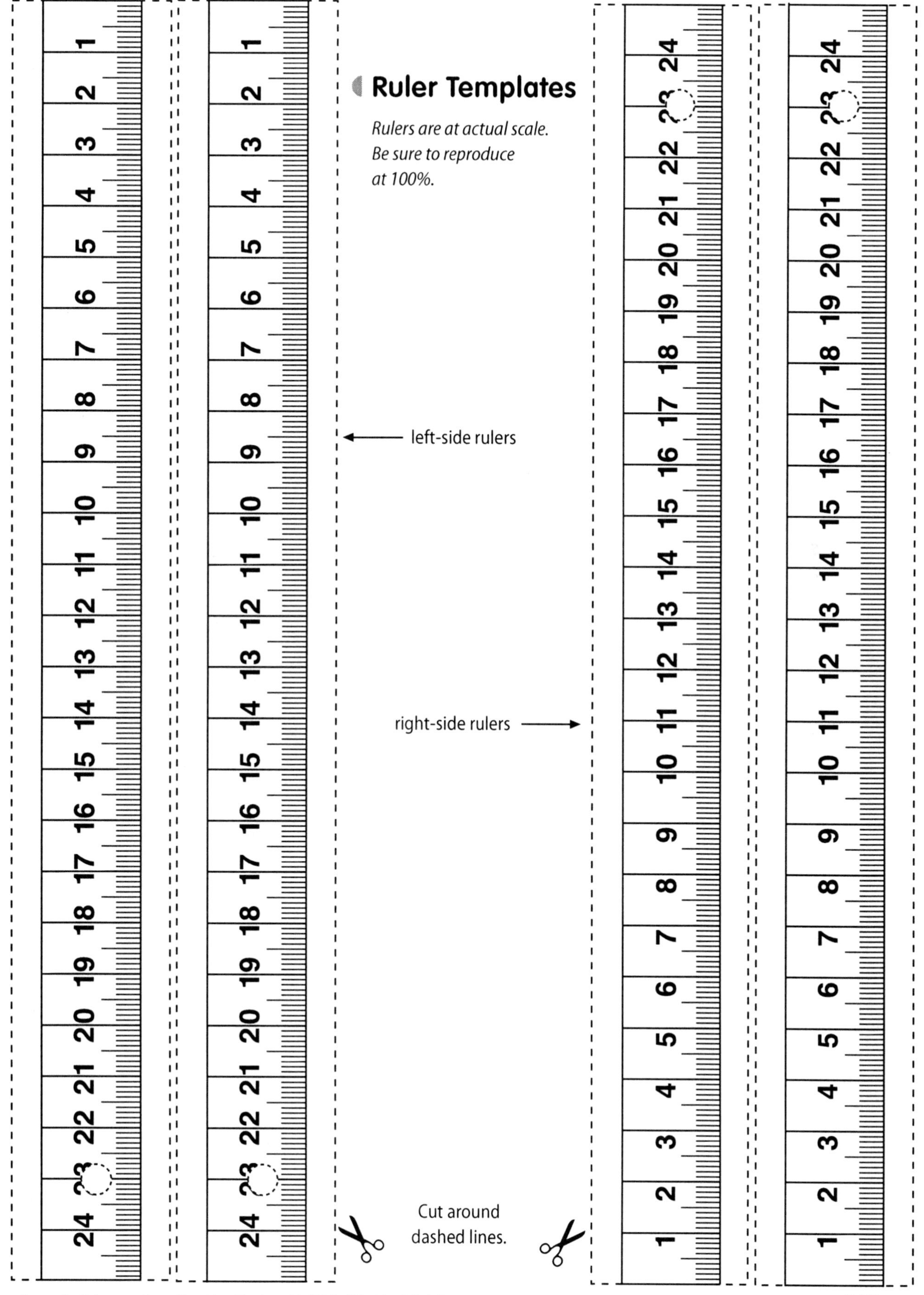

Levers at Work

Students explore a seesaw, wheelbarrow, and shovel to learn how first-class, second-class, and third-class levers work.

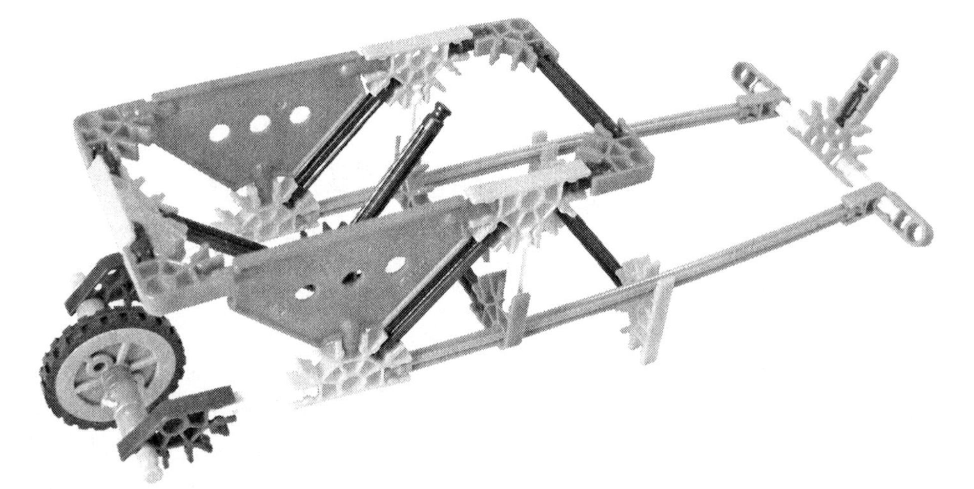

Grade Levels

Science activity appropriate for grades 7–9

Student Background

Students should have a basic understanding of how levers work. The activity "Seesaw Forces" in this book teaches this concept.

Time Required

Setup	10	minutes
Build K'NEX® models	30	minutes
Part A	30	minutes
Part B	30	minutes
Part C	30	minutes
Cleanup	10	minutes

Assessment time is not included.

Key Science Topics

- force
- mass
- mechanical advantage
- simple machines
- work

National Science Education Standards Overview

See *www.terrificscience.org/physicsez/* for details of how these standards relate to the activity.

Science as Inquiry

Abilities Necessary to Do Scientific Inquiry

5–8 *Develop descriptions, explanations, predictions, and models using evidence.*
5–8 *Think critically and logically to make the relationships between evidence and explanations.*
5–8 *Communicate scientific procedures and explanations.*
5–8 *Use mathematics in all aspects of scientific inquiry.*

9–12 *Use technology and mathematics to improve investigations and communications.*
9–12 *Formulate and revise scientific explanations and models using logic and evidence.*

Physical Science

5–8 *Motions and forces*

9–12 *Motions and forces*

Science and Technology

Abilities of Technological Design

5–8 *Design a solution or product.*
5–8 *Communicate the process of technological design.*

9–12 *Communicate the problem, process, and solution.*

Materials

For the Procedure

Activity Introduction, per group

- (optional) spring scale calibrated for 0–5 newtons

A spring scale that measures 0 to 5 newtons will enable students to see force differences much more easily than a spring scale designed to measure a larger range of force.

- (optional) 50 g, 100 g, and 200 g standard masses, slotted or with hooks

Part A, per group

- assorted K'NEX rods, connectors, and panels

For specifics, see K'NEX Parts Inventory at www.terrificscience.org/physicsez/ or www.terrificscience.org/supplies/.

- K'NEX assembly diagrams
- tape and index card
- 2 paper clips
- 2, 50 g masses
- spring scale calibrated for 0–5 newtons
- washers and spacers
- (optional) math balance

As an alternative to the K'NEX seesaw, you can use a math balance. See supply sources at www.terrificscience.org/supplies/.

Part B, per group

All materials listed for Activity Introduction, plus:

- assorted K'NEX rods, connectors, panels, and wheel

For specifics, see K'NEX Parts Inventory at www.terrificscience.org/physicsez/ or www.terrificscience.org/supplies/.

- metric ruler

Part C, per group

- toy shovel

Any plastic toy shovel will work, but choose one with the longest handle you can find.

- object to lift with shovel
- 1 or 2 metric rulers
- (optional) assorted K'NEX rods and connectors to make a hockey stick

For specifics, see K'NEX Parts Inventory at www.terrificscience.org/physicsez/ or www.terrificscience.org/supplies/.

For the Assessment

Assessment B, per group

- assorted objects such as craft sticks, rulers, blocks, mass sets, and K'NEX pieces

Safety and Disposal

No special safety or disposal procedures are required.

Getting Ready

- If needed, adjust spring scales so that they read 0 with nothing hanging from the hook. (Most models of spring scales have an adjustment knob at the top.)
- If your spring scales have a side showing grams, you may want to cover that side up so students do not confuse force with mass. (Spring scales actually measure force, not mass.)
- Print a set of K'NEX assembly diagrams for each group. (See *www.terrificscience.org/physicsez/*.) Since K'NEX pieces are color coded, the assembly diagrams need to be printed in color.

Depending on your schedule, you may want to have students build models in advance.

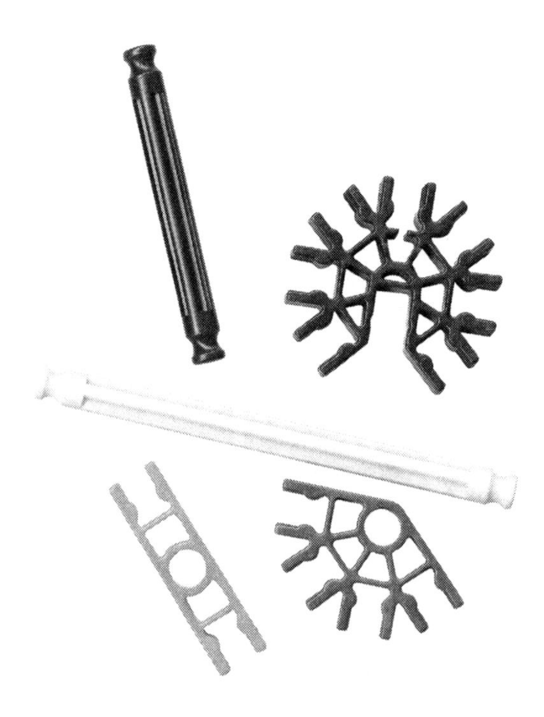

Procedure

This section provides teacher notes corresponding to each step of the student procedure. The procedure without teacher notes is included in the reproducible Student Notebook pages at the end of this activity and at www.terrificscience.org/physicsez/.

Student Procedure	Teacher Notes
Activity Introduction	Skip this introduction if your students are already familiar with mass, weight, and using a spring scale. Otherwise, through a discussion of students' ideas about mass and weight, make the following points: • Each of the items in the mass set has a different quantity of matter (mass). Two typical units of mass are gram and kilogram. • We experience differences in mass in terms of how light or heavy something feels. A particular mass feels lighter or heavier in our hands because of the size of the force gravity exerts on it. Weight is the measure of this force. Examples of weight units are newtons and pounds. See Explanation for a discussion of standard units.
Optional	• Have students pull on the spring scale with a force of 0.5, 1, 2, and 5 newtons. This step will give students practice using the correct units of measure from the spring scale. • Have students hang a 50 g mass on the spring scale and determine its weight in newtons. Repeat with the 100 g and 200 g masses. Students will discover that 50 g equals about 0.5 newton, 100 g equals about 1 newton, and 200 g equals about 2 newtons.
Part A: Explore a Seesaw	
① **Build the seesaw according to the K'NEX assembly diagrams. Explore the seesaw and observe how it works. A seesaw is a first-class lever.** **On another piece of paper, draw a diagram of your seesaw and label the fulcrum.**	Students will need color K'NEX assembly diagrams to build the model. (See *www.terrificscience.org/physicsez/*.) Very occasionally, the K'NEX seesaw will not be exactly balanced after you build it. (One side will be lower than the other.) If this happens, instruct students to tape a small piece of index card at the high end of the seesaw. If needed, trim or add to the card until the seesaw balances. See Sample Answers for student drawing example.

Student Procedure	Teacher Notes
② Use a paper clip to hang a 50 g mass from one end of the seesaw as shown. This mass is called the load. Add to your step 1 seesaw diagram by doing the following: ✎ Draw and label the load. ✎ The force of gravity acting on this mass is the load force (F_L). Draw and label an arrow showing the location and direction of the load force. ✎ Push on the other end of the seesaw until the load lifts. This push is known as the effort force (F_E). Draw and label an arrow showing the location and direction of the effort force. ✎ The load moment arm is the distance along the arm from the load force to the fulcrum. Draw and label a dashed line showing the location of the load moment arm. ✎ The effort moment arm is the distance along the arm from the effort force to the fulcrum. Draw and label a dotted line showing the location of the effort moment arm.	Help students label their drawings as needed. See Sample Answers for student drawing example. Note that the seesaw is shown in a balanced position. The moment arm distances are correct as described here only if the seesaw is balanced. Load force is sometimes referred to as resistance force. Mention this term if it is used in your classroom textbook or if students are familiar with the term from previous experience.
Class Discussion	Through a discussion of students' observations, bring out the following ideas: • The seesaw has two sides that pivot in the center. • The seesaw is an example of a first-class lever. The first-class lever configuration is: fulcrum F_L F_E • Have students help draw a large, labeled seesaw diagram based on their drawing from step 2. fulcrum load moment arm effort moment arm load load force (F_L) effort force (F_E)
③ Use a spring scale to determine the load force for a 50 g load. We can also call this load force the weight of a 50 g mass.	The load force for a 50 g load is 0.5 newton. Students need to understand that this means that gravity exerts a force of 0.5 newton on a mass of 50 g.

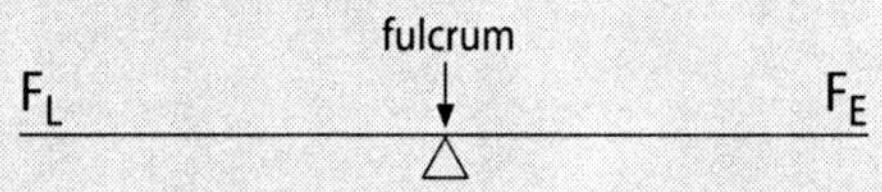

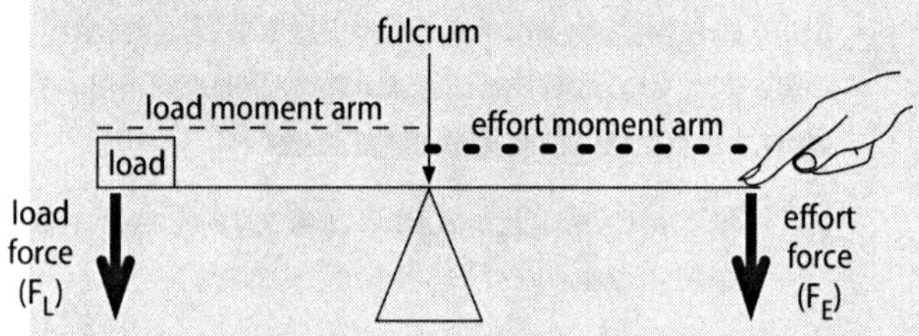

	Student Procedure	Teacher Notes
4	When the arms of the seesaw are equal in length, the effort force (push) needed to lift the load is equal to the load force. Push down on the seesaw and raise the 50 g load. ✎ What is the effort force of your push? ✎ What do you think would happen if you put a 50 g mass on the side of the seesaw that you just pushed down?	The effort force of the student's push is 0.5 newton. Some students may incorrectly answer 50 g. When asked to think about what would happen if a 50 g mass were added, students may realize that the seesaw will be level with equal mass on each side. If not, they will discover the answer in the next step.
5	Leave the 50 g mass on the seesaw. Using a paper clip, hang a different 50 g mass on the end of the seesaw you pushed down in step 4. ✎ What happens?	After adding 50 g to the side of the seesaw that was previously pushed, the seesaw should be level. Make sure students realize that putting a 50 g mass on the seesaw is just like using their hand to push down with a force of 0.5 newton. This understanding is critical to understanding later steps.
6	Remove both 50 g masses. In one of the arms of the seesaw, replace the two long yellow rods with two short green ones as shown below left. Since one side of the seesaw is now shorter than the other, the seesaw is no longer balanced. As shown at right, add the yellow rods you just removed to the short arm of the balance. Then, add washers and spacers as needed to make the seesaw balance. (This process is called "zeroing.")	This step creates a seesaw with unequal arms. Students should understand that the term "zeroing" is used to describe the process of adjusting an instrument before using it. If you are using a math balance instead of a K'NEX seesaw, you will need to instruct students to slightly revise steps 6–9 of the procedure. Instead of shortening one arm of the seesaw and hanging the load on the end of it, students using a math balance should hang the load about halfway between the fulcrum and the end of the arm.
7	Hang a 50 g load from the long arm. Push on the short arm until the load rises. Feel the resistance. ✎ Compared to the equal arm seesaw, is it now easier or harder to lift the load? ✎ Leave the 50 g load on the long arm. What do you think will happen if, instead of pushing on the short arm, you hang a 50 g load from it?	Students may feel that it is harder to lift the load this time. (More effort force is needed to raise the load.) Students will find this comparison easier if you keep one seesaw set up with equal arms. They can try the equal arm seesaw and then the unequal arm seesaw in rapid succession.
8	Now, try adding the 50 g load on the short arm. ✎ What happens? ✎ Based on what you just observed, was the force of your push in step 7 greater or less than 0.5 newton?	Students should observe that an effort force of 0.5 newton (provided by the mass on the short arm) does not raise the load. Students should be able to conclude that when they push the short arm and raise the load, the force of the push (effort force) must be greater than 0.5 newton.

	Student Procedure	Teacher Notes
9	Remove the load from the long arm. Leave the 50 g load on the short arm. Push on the long arm until the load rises. Feel the resistance. ✎ Compared to the equal arm seesaw, is it now easier or harder to lift the load? ✎ Leave the 50 g load on the short arm. What do you think will happen if, instead of pushing on it, you hang a 50 g load from the long arm?	Less effort force is needed to raise the load this time. When asked to think about what would happen if a 50 g mass were added, students may realize that the long arm will tip down. If not, they will discover the answer in the next step.
10	Now, try adding the 50 g load to the long arm. ✎ What happens? ✎ Based on what you just observed, was the force of your push in step 9 greater or less than 0.5 newton? Explain. ✎ Summarize what you learned in Part A by answering "yes" or "no" in each box. If necessary, repeat steps 4–10 and fill out the table as you complete each step.	Students should realize that an effort force of 0.5 newton (provided by the mass in the long arm) is more than enough force to raise the load. The long arm tips all the way to the table, rather than just balancing with the short arm. Therefore, the push required to raise the load must be less than 0.5 newton. Students should be able to explain that a first-class lever can make lifting a load easier to do. When the moment arm for the effort force is longer than the moment arm for the load force, the amount of force needed to lift the load is reduced. See Sample Answers for typical student answers to the questions in the data table.
	Part B: Explore a Wheelbarrow	
1	Build the wheelbarrow according to the K'NEX assembly diagrams. Explore the wheelbarrow and observe how it works.	Students will need color K'NEX assembly diagrams to build the model.
2	Anchor the wheel of the wheelbarrow in the holder as shown. Place a 100 g mass in the barrel of the wheelbarrow. As before, this mass is called the load. The force of gravity acting on this mass is the load force. Pull up on the wheelbarrow handles until the load lifts. This pull is the effort force. ✎ On another piece of paper, draw a diagram of the wheelbarrow. Show and label the load force and the effort force. Mark where you think the fulcrum is.	See Sample Answers for example diagram.

Student Procedure	Teacher Notes
Class Discussion	• Ask students to compare the wheelbarrow to the seesaw, identifying what is similar and what is different. Through discussion, help students realize that the wheelbarrow differs from the seesaw in that the fulcrum is on one end of the object, the load force (F_L) is in the center, and effort force (F_E) is on the other end. • Introduce the idea that the wheelbarrow is an example of a second-class lever. The second-class lever configuration is: fulcrum F_L F_E
③ Hold a 100 g mass in your hand to get a feeling for how much it weighs. Place the mass into the wheelbarrow and lift the handles. Compare the weight of the mass when you hold it in your hand to the effort force used to lift the mass in the wheelbarrow. Repeat with the 200 g and 300 g masses and record your observations.	Students should be able to feel that less effort force is needed to raise the load with the wheelbarrow.
④ Now you'll use a spring scale to measure what you felt in step 3. Record the load force for a 100 g load in the data table. What effort force do you think you'll need to lift the load? Record your prediction in the data table. Place the 100 g mass in the wheelbarrow. Lift the wheelbarrow with the spring scale as shown. Record the effort force in the data table.	See Sample Answers for example data.
⑤ Repeat step 4 with the 200 g and 300 g masses. Is the effort force when using the second-class lever more or less than the load force?	Predictions of the effort force needed to lift the load should improve as students see the pattern with each mass. When using a second-class lever, the effort force is less than the load force.

Student Procedure	Teacher Notes
(6) Put any load from 100 g to 300 g in the wheelbarrow and lift the handles to raise the load. ✎ Without lowering the wheelbarrow, measure and record how far up from the table you raised the load (d_L). Measure and record how far up from the table the ends of the wheelbarrow handles rose (d_E). ✎ Which travels a longer distance, the load or the handles?	When raising the load in a wheelbarrow, the handles travel a longer distance than the load travels.
Class Discussion	• Introduce or review the rule that "work in" must always equal "work out" and the idea that *work = force × distance*. So, the *force × distance* for the load must equal the *force × distance* for the effort. (See Explanation for further details.) • Help students relate this idea to the forces and distances they measured using the wheelbarrow. The relatively heavy load in the wheelbarrow is moved a short distance while the handles are moved a longer distance. The large load force times the small distance the load travels equals the small effort force (used to lift up the handle) times the large distance the handle travels when the wheelbarrow is lifted. The amount of work done on the load and the amount of work done by the effort are the same. • Have students calculate the "work in" and the "work out" using the equation below, where d_L is the distance moved by the load and d_E is the distance moved by the effort. $F_L \times d_L = F_E \times d_E$ Students should see that the work in and work out are equal.
(7) Determine the mechanical advantage (MA) of the wheelbarrow by dividing the load force you recorded in step 4 by its measured effort force ($MA = F_L/F_E$). ✎ Record the MA. Is the number you get greater than, equal to, or less than 1.0? What does this tell you about the effort force needed to use the wheelbarrow?	The MA of the wheelbarrow should be greater than 1.0, meaning that less force is required to lift the load with the wheelbarrow than without the wheelbarrow. (See Explanation.)

Student Procedure	Teacher Notes
Part C: Explore a Shovel	
① **Explore the toy shovel. Observe how it works.**	
② **Sit in a chair and hold the top of the shovel in one hand. Use the shovel to pick up an object from the floor.** **On another piece of paper, draw a diagram of the shovel. Show and label the load force and the effort force. Mark where you think the fulcrum is.**	See Sample Answers for example diagram.
Class Discussion	• Ask students to compare the shovel to the seesaw and wheelbarrow. Is the shovel a first-class or second-class lever? Through discussion, help students realize that the shovel differs from both the seesaw and the wheelbarrow. • The object being picked up is the load. The force of gravity acting on the object is the load force. When picking up the object with the shovel, the hand holding the top of the shovel becomes the fulcrum. The hand in the middle of the shovel provides the effort force. • Introduce the idea that the shovel is an example of a third-class lever. The rule for a third-class lever is that the fulcrum is on one end of the object, the effort force (F_E) is in the center, and the load force (F_L) is on the other end. When using the lever, the fulcrum at one end of the lever remains still as the middle of the lever (at the effort force) and the other end of the lever (at the load force) move in the same direction. The third-class lever configuration is: fulcrum F_E F_L

Student Procedure	Teacher Notes
③ Sit in a chair and hold the shovel in the starting position as shown below. Have a partner measure from the scoop to the floor and from the middle of the handle to the floor as shown below. Without moving the hand holding the top of the shovel, lift the object with the shovel. Have a partner measure the final distances of the scoop and the middle of the handle from the floor as shown. ✎ Record your measurements in the data table. Calculate the difference between ending and starting positions to determine the load force distance (d_L) and the effort force distance (d_E). ✎ Which travels a longer distance when you lift the shovel, the middle of the shovel or the scoop of the shovel? ✎ Give another example of a third-class lever. Explain its fulcrum, load force, effort force, and load.	If supplies are available, two metric rulers work better than one. Students can hold one ruler upright at the scoop and the other upright where the hand holds the middle of the handle. As an alternate, the two rulers can be taped to the wall. When the hand holding the top of the shovel (the fulcrum) remains nearly motionless, the hand lifting the middle of the shovel only moves a short distance, exerting a large force at a relatively slow speed. The end of the shovel moves faster and travels the longer distance. (See Sample Answers for example data.) Golf clubs, tennis rackets, and hockey sticks are examples of third-class levers.
Class Discussion	• Discuss with your class how forces and distances relate to third-class levers. (See Explanation.) • Ask students to think about what advantage you gain using a third-class lever. Ask them to visualize casting with a fishing rod, hitting a tennis or golf ball, or shoveling dirt from the ground into a truck. • Tell students that, unlike first- and second-class levers, third-class levers require a relatively larger force to move the load, not a smaller one. The trade-off is that the load moves over a longer distance than the effort and moves more quickly.
④ You calculated the mechanical advantage (MA) of the wheelbarrow in Part B by one method. MA can also be calculated by dividing effort force distance (d_E) by load force distance (d_L). Determine the MA of the shovel using the distances you measured in step 3 and the equation $MA = d_E/d_L$. ✎ Record the MA. Is the number you get greater than, equal to, or less than 1.0? What does this tell you about the effort force needed to use the shovel?	The MA of the shovel should be less than 1.0, meaning that greater effort force is needed to lift the load with the shovel than without the shovel. (See Explanation.)

Student Procedure	Teacher Notes
Optional	

Optional

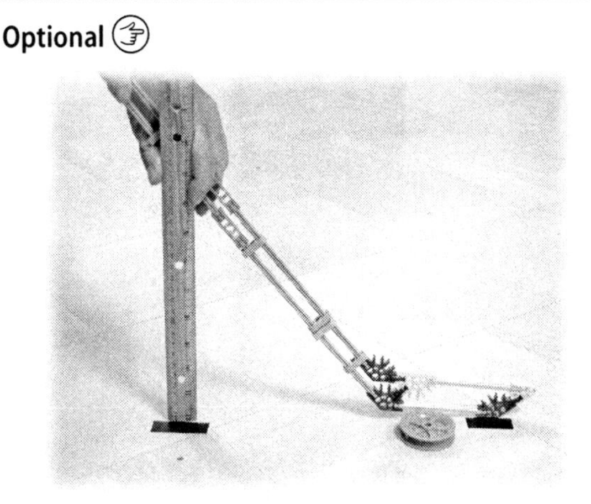

Place one piece of tape where the stick touches the floor and another piece directly below the hand providing the effort force.

After the swing, place tape to show the location of the bottom of the stick and the location of the hand providing the effort force.

To see an alternative example of a third-class lever, students can build a K'NEX hockey stick according to the color K'NEX assembly diagrams.

- Have one student hold the hockey stick while a partner marks the floor with tape as shown on left.
- Without moving the hand holding the top of the stick (the fulcrum), hit the puck and hold still at the end of the swing. Mark the floor with tape as shown on left.
- Measure and record the distances that the hand moved and the bottom of the stick moved. Ask students which travels a longer distance, the middle of the stick or the bottom hitting the puck. Have them calculate the MA of the hockey stick.

When the hand holding the top of the stick (the fulcrum) remains nearly motionless, the hand at the middle of the stick (which moves the stick forward) provides the effort force. This hand only moves a short distance but exerts a large force. The bottom of the stick moves in a longer arc than the hand providing the effort force. This means that the bottom of the stick moves faster than any other part of the stick. The bottom of the stick also travels a longer distance and exerts a relatively smaller force on the puck.

Assessment A

1 **In the physics department museum of Portugal's University of Coimbra, there is a trick balance with one of its arms measuring 21 cm long and the other arm measuring 17.5 cm long. The pans are also of unequal weights so that, when the pans are empty, the balance arms are level.**

Imagine that you have a bag of gold. You want to sell the gold based on its weight. How could someone use the balance to cheat you?

If the gold was placed on the side of the balance with the shorter arm (the 17.5-cm side), the gold would appear to weigh less than it actually does. The balance would behave like the seesaw did in step 9 of Part A. When one arm of the balance is shorter than the other and the load is placed at the end of the short arm, less effort force is needed to make the scale balance. Thus, less mass would have to be placed at the end of the longer arm to make the arm balance. Since less mass would be required to balance the gold, the gold would appear to weigh less than it actually does.

Student Procedure	Teacher Notes
Assessment B	
1 **Build an object or arrange a series of objects to incorporate both a first- and second-class lever. Your arrangement of levers should operate to accomplish a task.** **Draw your object or objects and label the fulcrum, load force, and effort force for each of the levers. Show the task each lever is designed to do.**	See Sample Answers for a design example.

Sample Answers

Where students follow the same procedure using the same materials, these answers are close to answers you can expect. Where students design their own experiment or model, students' results will vary.

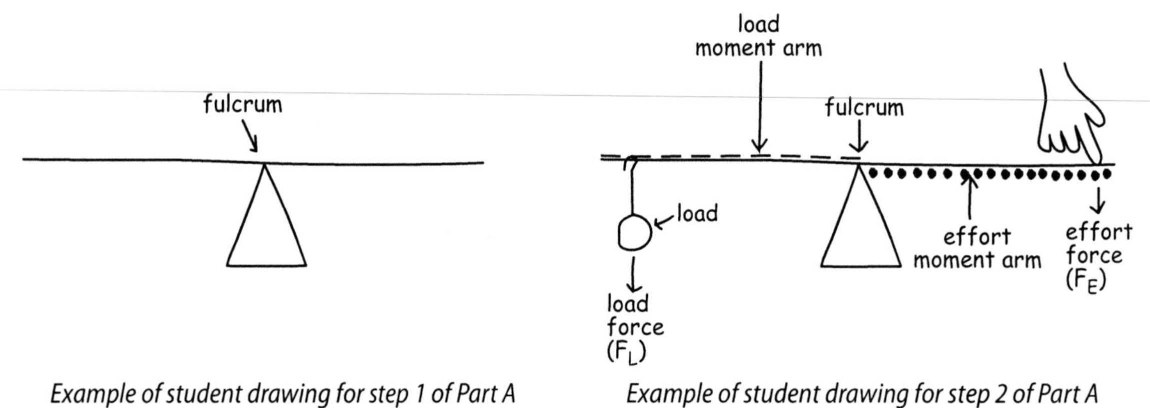

Example of student drawing for step 1 of Part A

Example of student drawing for step 2 of Part A

First-Class Lever Results					
Load			**Effort**		
Location of Load	**Mass**	**Load Force (Weight)**	**Less Than 0.5 Newton?**	**Equal to 0.5 Newton?**	**Greater Than 0.5 Newton?**
equal arm	50 g	0.5 newton	no	yes	no
long arm	50 g	0.5 newton	yes	no	no
short arm	50 g	0.5 newton	no	no	yes

Example of student answers for Part A

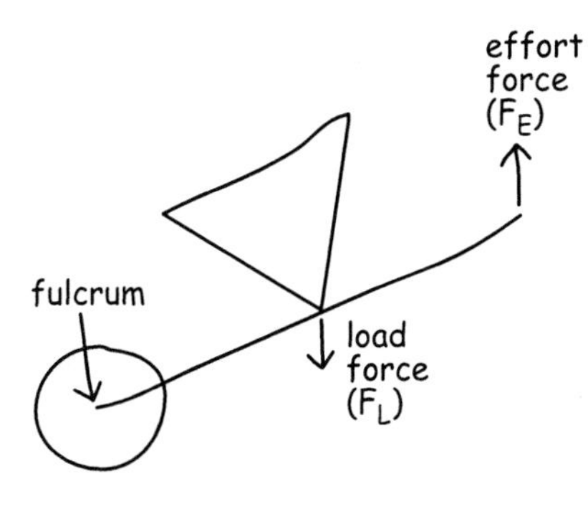

Example of student diagram for Part B

Second-Class Lever Data Table			
Mass Added	**Load Force (F_L) in Newtons**	**Predicted Effort Force in Newtons**	**Measured Effort Force (F_E) in Newtons**
100 g	1 newton	varies	0.6 newton
200 g	2 newtons	varies	1.0 newton
300 g	3 newtons	varies	1.2 newtons

Example of data for Part B

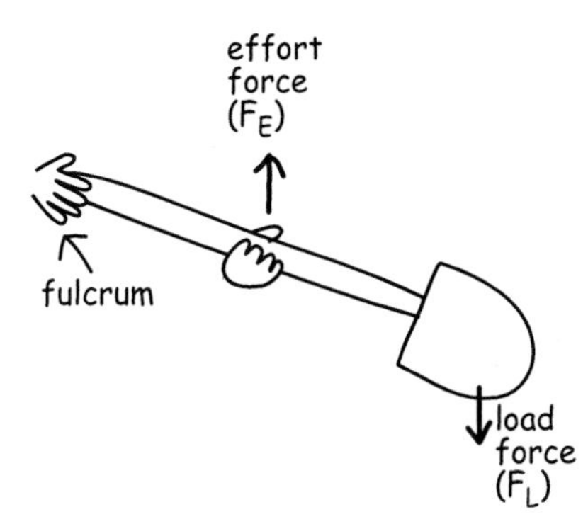

Example of student diagram for Part C

Third-Class Lever Data Table		
	Distance from Floor to Scoop	**Distance from Floor to Middle of Handle**
starting position	0 cm	27 cm
ending position	38 cm	43 cm
difference between ending and starting positions	$d_L = 38$ cm	$d_E = 16$ cm

Example of data for Part C

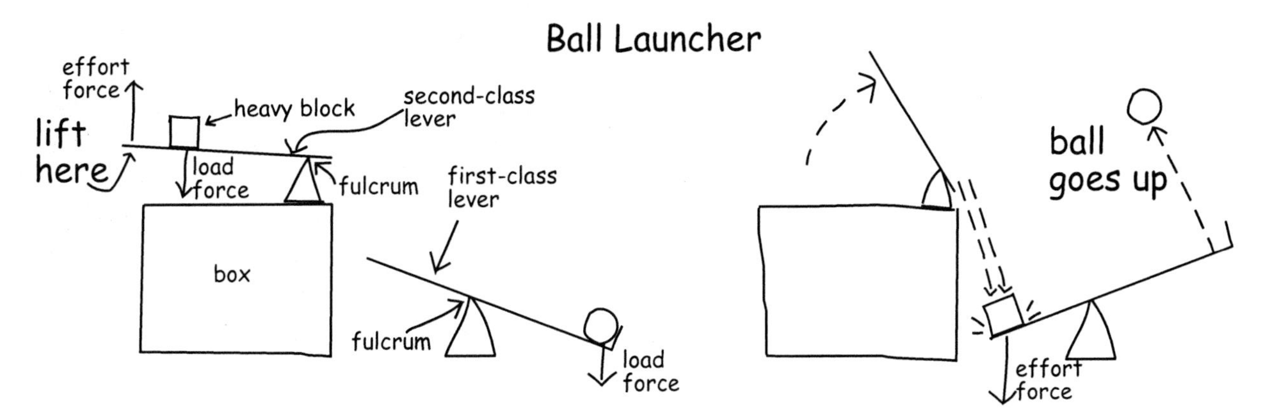

Example of student diagram for Assessment B

Explanation

This explanation is intended for teachers. Modify the explanation for students as needed.

Mass and Weight

Mass and weight are different concepts. Mass measures the quantity of matter in an object, and weight measures the force of gravity on an object. Two typical units of mass are gram (g) and kilogram (kg). Both of these are metric units. Weight has units of newtons in *Le Systême International d'Unités,* usually abbreviated as the SI (also called metric) system, and pounds in the United States Customary System (USCS). This lesson is not meant to teach the difference between mass and weight. Just be consistent in the use of the terms and their units. When using the word "mass," use the unit "g." When using the word "weight," use the unit "newton."

Newton's second law gives us a way to relate mass and weight. In general,

$$\text{force} = \text{mass} \times \text{acceleration}$$

Therefore,

$$\text{weight (newtons)} = \text{mass (kg)} \times \text{acceleration due to gravity (m/sec}^2\text{)}$$

The acceleration due to gravity is 9.8 m/sec^2 or approximately 10 m/sec^2. The following chart shows what some standard masses weigh.

Weights of Standard Masses	
Mass	**Approximate Weight**
50 g = 0.05 kg	0.5 newton
100 g = 0.1 kg	1 newton
200 g = 0.2 kg	2 newtons
500 g = 0.5 kg	5 newtons
1,000 g = 1.0 kg	10 newtons

Typical spring scales for use in science classrooms are often calibrated in both units. If you place a 500 g mass on the hook of a spring scale, one scale would tell you its mass (500 g) while the other side would tell you how much 500 g weighs (about 5 newtons). Spring scales really only measure weight. The mass is listed for convenience with the assumption that the spring scale will be used on Earth.

Three Classes of Levers

A first-class lever has the fulcrum (or pivot point) located on the lever arm between the load force (F_L) and the effort force (F_E). The seesaw is an example of a first-class lever. The first-class lever configuration is:

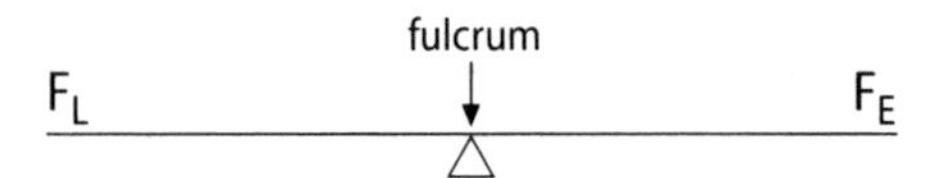

A second-class lever has the fulcrum located at one end of the lever arm, the load force is in the center, and effort force is on the other end. The wheelbarrow is an example of a second-class lever. The second-class lever configuration is:

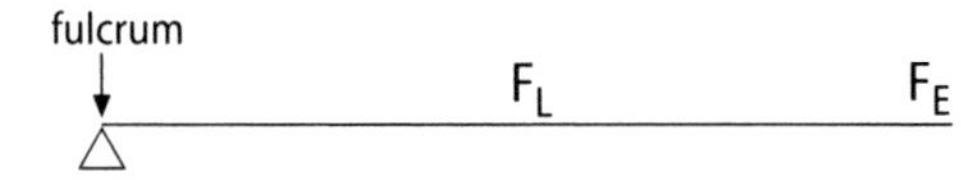

A third-class lever has the fulcrum located at one end of the lever arm, the effort force is in the center, and load force is on the other end. The shovel is an example of a third-class lever. The third-class lever configuration is:

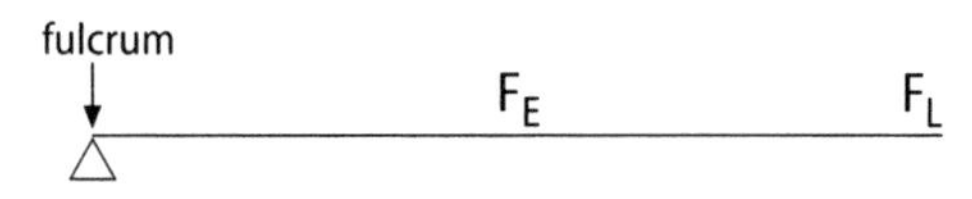

Work

How does a lever help you do work? The scientific definition of work is easiest to understand when the force is constant and always in the same direction as the object it displaces. In this case, work is the product of the force applied to an object and the distance the object moves.

$$\text{work} = \text{force} \times \text{distance}$$

Levers, like all other simple machines, do not reduce the amount of work needed to get a job done. Rather, a lever changes the amount of force needed to do the required or fixed amount of work. We can write the rule for work done by a lever or other simple machine as follows:

$$\text{work}_{in} = \text{work}_{out}$$

First-class levers help you do work by allowing you to apply a small force to accomplish a task that requires a large force. The trick is the distances over which the small force and the large force are applied. For example, it is nearly impossible to pull a nail out of a piece of wood just using your hand. However, if you use a claw hammer, you apply a small force to the handle of the hammer, moving the handle down through a large distance. The claw applies a large force to the nail and moves it up through a small distance, doing the same amount of work as your hand.

Another first-class lever is a car jack. None of us can lift the end of a car with our hands to change a flat tire. However, when we use a jack we can lift one end of a car. A relatively small force is applied downward on the handle, moving the handle through a large distance. The hook under the bumper of the car applies a large force but only moves up a very short distance. The same amount of work is done in each case.

Second-class levers (such as a wheelbarrow, nut cracker, garlic press, and bottle opener) also allow you to apply a relatively small force through a large distance, resulting in a large force moving through a small distance.

$$\text{(small force)} \times \text{(large distance moved)} = \text{(large force)} \times \text{(small distance moved)}$$

The advantage of using a third-class lever (such as a shovel, hockey stick, and fishing pole) is different than the advantage of using a first- or second-class lever. While "work in" must still equal "work out," with a third-class lever you actually exert more force to move the load rather than less. Rather than enabling you to do work with less effort, third-class levers enable you to increase the distance traveled by the load and increase its speed.

$$\text{(large distance moved)} \times \text{(small force)} = \text{(small distance moved)} \times \text{(large force)}$$

Torque

Just as forces cause levers to move up or down, torque causes them to rotate. As the various kinds of levers just discussed are moving up or down, they are also rotating around a fulcrum.

Although the concept of torque is not usually covered at this level, this activity introduces the term "moment arm," a measurement used to calculate torque. When the lever is horizontal, the moment arm is simply the distance from the force to the fulcrum. The torque is the force times the moment arm.

$$\text{torque} = \text{force} \times \text{moment arm}$$

If the torque on one side is greater than the torque on the other side, the lever will rotate so that the side with the larger torque is down. When a lever is balanced (not rotating), clockwise (CW) torque always equals counterclockwise (CCW) torque.

$$\text{torque}_{CW} = \text{torque}_{CCW}$$

The size of the torque produced by a force depends on how far from the fulcrum the force is applied. If the force is applied close to the fulcrum, the torque produced is small. The farther the force is from the fulcrum, the larger the torque. You can relate this to opening a door. If one pushes on the door near the hinges, it takes a very large force to rotate the door. If one pushes at the doorknob (farther away from the hinges), then a much smaller force will rotate the door into its open position.

Mechanical Advantage

Mechanical advantage (MA) is the factor by which a machine multiplies the force put into it. Mechanical advantage of a lever can be calculated from any of the following ratios:

- load force/effort force ($MA = F_L/F_E$)
- effort force distance/load force distance ($MA = d_E/d_L$)
- effort moment arm/load moment arm

Reference

K'NEX Industries, Inc. *Simple Machines: Levers and Pulleys Teacher's Guide;* Hatfield, PA, 2003.

Part A: Explore a Seesaw

① Build the seesaw according to the K'NEX® assembly diagrams. Explore the seesaw and observe how it works. A seesaw is a first-class lever.

On another piece of paper, draw a diagram of your seesaw and label the fulcrum.

② Use a paper clip to hang a 50 g mass from one end of the seesaw as shown. This mass is called the load. Add to your step 1 seesaw diagram by doing the following:

Draw and label the load.

The force of gravity acting on this mass is the load force (F_L). Draw and label an arrow showing the location and direction of the load force.

Push on the other end of the seesaw until the load lifts. This push is known as the effort force (F_E). Draw and label an arrow showing the location and direction of the effort force.

The load moment arm is the distance along the arm from the load force to the fulcrum. Draw and label a dashed line showing the location of the load moment arm.

The effort moment arm is the distance along the arm from the effort force to the fulcrum. Draw and label a dotted line showing the effort moment arm.

③ Use a spring scale to determine the load force for a 50 g load. We can also call this load force the weight of a 50 g mass.

④ When the arms of the seesaw are equal in length, the effort force (push) needed to lift the load is equal to the load force. Push down on the seesaw and raise the 50 g load.

What is the effort force of your push?

....

What do you think would happen if you put a 50 g mass on the side of the seesaw that you just pushed down?

....

⑤ Leave the 50 g mass on the seesaw. Using a paper clip, hang a different 50 g mass on the end of the seesaw you pushed down in step 4.

What happens?

....

⑥ Remove both 50 g masses. In one of the arms of the seesaw, replace the two long yellow rods with two short green ones as shown below left. Since one side of the seesaw is now shorter than the other, the seesaw is no longer balanced.

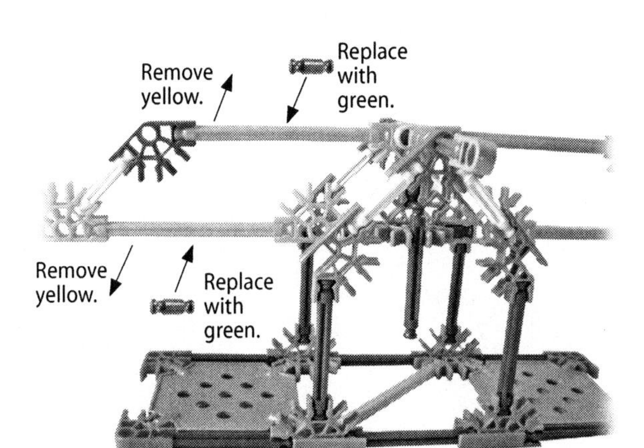

As shown at right, add the yellow rods you just removed to the short arm of the balance. Then, add washers and spacers as needed to make the seesaw balance. (This process is called "zeroing.")

⑦ Hang a 50 g load from the long arm. Push on the short arm until the load rises. Feel the resistance.

✎ Compared to the equal arm seesaw, is it now easier or harder to lift the load?

....

✎ Leave the 50 g load on the long arm. What do you think will happen if, instead of pushing on the short arm, you hang a 50 g load from it?

....

⑧ Now, try adding the 50 g load on the short arm.

✎ What happens?

....

✎ Based on what you just observed, was the force of your push in step 7 greater or less than 0.5 newton?

....

⑨ Remove the load from the long arm. Leave the 50 g load on the short arm. Push on the long arm until the load rises. Feel the resistance.

✎ Compared to the equal arm seesaw, is it now easier or harder to lift the load?

....

✎ Leave the 50 g load on the short arm. What do you think will happen if, instead of pushing on it, you hang a 50 g load from the long arm?

....

(10) Now, try adding the 50 g load to the long arm.

✎ What happens?

....

✎ Based on what you just observed, was the force of your push in step 9 greater or less than 0.5 newton? Explain.

....

....

✎ Summarize what you learned in Part A by answering "yes" or "no" in each box. If necessary, repeat steps 4–10 and fill out the table as you complete each step.

First-Class Lever Results

Load			Effort		
Location of Load	**Mass**	**Load Force (Weight)**	**Less Than 0.5 Newton?**	**Equal to 0.5 Newton?**	**Greater Than 0.5 Newton?**
equal arm	50 g	0.5 newton			
long arm	50 g	0.5 newton			
short arm	50 g	0.5 newton			

Part B: Explore a Wheelbarrow

(1) Build the wheelbarrow according to the K'NEX assembly diagrams. Explore the wheelbarrow and observe how it works.

(2) Anchor the wheel of the wheelbarrow in the holder as shown. Place a 100 g mass in the barrel of the wheelbarrow. As before, this mass is called the load.

The force of gravity acting on this mass is the load force. Pull up on the wheelbarrow handles until the load lifts. This pull is the effort force.

On another piece of paper, draw a diagram of the wheelbarrow. Show and label the load force and the effort force. Mark where you think the fulcrum is.

3. Hold a 100 g mass in your hand to get a feeling for how much it weighs. Place the mass into the wheelbarrow and lift the handles.

Compare the weight of the mass when you hold it in your hand to the effort force used to lift the mass in the wheelbarrow. Repeat with the 200 g and 300 g masses and record your observations.

....

....

....

4. Now you'll use a spring scale to measure what you felt in step 3.

Record the load force for a 100 g load in the data table.

What effort force do you think you'll need to lift the load? Record your prediction in the data table.

Place the 100 g mass in the wheelbarrow. Lift the wheelbarrow with the spring scale as shown. Record the effort force in the data table.

Second-Class Lever Data Table

Mass Added	Load Force (F_L) in Newtons	Predicted Effort Force in Newtons	Measured Effort Force (F_E) in Newtons
100 g			
200 g			
300 g			

5. Repeat step 4 with the 200 g and 300 g masses.

Is the effort force when using the second-class lever more or less than the load force?

....

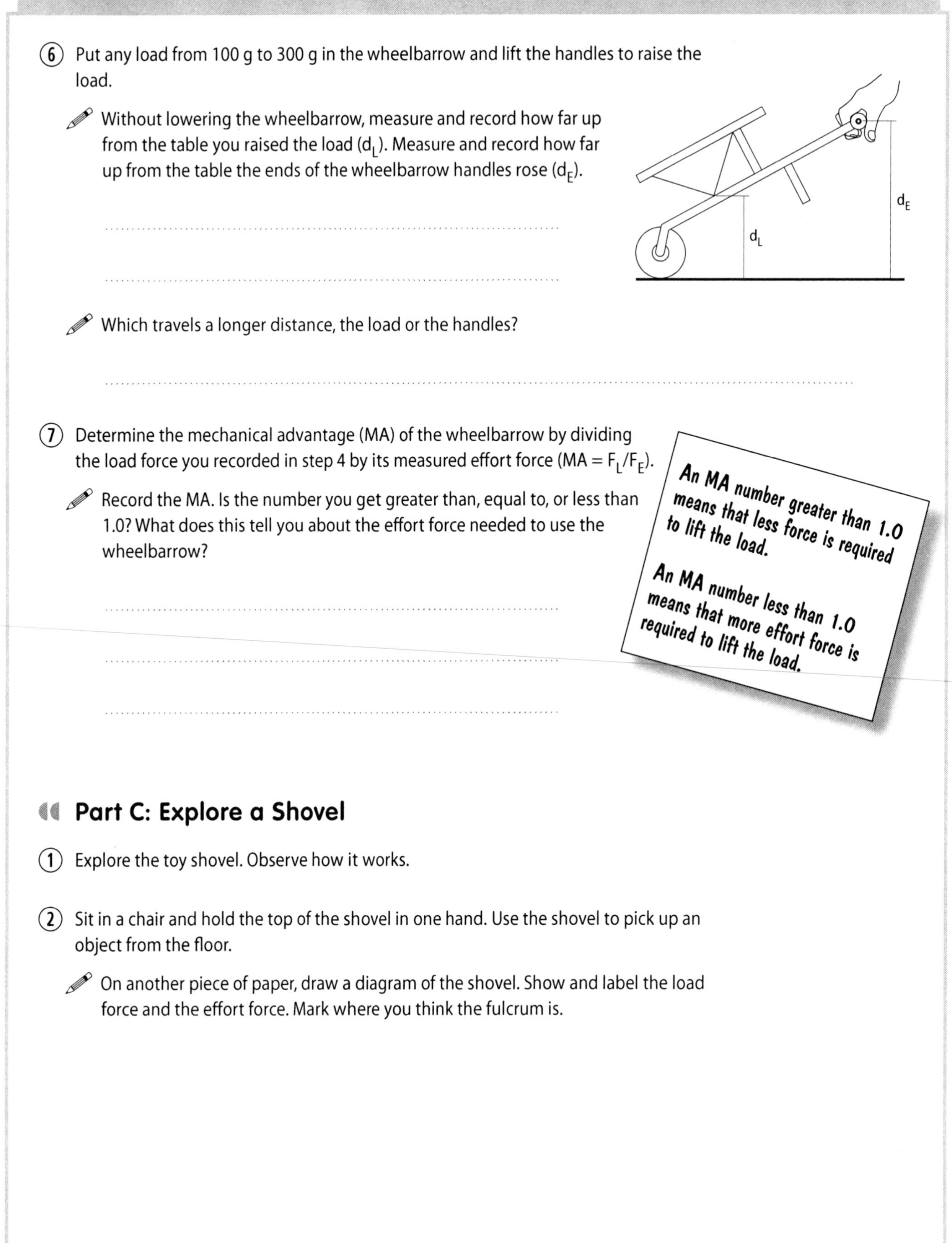

⑥ Put any load from 100 g to 300 g in the wheelbarrow and lift the handles to raise the load.

✎ Without lowering the wheelbarrow, measure and record how far up from the table you raised the load (d_L). Measure and record how far up from the table the ends of the wheelbarrow handles rose (d_E).

....

....

✎ Which travels a longer distance, the load or the handles?

....

⑦ Determine the mechanical advantage (MA) of the wheelbarrow by dividing the load force you recorded in step 4 by its measured effort force (MA = F_L/F_E).

✎ Record the MA. Is the number you get greater than, equal to, or less than 1.0? What does this tell you about the effort force needed to use the wheelbarrow?

....

....

....

An MA number greater than 1.0 means that less force is required to lift the load.

An MA number less than 1.0 means that more effort force is required to lift the load.

Part C: Explore a Shovel

① Explore the toy shovel. Observe how it works.

② Sit in a chair and hold the top of the shovel in one hand. Use the shovel to pick up an object from the floor.

✎ On another piece of paper, draw a diagram of the shovel. Show and label the load force and the effort force. Mark where you think the fulcrum is.

③ Sit in a chair and hold the shovel in the starting position as shown. Have a partner measure from the scoop to the floor and from the middle of the handle to the floor as shown. Without moving the hand holding the top of the shovel, lift the object with the shovel. Have a partner measure the final distances of the scoop and the middle of the handle from the floor as shown.

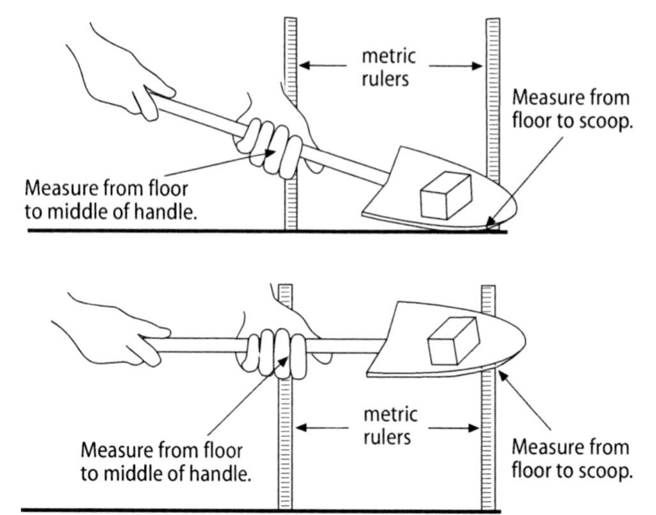

✏ Record your measurements in the data table. Calculate the difference between ending and starting positions to determine the load force distance (d_L) and the effort force distance (d_E).

Third-Class Lever Data		
	Distance from Floor to Scoop	**Distance from Floor to Middle of Handle**
starting position		
ending position		
difference between ending and starting positions	$d_L =$	$d_E =$

✏ Which travels a longer distance when you lift the shovel, the middle of the shovel or the scoop of the shovel?

....

✏ Give another example of a third-class lever. Explain its fulcrum, load force, effort force, and load.

....

....

④ You calculated the mechanical advantage (MA) of the wheelbarrow in Part B by one method. MA can also be calculated by dividing effort force distance (d_E) by load force distance (d_L). Determine the MA of the shovel using the distances you measured in step 3 and the equation $MA = d_E/d_L$.

✏ Record the MA. Is the number you get greater than, equal to, or less than 1.0? What does this tell you about the effort force needed to use the shovel?

....

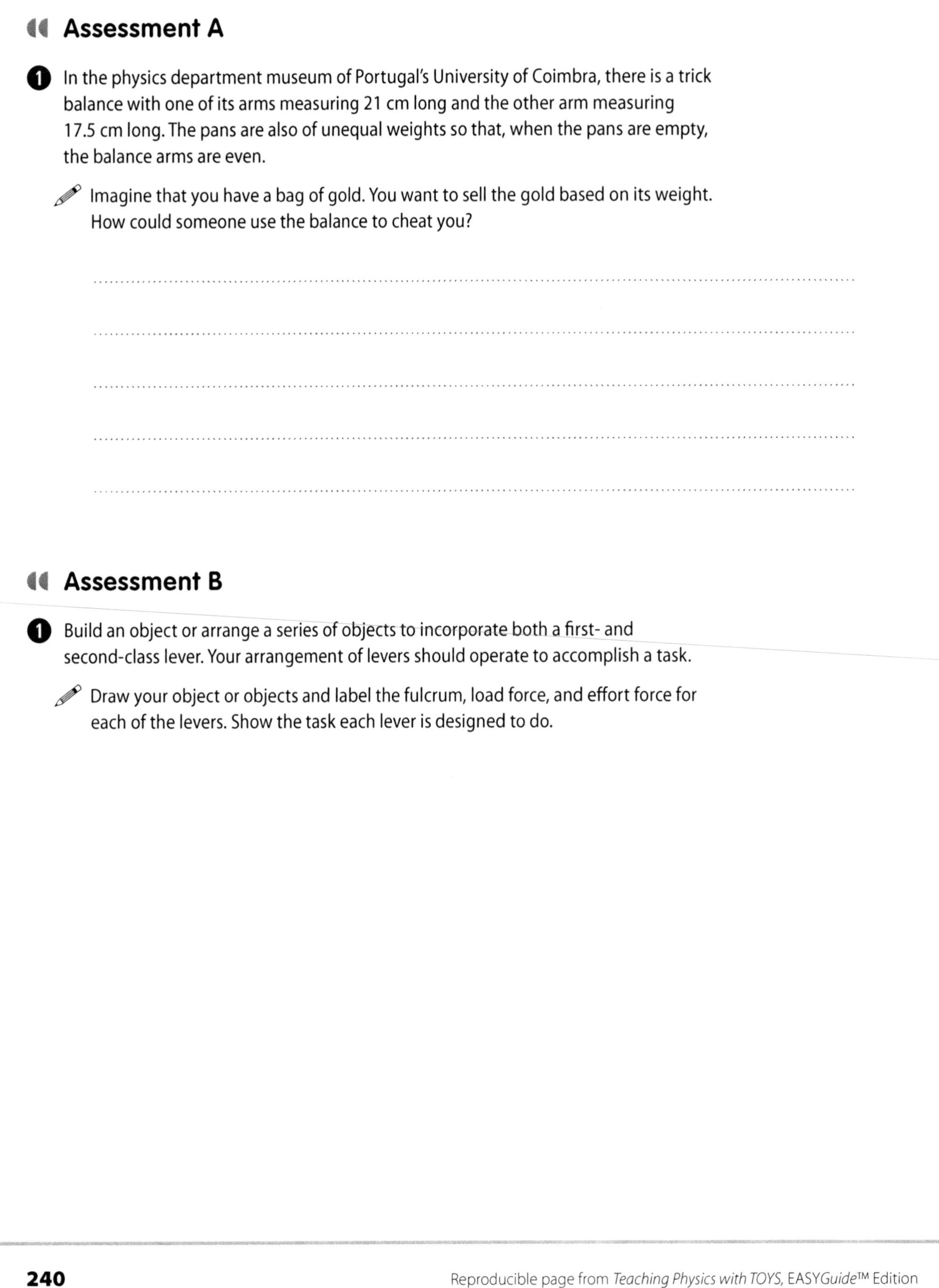

Assessment A

1 In the physics department museum of Portugal's University of Coimbra, there is a trick balance with one of its arms measuring 21 cm long and the other arm measuring 17.5 cm long. The pans are also of unequal weights so that, when the pans are empty, the balance arms are even.

Imagine that you have a bag of gold. You want to sell the gold based on its weight. How could someone use the balance to cheat you?

Assessment B

1 Build an object or arrange a series of objects to incorporate both a first- and second-class lever. Your arrangement of levers should operate to accomplish a task.

Draw your object or objects and label the fulcrum, load force, and effort force for each of the levers. Show the task each lever is designed to do.

Pulley Power Basics

Use flagpole and sailboat models to explore the basics of pulleys.

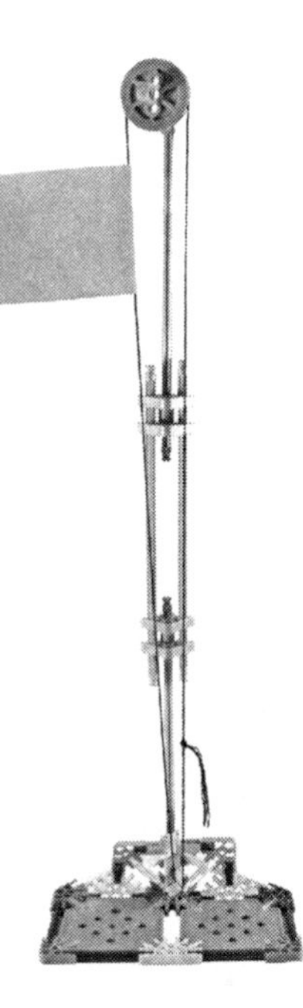

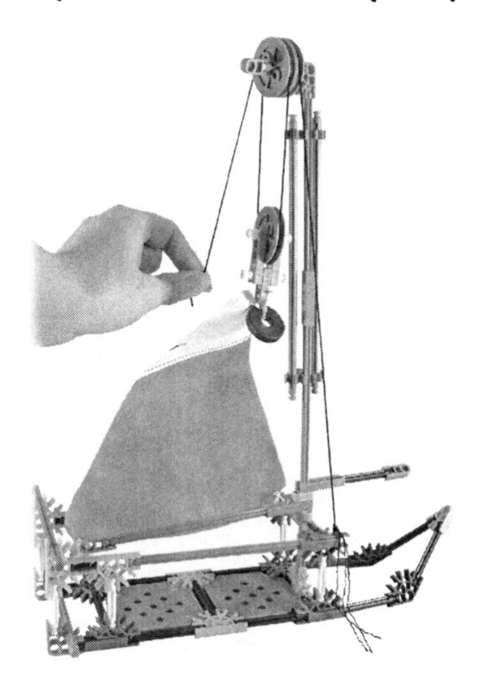

Grade Levels

Science activity appropriate for grades 4–8

Student Background

Students should have previous knowledge of how levers work or should have completed either "Levers at Work" or "Seesaw Forces" in this book before performing this activity.

Time Required

Setup	10	minutes
Build K'NEX® models	30	minutes
Part A	20	minutes
Part B	20	minutes
Cleanup	10	minutes

Assessment time is not included.

Key Science Topics

- forces
- simple machines
- work

National Science Education Standards Overview

See *www.terrificscience.org/physicsez/* for details of how these standards relate to the activity.

Science as Inquiry

Abilities Necessary to Do Scientific Inquiry

K–4 Communicate investigations and explanations.

5–8 Communicate scientific procedures and explanations.

Physical Science

K–4 Position and motion of objects

5–8 Motions and forces
5–8 Transfer of energy

Science and Technology

Abilities of Technological Design

K–4 Evaluate a product or design.

5–8 Evaluate completed technological designs or products.

Materials

As an alternative to using K'NEX pieces, students can use commercial pulleys and other assorted materials to create flagpole and sailboat models.

For the Procedure

Part A, per group

- assorted K'NEX rods, connectors, and pulleys

For specifics, see K'NEX Parts Inventory on www.terrificscience.org/physicsez/ or www.terrificscience.org/supplies/.

- K'NEX assembly diagrams
- scissors
- string

Braided cotton string works well.

- paper
- tape
- meterstick or measuring tape

Part B, per group

All materials listed for Part A, plus

- sail pattern
- hole punch
- (optional) Tyvek® instead of paper

Tyvek is a strong, flexible material that makes a good sail that won't tear. Used overnight mail envelopes are a good source.

- (optional) weights such as washers

Getting Ready

- Make one sail pattern per group by copying the pattern from *www.terrificscience.org/physicsez/*.
- Print a set of K'NEX assembly diagrams for each group. (See *www.terrificscience.org/physicsez/*.) Since K'NEX pieces are color coded, the assembly diagrams need to be printed in color.

Depending on your schedule, you may want to have students build models in advance.

Safety and Disposal

No special safety or disposal procedures are required.

Procedure

This section provides teacher notes corresponding to each step of the student procedure. The procedure without teacher notes is included in the reproducible Student Notebook pages at the end of this activity and at www.terrificscience.org/physicsez/.

	Student Procedure	Teacher Notes
	Part A: Flagpole Pulley System	
①	**Build the flagpole model according to the K'NEX assembly diagrams. Do not attach the string yet.**	Students will need color K'NEX assembly diagrams to build the model. (See *www.terrificscience.org/physicsez/*.)
②	**Cut a piece of string a little over 1 m long. Tape a small paper flag at about the middle of the string. Thread one end of the string through the gray connector at the bottom of the flagpole as shown. Pull the string through the gray connector so that the flag rests at the bottom of the flagpole.**	
③	**Hold the string just above the flag. Move your hand so that the flag moves to the top of the flagpole.** ✎ **Which way did you move your hand?**	Students will see that they must move their hands straight up to raise the flag to the top of the flagpole.
④	**Loop the string over the flagpole pulley and tie it as shown below. Start with the flag at the bottom again. Hold the string just below the pulley. Move your hand so that the flag moves to the top of the flagpole.** ✎ **Which way did you move your hand?** ✎ **A single fixed pulley behaves like a first-class lever. On another piece of paper, draw the pulley on the flagpole and identify the load, fulcrum, and effort.**	Students will see that they now move their hands downward to raise the flag. With prior experience with levers, students should be able to identify the flag as the load, the axle of the pulley as the fulcrum, and the force exerted on the rope as the effort. (See Sample Answers for student drawing example.) Load force is sometimes referred to as resistance force. Mention this term if it is used in your classroom textbook or if students are familiar with the term from previous experience.
	Class Discussion ☞	Explain that the flagpole is an example of a single fixed pulley. This type of pulley changes the direction of the force. The rope is pulled down and the flag goes up. Help the class identify the load, fulcrum, and effort when the single fixed pulley is used. (See the figure in Explanation.)

	Student Procedure	Teacher Notes
5	Compare the feeling of raising the flag with and without the pulley. ✎ Did one way seem easier than the other? ✎ Is the length of string you have to pull less than, more than, or equal to the distance that the flag moves up? ✎ Do you see any advantage to using the pulley? Why or why not? TIP: Think about a full-size flagpole.	There is no mechanical advantage to the use of a fixed pulley. The force required to raise the flag (effort force) is equal to the weight of the flag (load force). ("More Pulley Power" in this book explores this idea.) Students may feel that the effort is easier with the pulley because the effort force is applied downward in the same direction as gravity. However, the length of the string pulled is equal to the distance the flag moves up. A big advantage to using the pulley is that a load, such as a flag, can be raised a great height by pulling down on the rope from ground level. To raise a flag without a pulley, a person would have to stand on something at the same height as the top of the flagpole and pull up on the rope.
	Part B: Sailboat Pulley System	
1	Build the sailboat model according to the K'NEX assembly diagrams. Make a sail out of paper using the sail pattern provided by your teacher. Use a hole punch and string to attach the lower left corner of the sail as shown. Use the same method to attach the other two corners of the sail at the locations shown in the K'NEX assembly diagram.	Students will need color K'NEX assembly diagrams to build the model.
2	Observe the pulley system on the sailboat. Gently raise and lower the sail. ✎ How is the sailboat's pulley system different than the flagpole's pulley? How is it similar?	The flagpole uses only a fixed pulley. The sailboat uses a pulley system that is a combination of a fixed pulley and a movable pulley. The sailboat system is similar to the flagpole in that you still pull down on the rope to apply the effort force. Warn students to raise and lower the sails gently. The strings can easily slip from the pulleys as students raise and lower the sail, but they can quickly be put back in place. Adding weights to the top of the sails by attaching washers can help keep the strings in the grooves of the pulleys.

Student Procedure	Teacher Notes
③ **Gently raise and lower the sail a few more times. Devise a method to determine the amount of string you pull to completely raise the sail.** ✎ **Is the length of string you have to pull less than, more than, or equal to the distance that the sail moves up?** ✎ **How does this compare with what you observed with the flagpole in step 5 of Part A?**	Students should observe that they must pull a greater length of string than the distance the sail moves. With the flagpole, the length of string pulled was equal to the distance the flag moves.
④ **Review your work with the flagpole and sailboat models. The flagpole uses only a fixed pulley. The sailboat uses a pulley system that is a combination of a fixed pulley and a movable pulley.** ✎ **What do you think is the advantage of using a fixed and movable pulley together?**	The combined fixed and movable pulley reduces the amount of effort force needed to move the load. Because the paper flag and sail are such light loads, students cannot actually feel the difference in effort force and therefore may find answering this question challenging. The data students collect using heavier loads in "More Pulley Power" in this book will demonstrate the advantage of using multiple pulleys.
Class Discussion	• Review the idea that *work* = *force* × *distance* and that "work in" must always equal "work out." So, *force* × *distance* on the side of the load (the side with the sail) must equal *force* × *distance* on the side of the effort (the side you pull on). (See Explanation.) • Through discussion, help students to see that for a *greater distance* × *force* to equal a *smaller distance* × *force,* the forces cannot be equal. The force must be smaller on the side of the greater distance. Help students relate this idea to raising the sail. They pulled a greater length of string than the distance the sail went up. Therefore, they must have used less force to raise the sail than they would have used with only one pulley, where the length of string pulled equals the distance the load moves.

Student Procedure / Teacher Notes

Assessment

1. Many old Dutch houses were built narrow and high because owners' taxes were based on the width of their houses. For this reason, the houses have very narrow staircases. Each of these houses has a roof beam sticking out toward the street with a big hook on the end.

   Draw and label a diagram below showing how people can use this beam to move furniture into the house through the windows.

   Explain how your diagram works.

Teacher Notes:

If students seem confused by this scenario, encourage them to think about what they learned with the flagpole and sailboat models. If necessary, explain that the beam and hook are used to support a pulley. Diagrams should show a rope going through the beam's pulley, with one end of the rope tied to a piece of furniture and the other end held by a person. (See Sample Answers for student drawing example.)

Some students may draw the beam's pulley as a single fixed pulley that changes the direction of the applied force. This pulley permits people to pull down on the rope to lift the furniture. Other students may draw a combined fixed and movable pulley like the one used with the sailboat. This type of pulley system not only changes the direction of the applied force, but requires less effort force to lift the furniture.

Sample Answers

Where students follow the same procedure using the same materials, these answers are close to answers you can expect. Where students design their own experiment or model, students' results will vary.

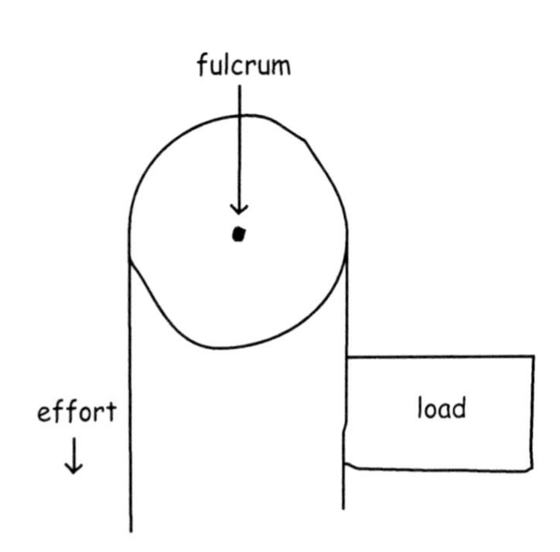

Example of student drawing for Part A

Example of student drawing for Assessment

Explanation

This section is intended for teachers. Modify the explanation for students as needed.

Fixed Pulley

A single fixed pulley is equivalent to a first-class lever, with the pulley's axle at the center acting as the fulcrum. (See figure below.) The load force is on one end of the string (located on one side of the pulley wheel) while the effort force is on the other end of the string (located on the other side of the pulley wheel). So, the load and effort on opposite ends of the string are just like the load and effort on opposite ends of the seesaw. (See Part A of the activity "Levers at Work" in this book for a more detailed lesson on first-class levers.)

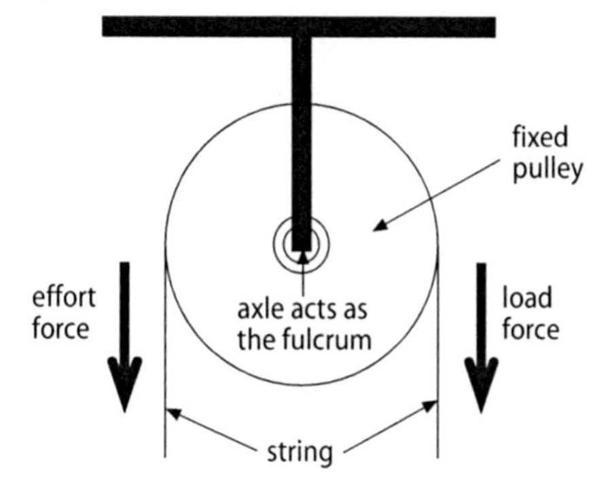

Fixed pulley as a first-class lever

A single fixed pulley changes the direction of the force, but there is no mechanical advantage to the use of a single fixed pulley. The distance the rope is pulled down on the effort side of the pulley equals the distance the flag moves up on the load side of the pulley. The force required to raise the flag (effort force) is equal to the weight of the flag (load force). Students may feel that the effort force is easier with the pulley because the effort force is applied downward in the same direction as gravity. A big advantage to using the pulley is that a load, such as a flag, can be raised a great height by pulling down on the rope from ground level. To raise a flag without a pulley, a person would have to stand on something at the same height as the top of the flagpole and pull up on the rope.

Movable Pulley

The sailboat uses a system that is a combination of a fixed pulley and a movable pulley. As in the flagpole, the fixed pulley serves to change the direction of the effort force so that you can pull downward on the rope. The movable pulley acts like a second-class lever to make the work of raising the sail easier. (See Part B of the activity "Levers at Work" in this book for a more detailed lesson on second-class levers.)

In a second-class lever, the fulcrum is at one end of the lever arm, the effort force is applied at the other end, and the load is between the fulcrum and effort force. The following figure illustrates where the fulcrum, load force, and effort force are located on a movable pulley.

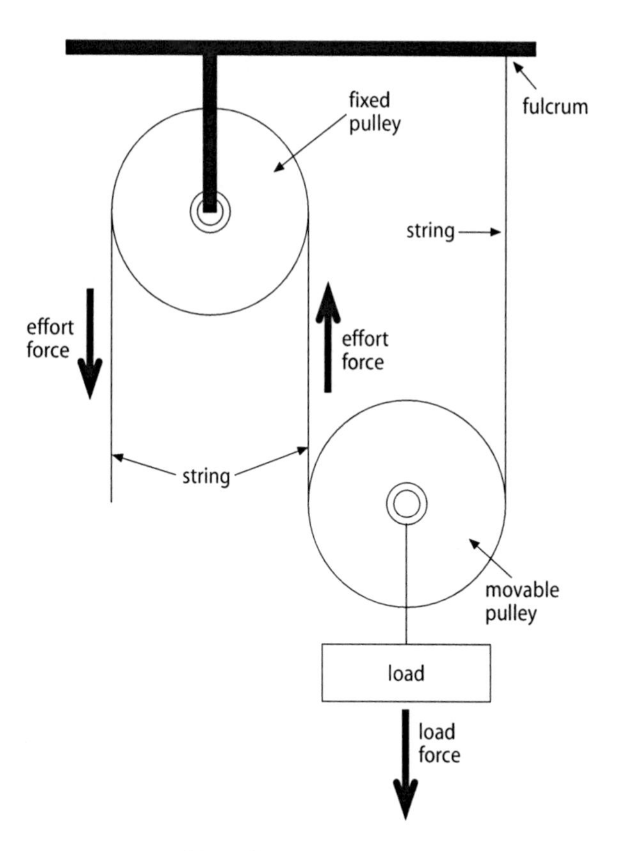

Movable pulley as a second-class lever

Work

The scientific definition of work is easiest to understand when the force is constant and always in the same direction as the object it displaces. In this case, work is the product of the force applied to an object and the distance the object moves. As in a wheelbarrow, the movable pulley reduces the effort force required to do the work of lifting the load. The pulley system does not change the amount of work needed to lift the sail; it just makes the work easier.

$$\text{work}_{\text{in}} = \text{work}_{\text{out}}$$

$$\text{force} \times \text{distance} = \text{force} \times \text{distance}$$

$$(\text{small force}) \times (\text{large distance moved}) = (\text{large force}) \times (\text{small distance moved})$$

For example,

$$(2 \text{ newtons}) \times (0.5 \text{ m}) = (4 \text{ newtons}) \times (0.25 \text{ m})$$
$$(1.0 \text{ newton m}) = (1.0 \text{ newton m})$$

To reduce the effort force required to raise the sail, you must pull a greater length of rope than the distance the sail moves. If you measure and compare these distances, you will find that the rope on the effort side moves twice as far as the load. Therefore, the mechanical advantage of this system is 2. (See "More Pulley Power" in this book for a discussion of mechanical advantage.)

Cross-Curricular Integration

Language arts:

- Read aloud or suggest that students read the following book:
  - *Sailboats, Flagpoles, Cranes: Using Pulleys as Simple Machines*, by Christopher Lampton (grades 3–5)
    This book explores the uses and benefits of pulleys in common applications.
  - *How Do You Lift a Lion?*, by Robert E. Wells (grades 4–7)
    Students learn basic physics concepts when they read about how levers, pulleys, and wheels can be used to lift heavy animals.

Reference

K'NEX Industries, Inc. *Simple Machines: Levers and Pulleys Teacher's Guide;* Hatfield, PA, 2003.

Part A: Flagpole Pulley System

① Build the flagpole model according to the K'NEX® assembly diagrams. Do not attach the string yet.

② Cut a piece of string a little over 1 m long. Tape a small paper flag at about the middle of the string. Thread one end of the string through the gray connector at the bottom of the flagpole as shown. Pull the string through the gray connector so that the flag rests at the bottom of the flagpole.

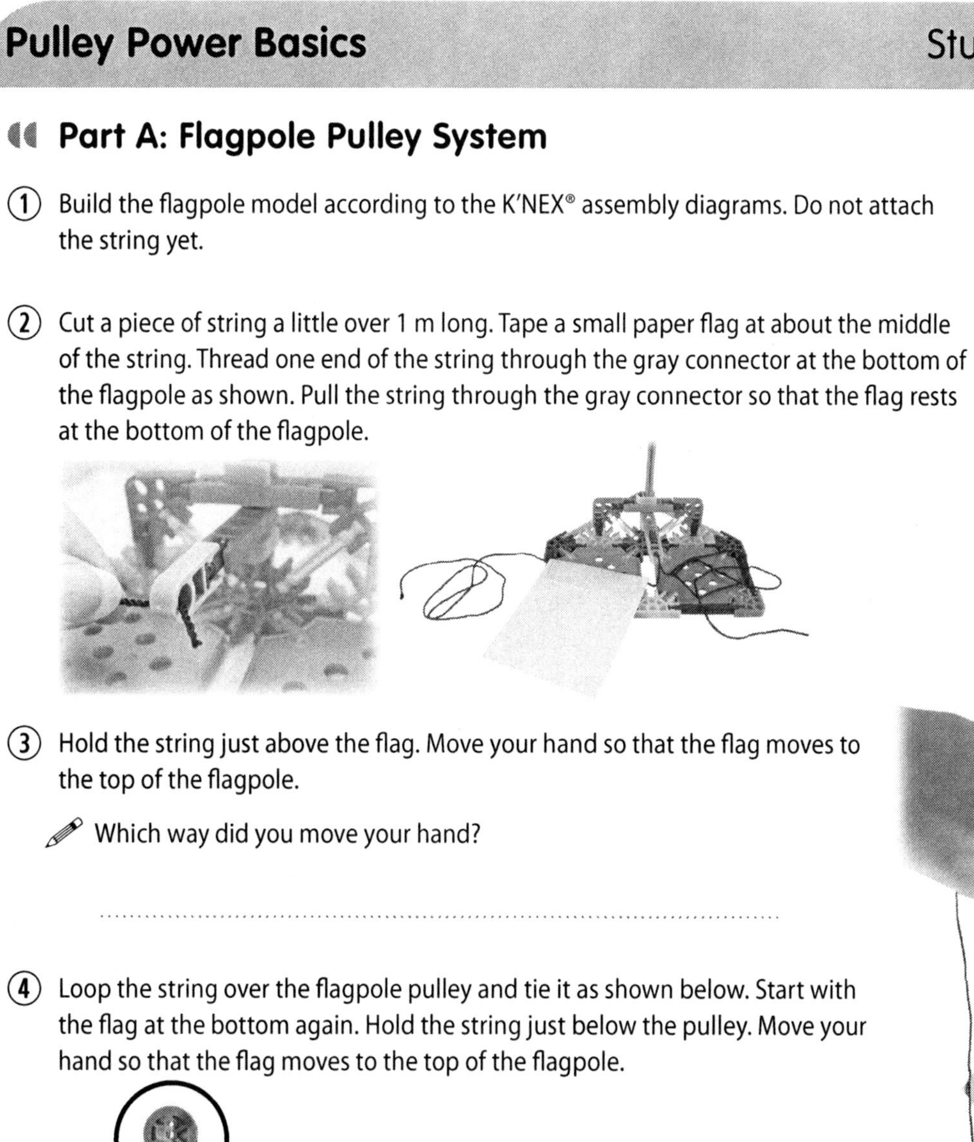

③ Hold the string just above the flag. Move your hand so that the flag moves to the top of the flagpole.

✎ Which way did you move your hand?

....

④ Loop the string over the flagpole pulley and tie it as shown below. Start with the flag at the bottom again. Hold the string just below the pulley. Move your hand so that the flag moves to the top of the flagpole.

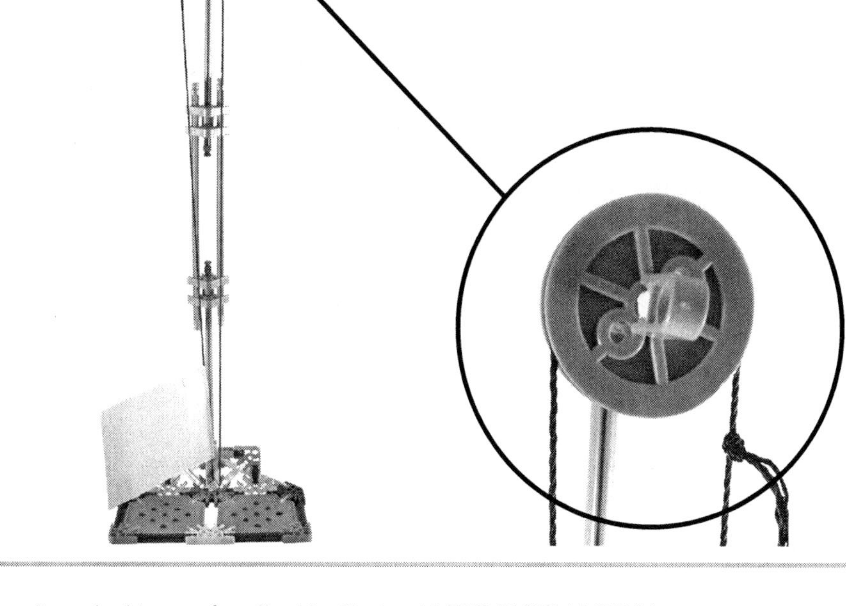

Which way did you move your hand?

A single fixed pulley behaves like a first-class lever. On another piece of paper, draw the pulley on the flagpole and identify the load, fulcrum, and effort.

5. Compare the feeling of raising the flag with and without the pulley.

Did one way seem easier than the other?

Is the length of string you have to pull less than, more than, or equal to the distance that the flag moves up?

Do you see any advantage to using the pulley? Why or why not?

TIP: Think about a full-size flagpole.

Part B: Sailboat Pulley System

1. Build the sailboat model according to the K'NEX assembly diagrams. Make a sail out of paper using the sail pattern provided by your teacher. Use a hole punch and string to attach the lower left corner of the sail as shown. Use the same method to attach the other two corners of the sail at the locations shown in the K'NEX assembly diagram.

2. Observe the pulley system on the sailboat. Gently raise and lower the sail.

How is the sailboat's pulley system different than the flagpole's pulley? How is it similar?

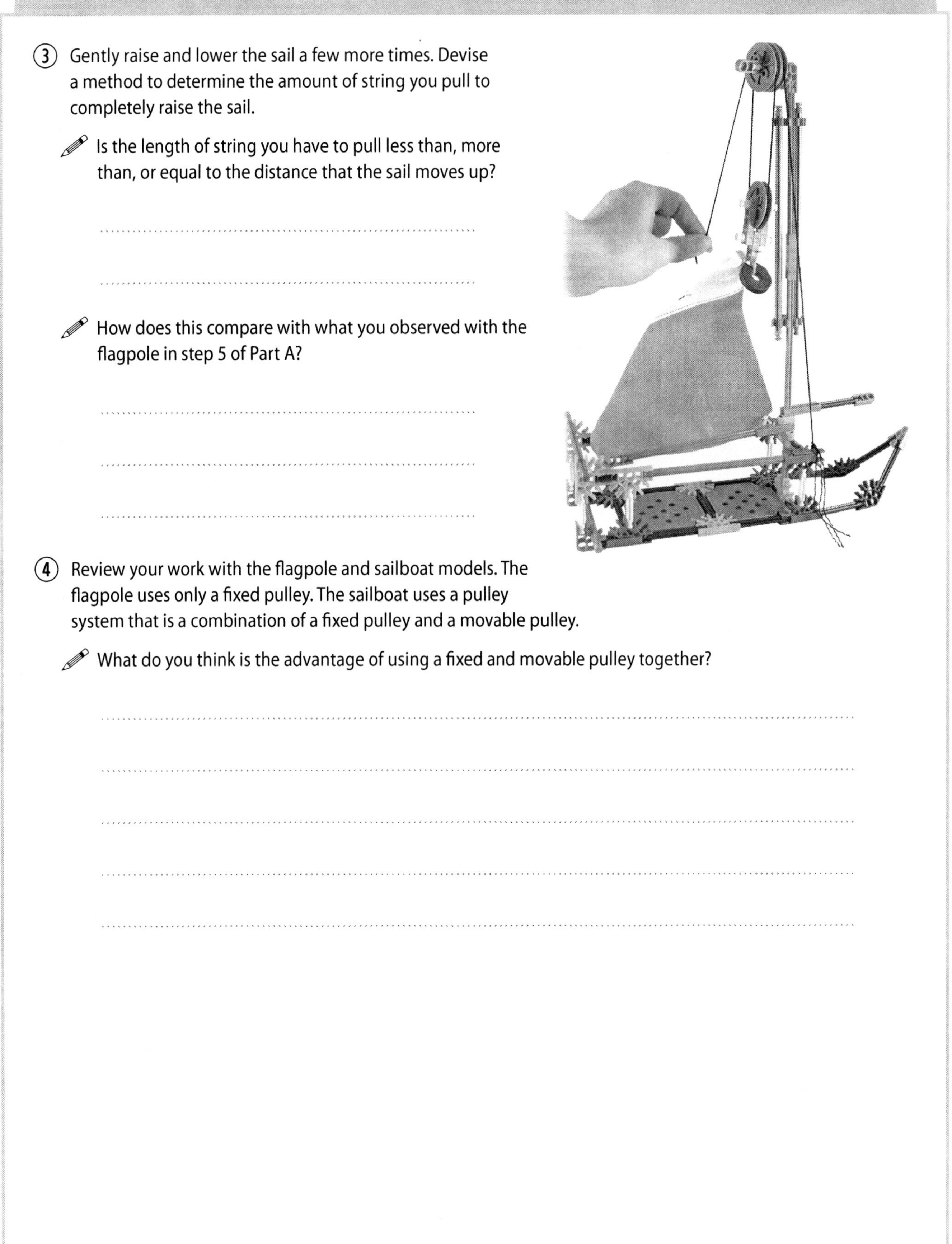

③ Gently raise and lower the sail a few more times. Devise a method to determine the amount of string you pull to completely raise the sail.

✎ Is the length of string you have to pull less than, more than, or equal to the distance that the sail moves up?

....

....

✎ How does this compare with what you observed with the flagpole in step 5 of Part A?

....

....

....

④ Review your work with the flagpole and sailboat models. The flagpole uses only a fixed pulley. The sailboat uses a pulley system that is a combination of a fixed pulley and a movable pulley.

✎ What do you think is the advantage of using a fixed and movable pulley together?

....

....

....

....

....

Assessment

1 Many old Dutch houses were built narrow and high because owners' taxes were based on the width of their houses. For this reason, the houses have very narrow staircases. Each of these houses has a roof beam sticking out toward the street with a big hook on the end.

Draw and label a diagram below showing how people can use this beam to move furniture into the house through the windows.

Explain how your diagram works.

More Pulley Power

Students explore different pulley arrangements to learn about mechanical advantage.

Grade Levels

Science activity appropriate for grades 6–9

Student Background

Students should be familiar with the use of a spring scale and should have completed the activity "Pulley Power Basics" in this book or similar material before performing this activity.

Time Required

Setup	10	minutes
Build K'NEX® models	30	minutes
Procedure	45	minutes
Cleanup	10	minutes

Assessment time is not included.

Key Science Topics

- force
- mechanical advantage
- simple machines
- work

National Science Education Standards Overview

See *www.terrificscience.org/physicsez/* for details of how these standards relate to the activity.

Science as Inquiry

Abilities Necessary to Do Scientific Inquiry

5–8 *Use appropriate tools and techniques to gather, analyze, and interpret data.*
5–8 *Think critically and logically to make the relationships between evidence and explanations.*
5–8 *Communicate scientific procedures and explanations.*
5–8 *Use mathematics in all aspects of scientific inquiry.*

9–12 *Use technology and mathematics to improve investigations and communications.*
9–12 *Communicate and defend a scientific argument.*

Physical Science

5–8 *Motions and forces*
5–8 *Transfer of energy*

9–12 *Motions and forces*

Science and Technology

Abilities of Technological Design

5–8 *Implement a proposed design.*
5–8 *Evaluate completed technological designs or products.*
5–8 *Communicate the process of technological design.*

9–12 *Implement a proposed solution.*
9–12 *Evaluate the solution and its consequences.*
9–12 *Communicate the problem, process, and solution.*

Materials

For the Procedure

Per group

- pulley tower and set of pulleys, such as
  - assorted K'NEX rods, connectors, and pulleys

  For specifics, see K'NEX Parts Inventory at www.terrificscience.org/physicsez/ or www.terrificscience.org/supplies/.
  - commercial pulleys (2 single wheels, 1 double wheel, 1 triple wheel, 2 two-wheel or double tandems), supports to hang them from (such as support stands, rods, and connectors), and a holder for the masses
- K'NEX assembly diagrams (if using K'NEX pieces)
- spool of pulley string
- 100 g, 200 g, and 500 g standard masses
- spring scale calibrated for 0–5 newtons

  A spring scale that measures 0 to 5 newtons will enable students to see force differences much more easily than a spring scale designed to measure a larger range of forces.
- metric ruler or meterstick

Safety and Disposal

No special safety or disposal procedures are required.

Getting Ready

- If needed, adjust spring scales so that they read 0 with nothing hanging from the hook. (Most models of spring scales have an adjustment knob at the top.)
- If using K'NEX pieces, print a set of K'NEX assembly diagrams for each group. (See *www.terrificscience.org/physicsez/*.) Since K'NEX pieces are color coded, the assembly diagrams need to be printed in color.

  Depending on your schedule, you may want to have students build models in advance.
- The procedure is written so that all groups of students make and experiment with all five pulley systems. As an alternative,
  - have groups share the pulleys by circulating the pulleys around the room,
  - set up five stations (each containing one type of pulley system) and have students circulate to each station, or
  - assign only one type of pulley system to each group of students. (Make sure each group uses the same load so data can be compared.)
- If stations are used, build the five block and tackle towers and five pulley systems prior to class.

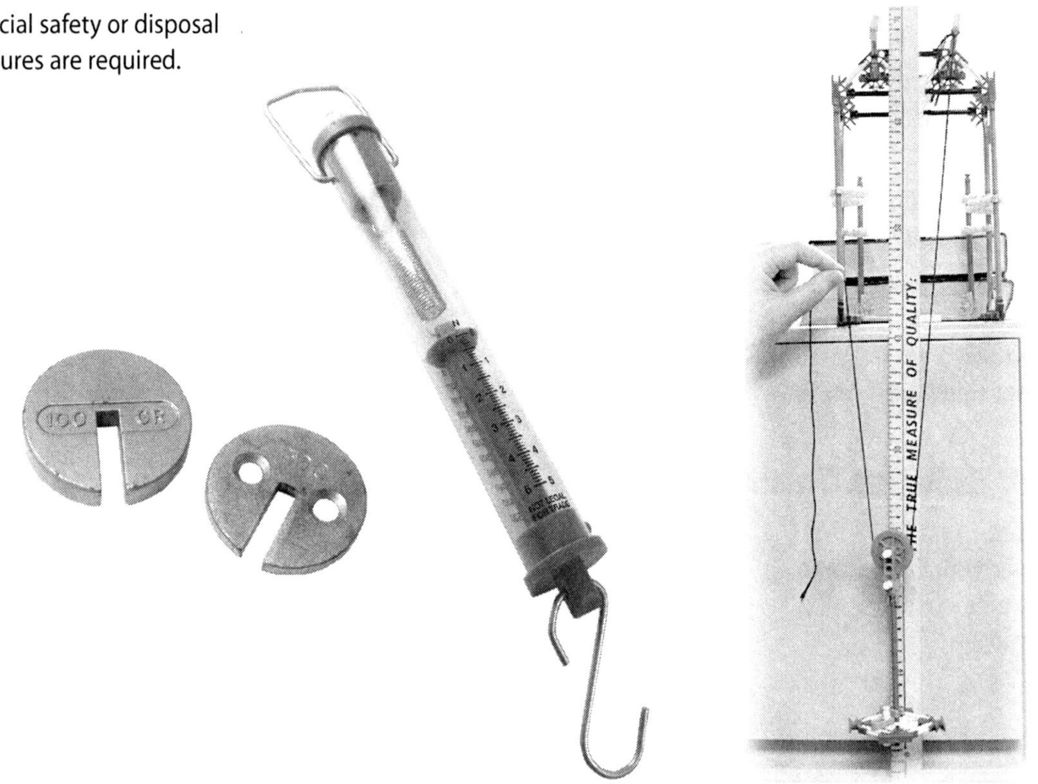

Procedure

This section provides teacher notes corresponding to each step of the student procedure. The procedure without teacher notes is included in the reproducible Student Notebook pages at the end of this activity and at www.terrificscience.org/physicsez/.

Student Procedure	Teacher Notes
Activity Introduction	• If students have done the activity "Pulley Power Basics" in this book, remind them of the flagpole and sailboat models they built. You may want to build a sample of each model as a reminder. (If your students have not done the activity, give them a chance to explore the function of the pulleys in the flagpole and sailboat models.) • Discuss other situations in which pulleys are commonly used and why people use them. Tell students that they will now work with pulleys again, this time measuring the forces involved with a spring scale. • If your students need a review of mass, weight, and using spring scales, have them do the Activity Introduction of "Levers at Work" in this book.
Towering Pulley System	
① **Prepare a pulley tower, set of pulleys, and weight support as instructed by your teacher.**	If using K'NEX pieces, students will need color K'NEX assembly diagrams to build the model. (See *www.terrificscience.org/physicsez/*.) If not using K'NEX pieces, instruct students as needed.
② **Select at least 200 g from the mass set to use as a load. Determine the weight of the load with the spring scale, making sure to include the weight support in your measurement. This total weight is the load force.** **Record the load force in Column A of the data table.**	Load force is sometimes referred to as resistance force. Mention this term if it is used in the classroom textbook or if students are familiar with the term from previous experience. See Sample Answers for example data.
③ **Attach the single fixed pulley to the tower. Run the string over the pulley, attach the load, and attach the spring scale as shown in the diagram. The exact appearance of your setup will vary depending on the materials you use.**	

Student Procedure	Teacher Notes
④ Complete Column A of the data table following the steps below: **a. Pull down on the spring scale and raise the load. The spring scale reading is the effort force. Record this effort force in Column A of the data table.** **b. Note the number of segments of string listed in the data table. Although you may see more string segments, only those that exert a force directly on the load have been counted.** **c. Raise the load two or three times. Compare the speed at which the load moves to the speed at which your hand moves (the effort). Record whether the load speed is faster than your hand speed, slower than your hand speed, or whether the two speeds are equal.** **d. Move your hand to lift the load. Measure the distance your hand moves from start to finish. Record this distance (d_E). Measure and record the corresponding distance the load moves (d_L).**	See Sample Answers for example data. Introduce or review the idea that a single fixed pulley functions like a first-class lever. (See Explanation in the "Pulley Power Basics" activity in this book.) Understanding the way that string segments are counted can be challenging. One simple rule of thumb is that you count the effort string segment if it is being pulled up but not down. However, looking at which string segments are actually exerting a force on the load is best. (See Explanation.) See the photograph below Getting Ready for a suggested setup for measuring distances.
⑤ Set up the single movable pulley as shown. **TIP: In this case, you will pull up rather than down to raise the load.** **Repeat step 4 with the same mass you chose in step 2, this time filling in Column B of the data table.**	See Sample Answers for example data.
⑥ Use the single fixed pulley data from Column A of your table to fill in the values for force and distance in the equations below. Be sure to convert the distance your hand moves from cm to m. Then, multiply to complete the equations. **What do you notice about the two answers you get when you multiply?**	See Sample Answers for example calculations. Students will see that the two answers are approximately the same.
⑦ Repeat step 6, this time with single movable pulley data from Column B. **What do you notice about the two answers you get when you multiply?**	See Sample Answers for example calculations. Students will see that the two answers are again approximately the same. You will use the calculations done in steps 6 and 7 during the class discussion after step 10.

	Student Procedure	Teacher Notes
8	Attach the top pulley of the combined fixed and movable pulley system to the tower, and set up the rest as shown. ✎ Repeat step 4 with the same mass you chose in step 2, this time filling in Column C of the data table.	See Sample Answers for example data.
9	Attach the top pulley of the linear block and tackle system to the tower, and set up the rest as shown. ✎ Repeat step 4 with the same mass you chose in step 2, this time filling in Column D of the data table.	See Sample Answers for example data.
10	Attach the top pulley of the parallel block and tackle system to the tower, and set up the rest as shown. ✎ Repeat step 4 with the same mass you chose in step 2, this time filling in Column E of the data table.	See Sample Answers for example data.
11	Based on the data you collected, answer the following questions: ✎ What happens to the number of segments of string as you add more pulleys to the system? ✎ What happens to the effort force as the number of segments increases beyond one? ✎ If you pull the string a fixed amount, what happens to the distance the load moves as you add segments to the load? ✎ What is the relationship between the effort force and distance you must pull the string to raise the load? ✎ How do the results of the linear block and tackle system and the parallel block and tackle system compare?	• The number of segments of string increases as more pulleys are added to the system. • The effort force becomes less as the number of segments increases. If you add a second segment, the effort force is about half of the load. If you add three more segments (totaling four), the effort force is about ¼ the load. • With one segment, the load moves the same distance as you pull. The load moves a shorter distance as you add segments. With two segments, the load moves half the distance that you pull. With four segments, the load moves ¼ the distance that you pull. • The more the effort force decreases, the further you have to pull the string to raise the load to a given height. • Both pulley systems give approximately the same results because the number of pulleys and segments of string is the same. Only the arrangement is different.

Student Procedure	Teacher Notes
Class Discussion	• Tell students that the answers they calculated in steps 6 and 7 are the values for work done with the pulley. Based on their calculations, have students suggest what the formula for work is. Through discussion, establish that *work = force × distance.* Equation A represents "work in" and Equation B represents "work out." Ask students to suggest a rule to describe the relationship between "work in" and "work out." Through discussion, establish that: $work_{in} = work_{out}$ • Ask students to share their ideas about the advantages of using pulleys. Ask if they see different advantages to any of the pulley arrangements they have tried. Encourage them to compare the single fixed versus single movable, single versus multiple pulleys, and linear versus parallel arrangements. (See Explanation.) • Introduce the concept of mechanical advantage. (See Explanation.)
(12) Use data previously collected in this activity to calculate mechanical advantage (MA) in two ways: Calculate and record the MA of all pulleys and pulley systems using the formula: MA = load force/effort force = F_L/F_E. Calculate and record the MA of all pulleys and pulley systems using a different formula: MA = effort distance/load distance = d_E/d_L. How does the MA of each pulley and pulley system compare with the number of segments of string in each pulley setup?	See Sample Answers for example calculations. Depending on the number of decimal places students use in their calculations, they may get slightly different answers for each method of calculating MA. You may want to explain to them the difference between theoretical and experimental results. Ask students what they think the theoretical MA values should be. With pulleys, MA equals the number of string segments that exert force directly on the load. (See Explanation.)
Assessment A	
(1) Suppose that you need to lift a 200,000 g object, but your arms can only produce 500 newtons. On another piece of paper, design and label a pulley system that would let you lift the object.	The 200,000 g object has a weight of approximately 2,000 newtons. Since the effort force in this situation can only be 500 newtons, students need to design a pulley system that is capable of lifting the object with ¼ the force required to lift it without the pulley system. Either a linear or parallel block and tackle pulley system will work to lift the 200,000 g object with an effort force of 500 newtons. (See Sample Answers for student drawing example.)

Student Procedure	Teacher Notes
Assessment B	
1 Observe the pulley system. The load is 1,000 g. **How much effort force is needed to raise the load?** **How far would you have to pull the string to lift the load 10 cm?**	The 1,000 g load has a weight of approximately 10 newtons. Using the combined fixed and movable pulley system, an effort force of 5 newtons will lift a 1,000 g load. The string will need to be pulled 20 cm in order to lift the load 10 cm.
Assessment C	
1 Review mechanical advantage. **Why is mechanical advantage a useful number?** **What is the mechanical advantage of the block and tackle system shown here?**	Mechanical advantage lets us figure out what force is needed to lift a particular object using a particular pulley system. For example, we can calculate the force it would take for a particular pulley system to lift a heavy object of known mass. Or, if we know how hard we can pull, we can calculate how much a particular pulley system can lift. This pulley system has a total of five segments of string. Since the free end of the string is to be pulled up, it counts when calculating the mechanical advantage. Therefore, the mechanical advantage is 5.

Sample Answers

Where students follow the same procedure using the same materials, these answers are close to answers you can expect. Where students design their own experiment or model, students' results will vary.

	Ⓐ	Ⓑ	Ⓒ	Ⓓ	Ⓔ
Note: Sample data here are based on a load force of 2.3 newtons (weight of a 200 g mass plus weight support) and a string pull distance of 30 cm.	**Single Fixed Pulley (1 Pulley)**	**Single Movable Pulley (1 Pulley)**	**Combined Fixed and Movable Pulley (2 Pulleys)**	**Linear Block and Tackle System (4 pulleys)**	**Parallel Block and Tackle System (4 pulleys)**
load force in newtons (F_L)	2.3 newtons	2.3 newtons	2.3 newtons	2.3 newtons	2.3 newtons
effort force in newtons (F_E)	2.3 newtons	1.2 newtons	1.2 newtons	0.6 newton	0.6 newton
number of segments of string	1	2	2	4	4
speed of load compared to speed of hand	equal	slower	slower	slower	slower
distance in cm that hand moves (d_E)	30 cm	30 cm	30 cm	30 cm	30 cm
distance in cm that load moves (d_L)	30 cm	15 cm	15 cm	7.5 cm	7.5 cm

Example data for different pulleys and pulley systems

A.	2.3 newtons (effort force)	× 0.3 m (distance hand moves)	=	0.7 newton m
B.	2.3 newtons (load force)	× 0.3 m (distance load moves)	=	0.7 newton m

Example calculations for step 6

A.	1.2 newtons (effort force)	× 0.3 m (distance hand moves)	=	0.4 newton m
B.	2.3 newtons (load force)	× 0.15 m (distance load moves)	=	0.4 newton m

Example calculations for step 7

Note: Sample data here are based on a load force of 2.3 newtons (weight of a 200 g mass plus weight support) and a string pull distance of 30 cm.	(A) **Single Fixed Pulley (1 Pulley)**	(B) **Single Movable Pulley (1 Pulley)**	(C) **Combined Fixed and Movable Pulley (2 Pulleys)**	(D) **Linear Block and Tackle System (4 pulleys)**	(E) **Parallel Block and Tackle System (4 pulleys)**
$MA = F_L/F_E$	1	1.9	1.9	3.8	3.8
$MA = d_E/d_L$	1	2	2	4	4

Example calculations for step 12

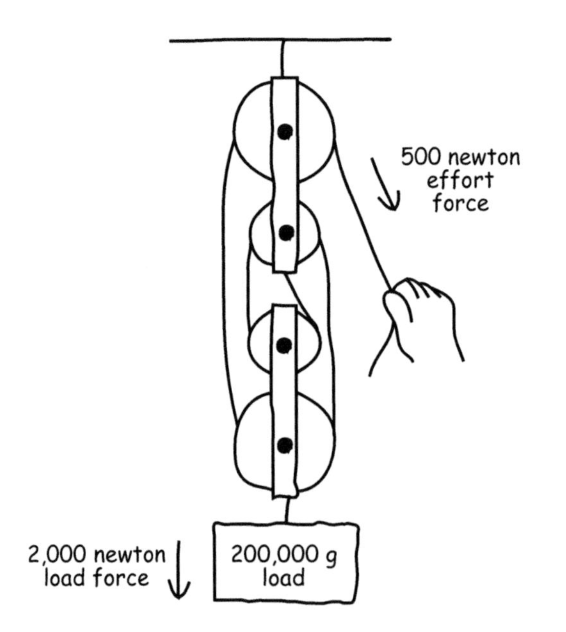

Example of student drawing for Assessment A

Explanation

This section is intended for teachers. Modify the explanation for students as needed.

Mass and Weight

Correctly using terms, units, and tools related to mass and weight is an important part of this activity. Mass measures the quantity of matter in an object, and weight measures the force of gravity on an object. The units are different as well. Two typical units of mass are grams (g) and kilograms (kg). Two typical units of weight are newtons and pounds. In this activity, when talking about mass, use the unit "gram". When talking about weight, use the unit "newton."

Newton's second law gives us a way to relate mass and weight. In general,

$$\text{force} = \text{mass} \times \text{acceleration}$$

Therefore,

$$\text{weight (newtons)} = \text{mass (kg)} \times \text{acceleration due to gravity (m/sec}^2\text{)}$$

The acceleration due to gravity is 9.8 m/sec^2 or approximately 10 m/sec^2. The following table shows the weight of some standard masses.

Weights of Standard Masses	
Mass	Approximate Weight
50 g = 0.05 kg	0.5 newton
100 g = 0.1 kg	1 newton
200 g = 0.2 kg	2 newtons
500 g = 0.5 kg	5 newtons
1,000 g = 1.0	10 newtons

Spring scales for use in science classrooms are often calibrated in both units of mass (grams) and units of weight (newtons). However, the spring scale is only appropriate for measuring weight, not mass. A balance is the appropriate tool for measuring mass.

A good way to help students visualize the appropriate use of these tools is to imagine a spring scale and a double pan balance operating on Earth compared to operating on the moon.

- On Earth, the spring scale reads 1 newton with a 100 g object attached to the bottom.
- On Earth, if the balance has a 100 g mass in the pan on the left side, then adding a 100 g mass to the pan on the right side causes the arms to balance.
- If we took the spring scale to the moon, it would read approximately 0.16 newton with the same object attached to the bottom. Since the spring scale measures weight and weight is a function of gravity, the weight of the object on the moon is different than the weight of the object on Earth.
- If we took the balance to the moon with a 100 g mass on the left side, adding a 100 g mass to the pan on the right side would still cause the arms to balance. Mass is an inherent property of the object and is not changed by gravity.

What Pulleys Do

This activity begins with fixed and movable pulleys similar to those explored in "Pulley Power Basics." (See "Pulley Power Basics" in this book for a discussion of how fixed and movable pulleys do work and function as first- and second-class levers.) In this activity, students make quantitative measurements that help them understand the concept of mechanical advantage.

Aside from changing the direction of the force, we use pulleys to make the work we do easier. For example, imagine trying to lift a heavy box off the ground. Lifting the box would be much easier if you could lift it using less force. A block and tackle pulley system allows you to lift the box using ¼ of the force required to lift it using your arms alone. In other words, the block and tackle system multiplies the force you exert by a factor of four. This factor is the mechanical advantage of the pulley system. The trade-off is that you have to exert this force over four times the original distance, because there is no free lunch where work is concerned. You have to make up for using less force by using it over a larger distance.

Counting String Segments

One challenge of working with pulleys is understanding how to count string segments. These counts are one way to calculate mechanical advantage, discussed in the next section. To correctly count string segments in pulley systems, we must think about which string segments exert a force directly on the load. When a string passes over a pulley, we can think of each segment as exerting a separate force. While we may teach students the simple rule to count the effort string segment if it is being pulled up but not down, looking at which string segments are actually exerting a force on the load better explains what is happening. Considering these forces also allows us to evaluate more complex systems.

- In the single fixed pulley system, the load is attached to the string, and one string segment directly exerts a force on the load.
- In the case of the single movable pulley, the load attaches to the pulley, not to the string, making the pulley itself part of the load. The segments of the string on the left and right of the pulley each exert a force on the pulley, so we say that this pulley has two string segments. (For simplicity in the procedure, we ignored the weight of the pulley. The load force is actually the weight of the pulley plus the weight of the load.)
- For the combined fixed and movable pulley system, we again have two string segments supporting the lower pulley.
- In the linear block and tackle system shown on the following page, the load attaches directly to the lower set of pulleys (pulleys 3 and 4), so those pulleys are part of the load. Four string segments pull up on the lower set of pulleys, so only these segments are counted: two for pulley 3 and two for pulley 4, making a total of four string segments.
- For the parallel block and tackle system, the load again attaches directly to the lower set of pulleys. Again, four string segments pull up on the lower set of pulleys, so we say this system has four strings.

Mechanical Advantage

Mechanical advantage (MA) is the factor by which a machine multiplies the force put into it. Mechanical advantage can be calculated as the ratio of load force divided by effort force: $MA = F_L/F_E$. It can also be calculated as the ratio of effort distance divided by the load distance: $MA = d_E/d_L$. Interestingly, in the case of pulleys, mechanical advantage can also be directly observed by looking at the number of string segments that exert force directly on the load.

In all practical pulley systems, the actual mechanical advantage is, of course, less that the ideal mechanical advantage because of friction. The amount of friction increases with the number of pulleys used, the load used, and the type of string used. In the procedure, we tried to minimize this deviation by suggesting a load of at least 200 g.

- Imagine a single fixed pulley with a load on one end of the string. (See figure below.) The person holding the other end must apply a force equal to the weight of the load to keep the load hanging in the air. The load force equals the effort force, so the mechanical advantage of this system is 1.

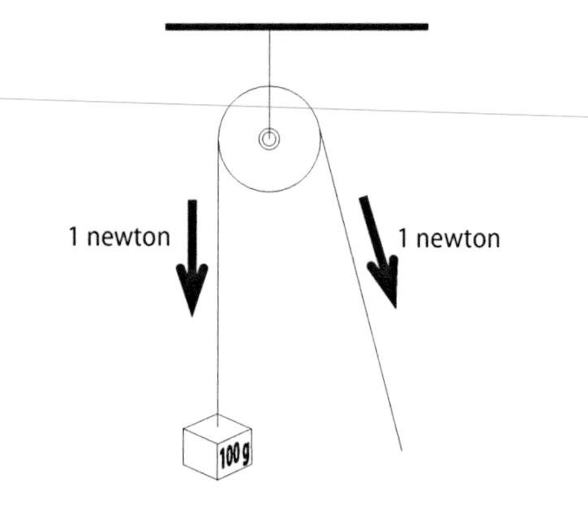

Single fixed pulley

- Now, imagine a single movable pulley with a load attached. (See figure at top of next page.) One end of the string is attached to the pulley tower. The other end is held up by a person. This time, the load is supported by two segments of string—the segment attached to the pulley tower and the segment held up by the person. Each segment of the string is supporting the weight of the load equally, so each segment carries only half the weight. To keep the load hanging in the air, the person needs to apply a force only half the weight of the load. The load force (weight of the load) is divided by the effort force, which is ½ the weight, so the mechanical advantage of this system is 2.

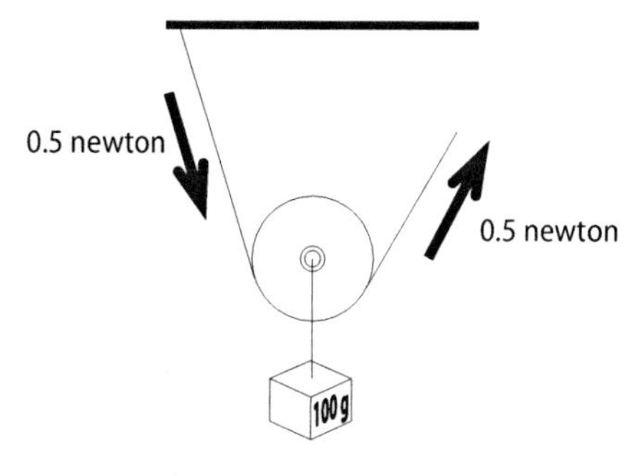

Single movable pulley

- The block and tackle system uses four pulleys. The load is in the middle directly below the large bottom pulley. (See figure below.) Four segments of string share the load—two around pulley 3 and two around pulley 4. This arrangement functions as if the load were suspended by four strings, one at each corner, each supporting ¼ of the load. (See figure below.) To keep the load hanging in the air, a person using the block and tackle system needs to apply a force equal to the force on one of these four string segments. The load force (weight of the load) is divided by the effort force, which is ¼ the weight, so the mechanical advantage of this system is 4.

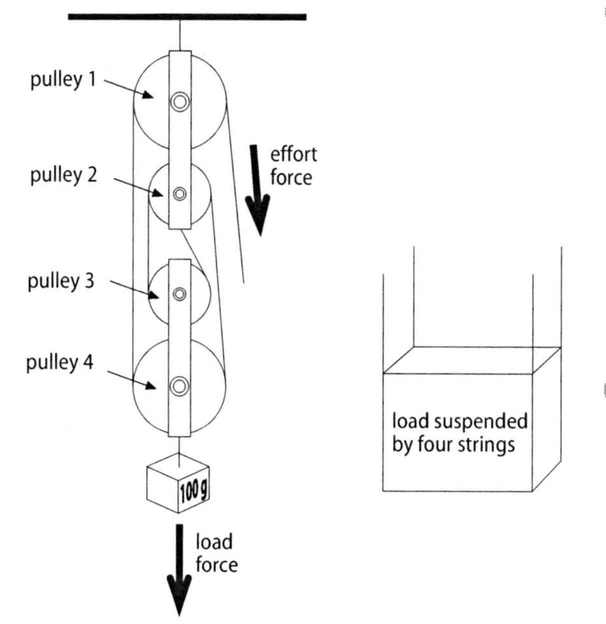

Block and tackle system

Work

The amount of work needed to lift a load using movable pulleys is not any different than lifting the same load using the single fixed pulley. The trade off comes when the effort force string is pulled. Consider the block and tackle system shown at bottom left. To lift the load a distance (d), the effort force string segment must be pulled four times the distance d. This way all four string segments (1 through 4) get shortened (or move) the same amount.

$$\text{work}_{\text{in}} = \text{work}_{\text{out}}$$

$$\text{effort force} \times 4d = \text{load force (4 times effort force)} \times d$$

or

$$\text{effort force (}\tfrac{1}{4}\text{ load force)} \times 4d = \text{load force} \times d$$

The mechanical advantage of this pulley system is 4 to 1. To easily calculate the mechanical advantage of a pulley system, count the total number of string segments that exert a force directly on the load. If the free end of the string pulls up, include it in your count. If the free end pulls down, don't count it. The resulting number is the mechanical advantage of the pulley system.

Cross-Curricular Integration

Language arts:

- Read aloud or suggest that students read the following book:
  - *Sailboats, Flagpoles, Cranes: Using Pulleys as Simple Machines*, by Christopher Lampton (grades 3–5)
    This book explores the uses and benefits of pulleys in common applications.

Reference

K'NEX Industries, Inc. *Simple Machines: Levers and Pulleys Teacher's Guide;* Hatfield, PA, 2003.

Towering Pulley System

1. Prepare a pulley tower, set of pulleys, and weight support as instructed by your teacher.

2. Select at least 200 g from the mass set to use as a load. Determine the weight of the load with the spring scale, making sure to include the weight support in your measurement. This total weight is the load force.

   ✎ Record the load force in Column A of the data table.

	A Single Fixed Pulley (1 Pulley)	B Single Movable Pulley (1 Pulley)	C Combined Fixed and Movable Pulley (2 Pulleys)	D Linear Block and Tackle System (4 pulleys)	E Parallel Block and Tackle System (4 pulleys)
load force in newtons (F_L)					
effort force in newtons (F_E)					
number of segments of string	1	2	2	4	4
speed of load compared to speed of hand					
distance in cm that hand moves (d_E)					
distance in cm that load moves (d_L)					

3. Attach the single fixed pulley to the tower. Run the string over the pulley, attach the load, and attach the spring scale as shown in the diagram. The exact appearance of your setup will vary depending on the materials you use.

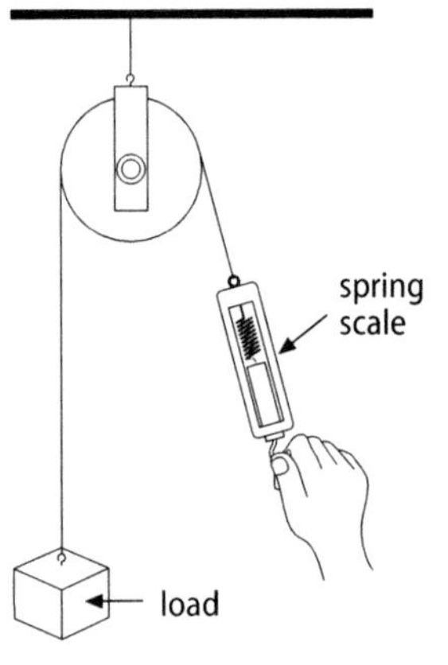

4. Complete Column A of the data table following the steps below:
   a. Pull down on the spring scale and raise the load. The spring scale reading is the effort force. Record this effort force in Column A of the data table.
   b. Note the number of segments of string listed in the data table. Although you may see more string segments, only those that exert a force directly on the load have been counted.
   c. Raise the load two or three times. Compare the speed at which the load moves to the speed at which your hand moves (the effort). Record whether the load speed is faster than your hand speed, slower than your hand speed, or whether the two speeds are equal.
   d. Move your hand to lift the load. Measure the distance your hand moves from start to finish. Record this distance (d_E). Measure and record the corresponding distance the load moves (d_L).

⑤ Set up the single movable pulley as shown.

TIP: In this case, you will pull up rather than down to raise the load.

Repeat step 4 with the same mass you chose in step 2, this time filling in Column B of the data table.

⑥ Use the single fixed pulley data from Column A of your table to fill in the values for force and distance in the equations below. Be sure to convert the distance your hand moves from cm to m. Then, multiply to complete the equations.

A. ______ newton × ______ m = ______ newton m
(effort force) (distance hand moves)

B. ______ newton × ______ m = ______ newton m
(load force) (distance load moves)

What do you notice about the two answers you get when you multiply?

....

⑦ Repeat step 6, this time with single movable pulley data from Column B.

A. ______ newton × ______ m = ______ newton m
(effort force) (distance hand moves)

B. ______ newton × ______ m = ______ newton m
(load force) (distance load moves)

What do you notice about the two answers you get when you multiply?

....

⑧ Attach the top pulley of the combined fixed and movable pulley system to the tower, and set up the rest as shown.

Repeat step 4 with the same mass you chose in step 2, this time filling in Column C of the data table.

⑨ Attach the top pulley of the linear block and tackle system to the tower, and set up the rest as shown.

✎ Repeat step 4 with the same mass you chose in step 2, this time filling in Column D of the data table.

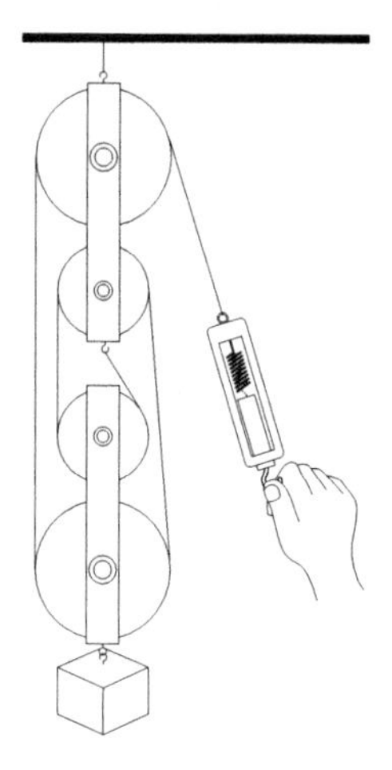

⑩ Attach the top pulley of the parallel block and tackle system to the tower, and set up the rest as shown.

✎ Repeat step 4 with the same mass you chose in step 2, this time filling in Column E of the data table.

⑪ Based on the data you collected, answer the following questions:

✎ What happens to the number of segments of string as you add more pulleys to the system?

....

....

✎ What happens to the effort force as the number of segments increases beyond one?

....

....

✎ If you pull the string a fixed amount, what happens to the distance the load moves as you add segments to the load?

....

....

....

✎ What is the relationship between the effort force and distance you must pull the string to raise the load?

....

....

✎ How do the results of the linear block and tackle system and the parallel block and tackle system compare?

....

....

⑫ Use data previously collected in this activity to calculate mechanical advantage (MA) in two ways:

✎ Calculate and record the MA of all pulleys and pulley systems using the formula: MA = load force/effort force = F_L/F_E

✎ Calculate and record the MA of all pulleys and pulley systems using a different formula: MA = effort distance/load distance = d_E/d_L

	Ⓐ Single Fixed Pulley (1 Pulley)	Ⓑ Single Movable Pulley (1 Pulley)	Ⓒ Combined Fixed and Movable Pulley (2 Pulleys)	Ⓓ Linear Block and Tackle System (4 pulleys)	Ⓔ Parallel Block and Tackle System (4 pulleys)
MA = F_L/F_E					
MA = d_E/d_L					

✎ How does the MA of each pulley and pulley system compare with the number of segments of string in each pulley setup?

....

....

....

....

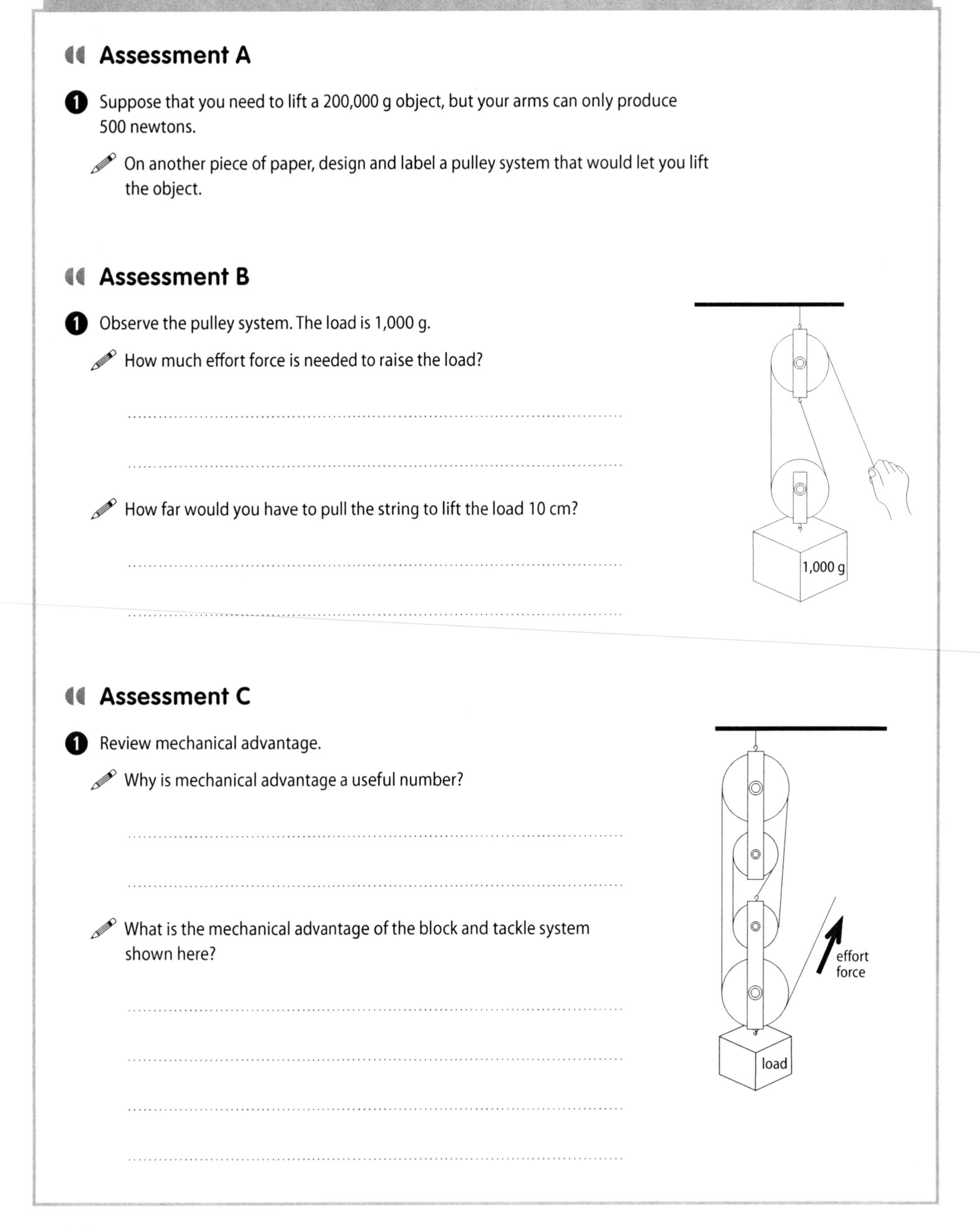

Assessment A

1. Suppose that you need to lift a 200,000 g object, but your arms can only produce 500 newtons.

   On another piece of paper, design and label a pulley system that would let you lift the object.

Assessment B

1. Observe the pulley system. The load is 1,000 g.

   How much effort force is needed to raise the load?

   ....

   ....

   How far would you have to pull the string to lift the load 10 cm?

   ....

   ....

Assessment C

1. Review mechanical advantage.

   Why is mechanical advantage a useful number?

   ....

   ....

   What is the mechanical advantage of the block and tackle system shown here?

   ....

   ....

   ....

   ....

Static Cling

Gelatin powder rises into the air as students investigate static electricity.

Grade Levels

Science activity appropriate for grades 4–8

Student Background

Students should know that matter is composed of small particles.

Time Required

Setup	15	minutes
Part A	15	minutes
Part B	15	minutes
Part C	15	minutes
Cleanup	5	minutes

Assessment time is not included.

Key Science Topics

- attractive and repulsive forces
- static electricity

National Science Education Standards Overview

See *www.terrificscience.org/physicsez/* for details of how these standards relate to the activity.

Science as Inquiry

Abilities Necessary to Do Scientific Inquiry

K–4 Use data to construct a reasonable explanation.
K–4 Communicate investigations and explanations.

5–8 Develop descriptions, explanations, predictions, and models using evidence.
5–8 Think critically and logically to make the relationships between evidence and explanations.
5–8 Communicate scientific procedures and explanations.

Physical Science

K–4 Light, heat, electricity, and magnetism

Materials

For the Procedure

Activity Introduction, per class

- inflated balloon
- (optional) piece of wool cloth

Part A, per group

- 2 inflated balloons
- 2 short pieces of string
- piece of wool cloth

Part B, per group

All materials listed for Part A, plus:

- transparent tape (the type that looks frosted rather than clear)

Some brands of tape work better than others.

Part C, per group

All materials listed for Part A, plus:

- another balloon
- tissue paper
- unflavored gelatin powder
- paper plate

For the Assessment

Assessment A, per group

- plastic food wrap (such as Glad® Cling Wrap)
- paper towel
- assorted objects (such as tissue paper pieces, black pepper, oven-toasted rice cereal, rubber bands, dried rice, salt, paper clips, sugar, flour, and glitter)

Safety and Disposal

For health purposes, it is important that two students not rub their hair on the same balloon. No special disposal procedures are required.

Getting Ready

This activity must be performed when humidity is low.

- Inflate three balloons per group and one for the Activity Introduction.
- Practice charging a balloon by rubbing it on your hair, a wool cloth, or your clothes. Prepare for the Activity Introduction by making sure that the charged balloon sticks to the classroom wall or another classroom object.
- Test the transparent tape for Part B in advance to make sure that the procedure works for the brand of tape you selected.

Procedure

This section provides teacher notes corresponding to each step of the student procedure. The procedure without teacher notes is included in the reproducible Student Notebook pages at the end of this activity and at www.terrificscience.org/physicsez/.

	Student Procedure	Teacher Notes
	Activity Introduction	Rub a balloon on your hair, with a wool cloth, or on your clothes, then "stick" the balloon to the wall or another classroom object. Tell students that they will talk about why the balloon sticks after they have explored some on their own.
	Part A: Charging Balloons	
1	**Tie separate short pieces of string to two balloons. Hold your balloons by the strings and bring the balloons close together.** **What happens?**	Students observe that nothing happens when the two balloons come close together.
2	**Now, rub your balloons on your hair, with a piece of wool cloth, or on your clothes. Listen carefully.** **Do you hear any sound other than the rubbing noise? If so, describe it.**	Often, rubbing on hair charges the balloon best. However, for health purposes, it is important that two students not rub their hair on the same balloon. Also, natural oil and hair care products in the hair may prevent the balloon from charging. In this case, wool cloth or clothing will work. The crackling that the students should hear is static electricity or "mini-thunder," which is similar to the physics of the thunder they hear in a storm, but on a much smaller scale. Some students may ask if the snap, crackle, pop sounds they hear when they pour milk on rice cereal are caused by static electricity. The answer is no. The air rushes out of the holes in the cereal when milk is poured, causing the crackling sounds.
3	**Charge your balloons again by rubbing them as in step 2. Hold your charged balloons by the strings and bring the two balloons close together.** **What happens?**	Students should observe that the balloons repel each other.

	Student Procedure	Teacher Notes
4	**Rub one of the balloons you used in step 3 with a wool cloth. Holding one corner of the cloth, put the cloth very near (but not touching) that balloon.** **Describe what happens.**	Students observe that the cloth is attracted to the balloon.
	Class Discussion	• Review group results with the class. • Explain that objects having no net charge are called neutral. Two neutral objects do not attract or repel each other. In step 1, before the balloons were rubbed with the cloth, the two balloons were neutral and nothing happened when you put them close together. • Talk about static electricity and electric force. Explain that charge is a property of an object just like mass. (See Explanation.) • Remind students that objects with the same charges repel each other. (Two negatively charged objects repel and two positively charged objects repel.) After the two balloons are rubbed in step 3, the balloons are both negatively charged and repel each other. • Explain that the rubber balloon pulls electrons from the wool cloth, leaving the balloon negatively charged and the cloth positively charged. Since negative and positive charges attract, the balloon and the cloth stick together in step 4.

Student Procedure	Teacher Notes
Part B: Charging Tape	
① Prepare two pieces of tape as follows: a. Cut off a piece of tape about 15 cm long. Fold about 3 cm of each end back onto itself to make tabs. Stick the tape to a clean, dry area on the desk. Write the letter "B" (for bottom piece) on one of the tabs. b. Repeat the same process, except stick the second tape directly on top of the B tape. Rub the tape pieces to firmly stick them together. Write the letter "T" (for top piece) on a tab of the second tape. c. Use the tab of the B tape to slowly pull the two-tape set from the desk, keeping the two pieces of tape stuck together. Gently tap your fingers down both sides of the tape until the tape set is no longer attracted to neutral objects like the desk or your arm. d. Now hold the B and T tabs and pull the two pieces of tape apart as fast as you can. (Keep the two pieces separated.)	Sometimes this procedure does not produce oppositely charged tape. If this happens, have students try again.
② Have another group member charge the balloon as in Part A and move it close to each piece of tape, one at a time. Try not to let either tape piece actually touch the balloon. ✎ What happens to the balloon and the B tape? Explain why. ✎ What happens to the balloon and the T tape? Explain why.	One piece of tape will be attracted to the balloon, meaning that the tape and balloon have different charges. The other tape will be repelled, meaning that the tape and balloon have the same charge. Students could use what they learned about the balloon's charge in Part A to identify the charge of each piece of tape. If different groups have different brands of tape, they won't necessarily get the same answer as to whether the B tape is positive or negative.
③ Repeat step 1, but this time make two sets of tapes. ✎ What happens when you move the two T tapes close to each other and why? ✎ What happens when you move the two B tapes together and why? ✎ What happens when you move one T tape and one B tape together and why?	Since the B and T tapes are oppositely charged, the two T tapes repel, the two B tapes repel, and the T and B tapes are attracted to each other. Students could use what they learned in step 2 to identify the charge of each piece of tape.

Student Procedure	Teacher Notes
Part C: Charging Other Objects	**Part C briefly introduces the idea of induction, but does not offer a complete explanation. Awareness of this concept is appropriate for older students. For younger students, Part C can be done simply as an observation of the effects of static electricity, without discussing induction.**
① **Tear tissue paper into small pieces and place the pieces on the desk in front of you. Charge a balloon as in Part A and lower it over the pieces of tissue.** ✎ **What happens?**	Some of the small tissue pieces jump up and attach to the charged balloon. If you plan to discuss induction, ask students to give a possible explanation for the behavior of the tissue. They will probably say that, since the balloon is negatively charged, that the tissue must have a positive charge. Pique their interest by telling them that the tissue is actually neutral.
② **Place an uncharged balloon (one you have never charged) over unflavored gelatin powder spread out on a paper plate.** ✎ **What happens?** ✎ **Now charge your balloon, try again, and record what happens.**	To ensure students use a neutral balloon for this step, be sure they use a balloon that has never been charged. Students observe that gelatin (which is neutral) is not attracted to the uncharged (neutral) balloon. However, the negatively charged balloon attracts the neutral gelatin powder.
Class Discussion	• Ask students to relate the balloon's sticking on the wall (from the Activity Introduction) to what they have observed in Part C. • Explain to older students that charged objects (either negatively charged or positively charged) can attract neutral objects such as tissue paper and gelatin. • When a charged object is brought near a neutral object, the neutral object may behave like a charged object because of a temporary change in the electrical balance within the material. This process is called induction. (More details about induction are usually not appropriate for students before high school. See Explanation.)

	Student Procedure	Teacher Notes
	Assessment A	
1	Lay a piece of plastic food wrap flat on the desk. Use a paper towel to vigorously rub the food wrap. Pull on one corner to lift the food wrap off the desk. ✎ What happens and why?	When students rub the food wrap with a paper towel, electrons are transferred so the food wrap becomes charged and clings to the hand lifting it.
2	Repeat step 1, but this time grasp the food wrap on opposite sides and lift up. ✎ What happens and why?	The two sides of the food wrap hang down without touching, making a shape like a tent or upside down letter V. The wrap becomes charged when rubbed with the paper towel, and since like charges repel, the two halves push away from each other.
3	Think about what would happen if the charged food wrap was put just above (but not touching) different objects lying on the table. ✎ On another piece of paper, make a data table listing the objects you are going to test. Write down your predictions. ✎ Test your objects and record the results in the data table.	See Sample Answers for example results. Answers will vary depending on what objects are used. Some objects (such as pieces of tissue paper) will be attracted to the food wrap and will lift off the table. Other objects (such as paper clips) will not move because they have too much weight. (The force of gravity is stronger than the electric force.)
	Assessment B	**Use for older students if discussing induction.**
1	Look at the picture at right. ✎ What makes the foil ball neutral?	The foil ball has the same number of protons as electrons, so it is neutral.
2	Look at the picture at right. ✎ Why are the negative charges clustered on the side of the foil ball away from the negatively charged balloon?	The electrons in the foil ball are repelled by the electrons stuck to the balloon, so the foil ball's electrons move to the side away from the balloon.
3	Look at the picture at right. ✎ Why does the foil ball swing towards the balloon now?	Since there are now fewer electrons on the side of the foil ball closest to the balloon, the side closest to the balloon has a positive charge. So, the foil ball is attracted to the balloon.

Sample Answers

Where students follow the same procedure using the same materials, these answers are close to answers you can expect. Where students design their own experiment or model, students' results will vary.

Object	Attracted to the Charged Food Wrap?			
	Predicted Results		Actual Results	
	Yes	No	Yes	No
tissue paper pieces	varies	varies	X	
black pepper	varies	varies	X	
rice cereal	varies	varies	X	
rubber bands	varies	varies		X
dried rice	varies	varies		X
salt	varies	varies	X	
paper clips	varies	varies		X

Example data for Assessment A

Explanation

This section is intended for teachers. Modify the explanation for students as needed.

Attractive and Repulsive Forces

All matter is made up of small particles. These particles are made up of parts, some of which have electric charges—protons that have a positive charge and electrons that have a negative charge. For some types of matter, it is relatively easy to build up a static charge. For example, when rubbing a balloon against wool cloth, the friction causes some of the electrons from the cloth to be transferred to the balloon. After this exchange, the balloon becomes negatively charged with an excess of electrons, and the cloth is positively charged with a shortage of electrons. The attraction of the balloon to the cloth is caused by an electric force known as the Coulomb force (named after Charles Coulomb). Coulomb, an 18th century physicist, developed the theory describing attraction and repulsion between objects of the same and opposite electrical charges.

Objects with the same net charge repel each other; two negatively charged objects repel, as do two positively charged objects. For example, two balloons that have been rubbed with a wool cloth become negatively charged and repel each other.

Objects with opposite charges attract each other. Since the cloth becomes positively charged after rubbing it with the balloon, the cloth and balloon will attract each other.

Induction

Objects can also attract each other when one is charged and the other is not charged (is neutral). This process, called induction, happens because a charged object causes a distortion in the electrical balance within the neutral material.

When the neutral object is metal, a charged object brought near the metal repels or attracts electrons in the metal, causing these electrons to move to another area in the metal. When the electrons move away, a positive region is created in the metal. This positive region is closer to the original charged object than the negative region so there is a net attractive force. (Keep in mind that only the electrons move around; the protons are stationary.) The figure below shows what happens when a positively charged object is brought near a neutral metal object such as a foil ball. Assessment B shows what happens when a negatively charged object is brought near a neutral foil ball. In both cases, the objects attract the metal. Note that we have exaggerated the movement of the electrons in the illustrations. The actual distance the electrons move is less than a nanometer.

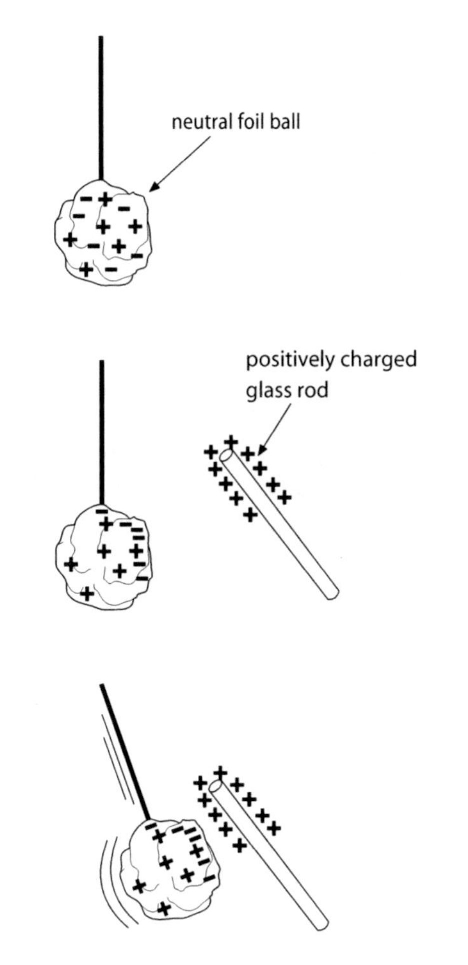

When the neutral object is nonmetal, a charged object brought near the nonmetal causes the molecules that make up the nonmetal material to polarize temporarily. In other words, each molecule temporarily forms a positive area and a negative area. This is what happens to the tissue paper and gelatin powder. These objects are lifted by the negatively charged balloon because the electrical attraction force is greater than the weight of the objects. (See the drawings in Assessment B.) Similarly, all of the objects picked up by the plastic wrap do so because they are charged by induction and their weights are less than the strength of the electrical force.

Physical Makeup Determines Behavior

The nature of a material determines what charge it will acquire when rubbed against another material. For example, if you rub a balloon with a wool cloth, the balloon becomes negatively charged because rubber and latex materials have a stronger tendency to become negatively charged than wool does. The friction between the wool and the latex of the balloon moves the electrons from the wool to the balloon. A glass rod rubbed with silk will become positively charged because silk has a stronger tendency to become negatively charged than glass. The same happens when a piece of acetate (overhead projector transparency) is rubbed with cotton. The friction between the cotton and the acetate moves the electrons from the acetate to the cotton.

The behavior of the tape in this activity depends on the adhesive on its surface. When the two pieces of tape are stacked together, the acetate side of the bottom piece is facing the glue side of the top piece. When the two pieces of tape are ripped apart, they usually become charged with opposite charges. If the two oppositely charged pieces of tape are brought near each other, they attract. Bring a negatively charged balloon near them, and the positively charged piece of tape is attracted to the balloon while the negatively charged piece of tape is repelled.

Why does one tape become negatively charged and one become positively charged? When the tape is ripped apart, the molecules that make up the adhesive become sheared in such a way that those in some regions of the tape have an excess of electrons (resulting in a negative charge) and those in other regions have a deficit of electrons (resulting in a positive charge). These negative and positive regions are scattered over the surface of both pieces of tape. Because the tape is not a conductor, excess electrons on one part of the tape don't move to the locations where there is a deficit of electrons. If one piece of tape has more negative regions than positive ones, it will end up with a net negative charge. If one piece of tape has more positive regions than negative, it will end up with a net positive charge. The result is that the piece of tape with the net negative charge will be repelled by the balloon and the piece with the net positive charge will be attracted to the balloon. Occasionally, the negative and positive regions on a piece of tape will balance each other out and the tape will not gain a net charge.

Another possibility is that both pieces of tape end up with the same charge because of interaction between the bottom tape and the table. Remember that before ripping the two pieces of tape apart as discussed above, the two pieces of tape are lifted as a unit from the table. This action can cause the lower surface of the bottom tape to become positively charged. These charges, in combination with the charges acquired when the tapes are ripped apart, can result in two pieces of tape with the same net charge. That is why students are instructed to try to remove any charge the tape acquired from the table before pulling the two tapes apart.

Cross-Curricular Integration

Art:

- Create a static electricity toy. Cut tissue paper into small seasonal shapes (such as snowflakes for winter, leaves for fall, and flowers for spring). Place the shapes in a clear plastic (deli) container with a lid. Hold a charged balloon over the lid and watch the tissue paper shapes dance. Have the students write an instruction manual for using the new toy. This manual should include drawings and a simple explanation about what happens.
- Cut a ghost shape out of newspaper. Make a face with washable markers. Hold the ghost against a metal piece of furniture or a wall. (Drywall works fine, but cinder block doesn't work.) Rub the ghost hard with the wooden side of a pencil, making sure to rub the ghost all over. Let go and the ghost will stick because of static electricity.

References

Barhydt, F. *Science Discovery Activities Kit;* The Center for Applied Research in Education: West Nyack, NY, 1982.

Chabay, R.; Sherwood, B. *Electric and Magnetic Interactions*; John Wiley and Sons: New York, 1995.

Hewitt, P. *Conceptual Physics,* 9th ed.; Addison Wesley: San Francisco, 2002; pp 412–432.

Paulu, N. *Helping Your Child Learn Science;* U.S. Department of Education: Washington, D.C., June 1991.

Sarquis, M.; Kibbey, B.; Smyth, E. *Science Activities for Elementary Classrooms;* Flinn Scientific: Batavia, IL, 1989.

Tolman, M.; Morton, J. *Physical Science Activities for Grades 2–8;* West Nyack, NY, Parker: 1986.

Utah Education Network Website. Static Electricity. http://www.uen.org (accessed February 19, 2005).

Part A: Charging Balloons

① Tie separate short pieces of string to two balloons. Hold your balloons by the strings and bring the balloons close together.

What happens?

② Now, rub your balloons on your hair, with a piece of wool cloth, or on your clothes. Listen carefully.

Do you hear any sound other than the rubbing noise? If so, describe it.

③ Charge your balloons again by rubbing them as in step 2. Hold your charged balloons by the strings and bring the two balloons close together.

What happens?

④ Rub one of the balloons you used in step 3 with a wool cloth. Holding one corner of the cloth, put the cloth very near (but not touching) that balloon.

Describe what happens.

Part B: Charging Tape

① Prepare two pieces of tape as follows:

a. Cut off a piece of tape about 15 cm long. Fold about 3 cm of each end back onto itself to make tabs. Stick the tape to a clean, dry area on the desk. Write the letter "B" (for bottom piece) on one of the tabs.
b. Repeat the same process, except stick the second tape directly on top of the B tape. Rub the tape pieces to firmly stick them together. Write the letter "T" (for top piece) on a tab of the second tape.
c. Use the tab of the B tape to slowly pull the two-tape set from the desk, keeping the two pieces of tape stuck together. Gently tap your fingers down both sides of the tape until the tape set is no longer attracted to neutral objects like the desk or your arm.
d. Now hold the B and T tabs and pull the two pieces of tape apart as fast as you can. (Keep the two pieces separated.)

② Have another group member charge the balloon as in Part A and move it close to each piece of tape, one at a time. Try not to let either tape piece actually touch the balloon.

What happens to the balloon and the B tape? Explain why.

....

....

What happens to the balloon and the T tape? Explain why.

....

....

③ Repeat step 1, but this time make two sets of tapes.

What happens when you move the two T tapes close to each other and why?

....

....

What happens when you move the two B tapes together and why?

....

....

What happens when you move one T tape and one B tape together and why?

Part C: Charging Other Objects

1. Tear tissue paper into small pieces and place the pieces on the desk in front of you. Charge a balloon as in Part A and lower it over the pieces of tissue.

   What happens?

2. Place an uncharged balloon (one you have never charged) over unflavored gelatin powder spread out on a paper plate.

   What happens?

   Now charge your balloon, try again, and record what happens.

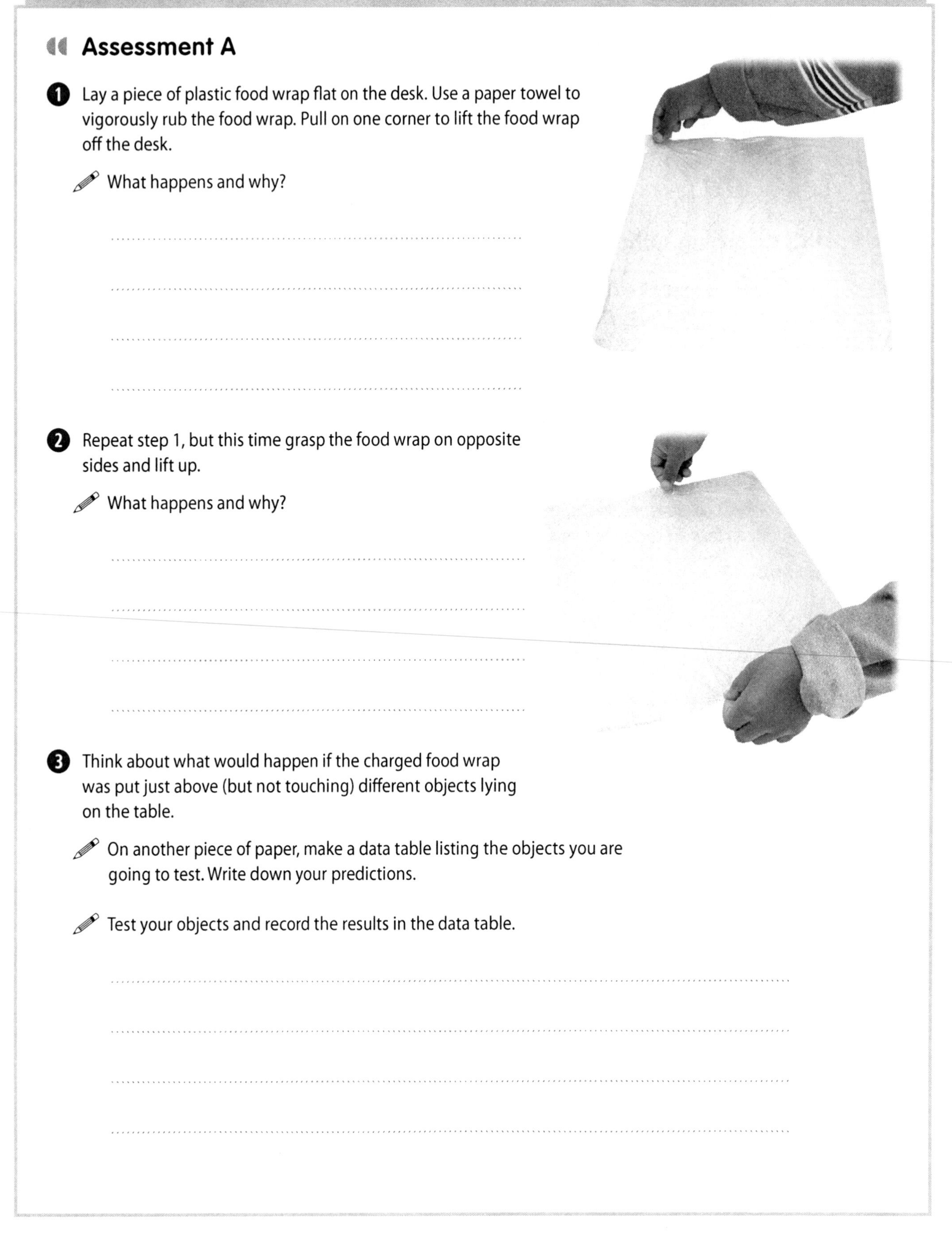

Assessment A

1. Lay a piece of plastic food wrap flat on the desk. Use a paper towel to vigorously rub the food wrap. Pull on one corner to lift the food wrap off the desk.

   What happens and why?

2. Repeat step 1, but this time grasp the food wrap on opposite sides and lift up.

   What happens and why?

3. Think about what would happen if the charged food wrap was put just above (but not touching) different objects lying on the table.

   On another piece of paper, make a data table listing the objects you are going to test. Write down your predictions.

   Test your objects and record the results in the data table.

Assessment B

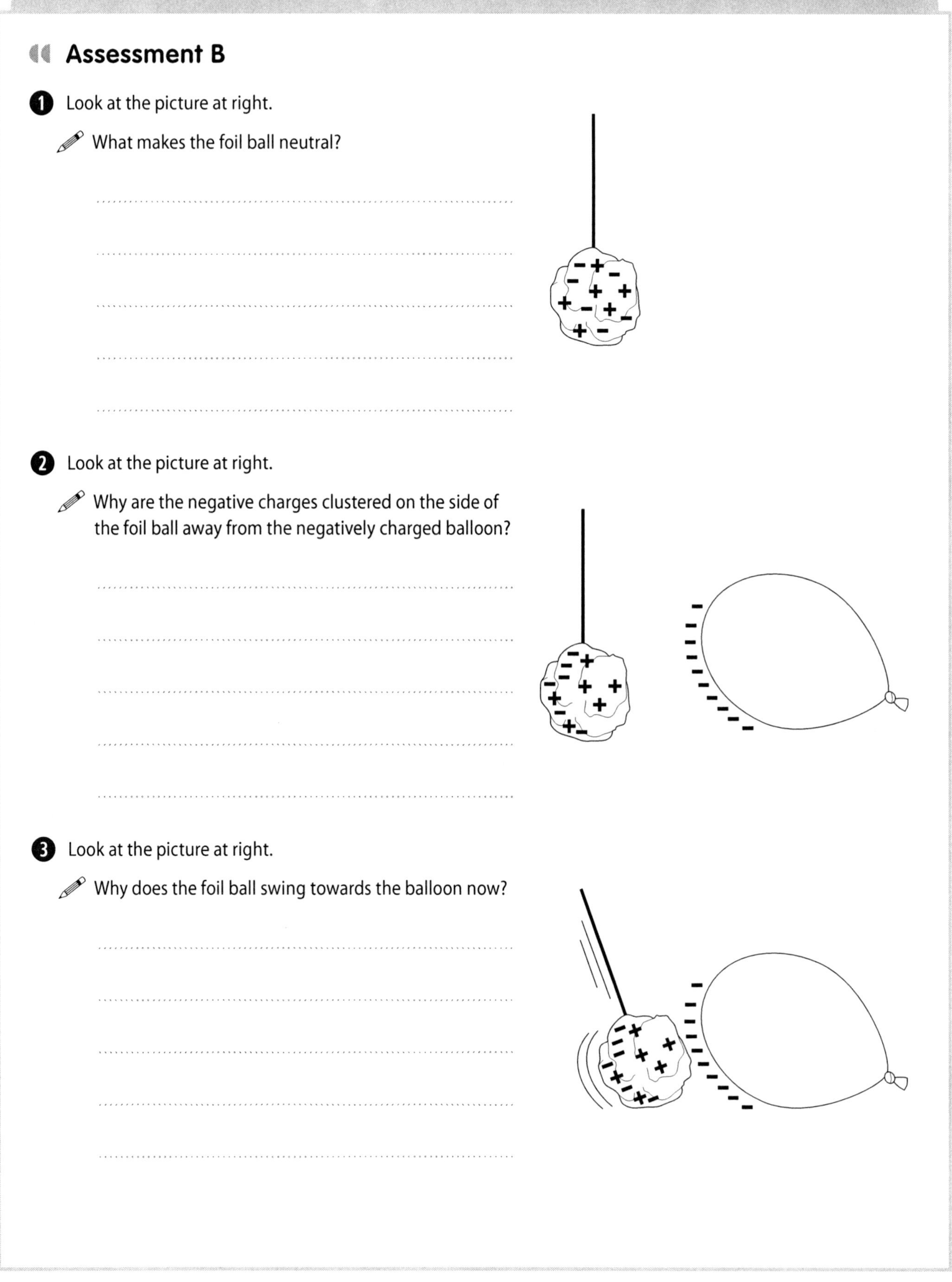

1. Look at the picture at right.

   What makes the foil ball neutral?

2. Look at the picture at right.

   Why are the negative charges clustered on the side of the foil ball away from the negatively charged balloon?

3. Look at the picture at right.

   Why does the foil ball swing towards the balloon now?

Magnet Cars

How do magnets change the way toy cars act?

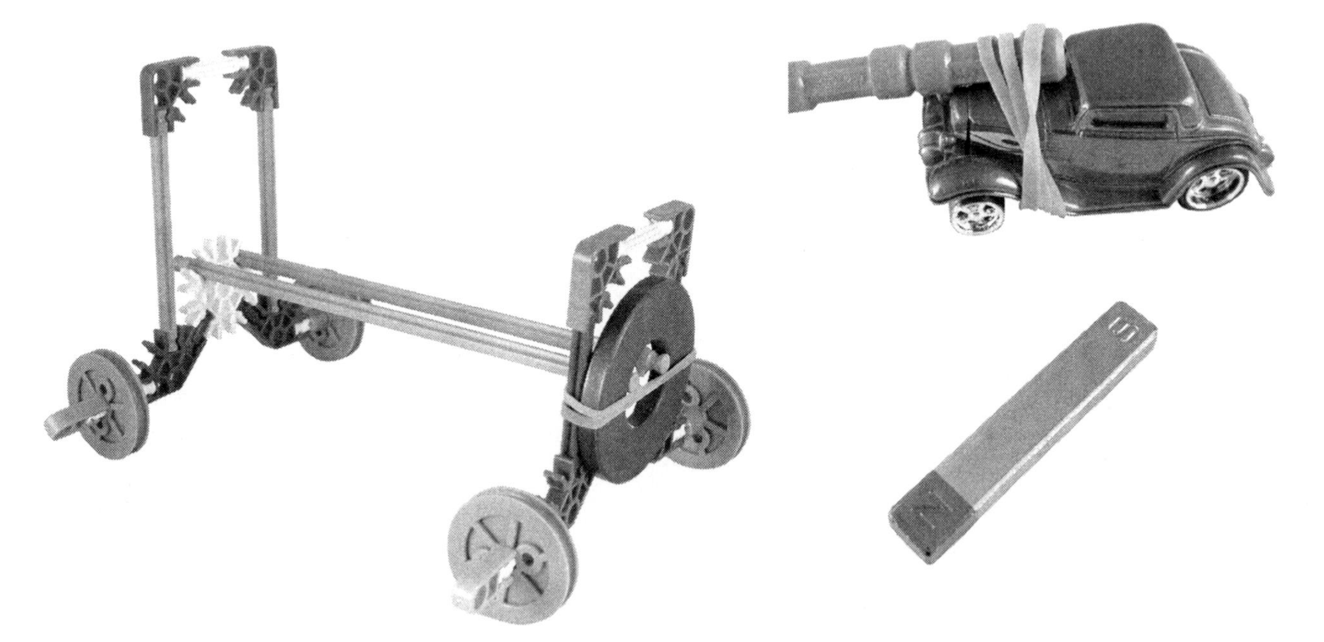

Grade Levels

Science activity appropriate for grades 3–5

Student Background

Students require no particular background preparation for this activity.

Time Required

Setup	10	minutes
Procedure	30	minutes
Cleanup	10	minutes

Assessment time is not included.

Key Science Topics

- attractive and repulsive forces
- magnets
- north and south poles

National Science Education Standards Overview

See *www.terrificscience.org/physicsez/* for details of how these standards relate to the activity.

Science as Inquiry

Abilities Necessary to Do Scientific Inquiry

K–4 Communicate investigations and explanations.

5–8 Develop descriptions, explanations, predictions, and models using evidence.

5–8 Communicate scientific procedures and explanations.

Physical Science

K–4 Position and motion of objects

K–4 Light, heat, electricity, and magnetism

5–8 Motions and forces

Science and Technology

Abilities of Technological Design

K–4 Communicate a problem, design, and solution.

5–8 Design a solution or product.

5–8 Communicate the process of technological design.

Materials

For Getting Ready

Per class

- (optional) factory-labeled magnet indicating N (for north) and S (for south)

For the Procedure

Activity Introduction, per class

- assorted magnet-based toys or magnets

For supply source suggestions, see www.terrificscience.org/supplies/.

Per group

- 2 rod- or bar-shaped magnets marked with N and S as described in Getting Ready

Magnetic poles are easier to identify and mark on rod- and bar-shaped magnets.

- 2 toy cars, such as
  - simple cars (such as Hot Wheels® and Matchbox®)
  - cars made from assorted K'NEX® rods, connectors, and wheels
- 2 rubber bands or adhesive tape
- assorted magnets having different shapes

Parts from some magnet-based toys may be used.

- materials to make a game

For the Assessment

All materials listed for the Procedure, with additional:

- toy cars
- rubber bands, adhesive tape, or string
- assorted magnets

Safety and Disposal

No special safety or disposal procedures are required.

Getting Ready

If the rod- or bar-shaped magnets are not marked N (for north) and S (for south), use a factory-labeled magnet to test the nonmarked magnets. Mark an N on the side or end of each magnet that is attracted to the south side of the factory-labeled magnet. Mark an S on the other side or end.

Procedure

This section provides teacher notes corresponding to each step of the student procedure. The procedure without teacher notes is included in the reproducible Student Notebook pages at the end of this activity and at www.terrificscience.org/physicsez/.

	Student Procedure	Teacher Notes
	Activity Introduction	Give students an opportunity for free exploration with magnet-based toys or assorted magnets. This play experience provides a foundation for the rest of the activity.
	Explore Magnets	
1	**Play with the two magnets and watch what they do when you put them close to each other. Try arranging them in different positions to see what happens.** ✎ **On another piece of paper, draw the two magnets in all the ways they stick together. Label the north (N) and south (S) poles on your drawing.** ✎ **Now draw the two magnets in all the ways they push away from each other. Label the north and south poles on your drawing.**	Magnets may attract or repel each other end to end, side by side, or one on top of the other. In all cases, students will see that opposite magnetic poles attract one another and like magnetic poles repel. Students reinforce these concepts by drawing the position of the magnets when they stick together and when they push away from each other. (See Sample Answers for example student drawing.)
2	**If you are building your own cars, use K'NEX pieces to make two cars that can roll.**	
3	**Roll your cars and watch what happens. Observe what happens when the cars are close together.**	Give the students time to play with the cars. Make sure they notice that two cars driven close to each other don't affect each other in any way. Cars without magnets will travel in a straight line when pushed.
4	**Use rubber bands or tape to attach one magnet to each car so that the two cars can stick together. Now roll the cars around and observe what the cars do when you drive them close to each other.** ✎ **Compare how the cars with magnets behave when near each other to how the cars without magnets behave in step 3.**	Let students decide where to position the magnets on the cars. Encourage students to reposition the magnets if the cars do not stick together on the first try. (See Sample Answers for example car and magnet arrangement.) Students should notice that the cars, when rolled near each other, will stick together if the magnets are positioned correctly. Introduce the terms "pull towards" and "attract."

	Student Procedure	Teacher Notes
5	Use rubber bands or tape to attach one magnet to each car so that the two cars can push each other away. Now roll the cars around and notice what the cars do when you drive them close to each other. ✎ Compare how these cars behave when near each other to how the cars behave in step 4.	Let students decide where to position the magnets on the cars. Encourage students to reposition the magnets if the cars do not push away from each other on the first try. Students should notice that the cars, when rolled near each other, will push away from each other if the magnets are positioned correctly. Introduce the terms "push apart" and "repel."
6	Play with a variety of differently shaped magnets. Observe how they attract and repel each other as you change the positions of the north and south poles. Try attaching these magnets to cars in different ways. Observe how the cars behave. ✎ Record your observations.	Students should observe how differently shaped magnets attract and repel each other. For example, in a ring-shaped magnet, instead of the poles being at two ends (as in a bar magnet), one flat face of the ring is the north pole and the other flat face is the south. Students should be able to apply what they have learned to create magnet cars with different magnets.
7	Invent and play a game that uses magnet cars. ✎ Write the rules of the game. ✎ On another piece of paper, draw and label a diagram of the game. Show the magnets and label their poles. ✎ Explain how magnetic pushing and/or pulling are part of the game.	Students' games should make use of magnetic attraction and/or repulsion. For example, a reasonable magnet car game uses two sets of cars having magnets set up to repel. Two people set their racing cars on the starting line. On "Go," students push their second, driver cars forward towards the racing cars. Magnetic repulsion causes the racing cars to move forward. The first racing car over the finish line wins. (See Sample Answers for a student drawing example for this game.) Students should be able to recognize and explain how magnetic attraction and/or repulsion is a part of their games.
	Assessment A	
1	Work with your group to make a long train of cars that all stick together. ✎ Why does your train stick together? ✎ On another piece of paper, draw a diagram of your train and label its parts. Label the north and south poles of the magnets.	Students should apply the skills they learned to make a line of cars that stick together. Let students decide where to position the magnets on the cars. Younger students may need help using the rubber bands, tape, and string. (See Sample Answers for an example of a car train student drawing.)

Student Procedure	Teacher Notes
Assessment B	
1 A few places in the world have a special type of train called a maglev. The name comes from the words "magnetic" and "levitation." Magnets lift and propel these trains down the track. ✎ What properties of magnets make such a train possible? ✎ Draw your idea of a maglev train, showing the position of the magnets and labeling the poles.	Maglev trains have large magnets on their undercarriage. Magnetized coils running along the tracks repel the train's magnets, causing the train to levitate above the track. Electric current supplied to the coils along the track change the polarity of the magnetic field in front of the train, pulling the train forward. (For more information, look up maglev trains on the Internet.) Although students will not know the details about maglev trains, they should deduce that the repulsion of opposite magnetic poles enables the train to levitate above the track. Without further background information, students probably will not be able to deduce the mechanism for the train's forward motion. However, they may realize that magnetic attraction plays a role.

Sample Answers

Where students follow the same procedure using the same materials, these answers are close to answers you can expect. Where students design their own experiment or model, students' results will vary.

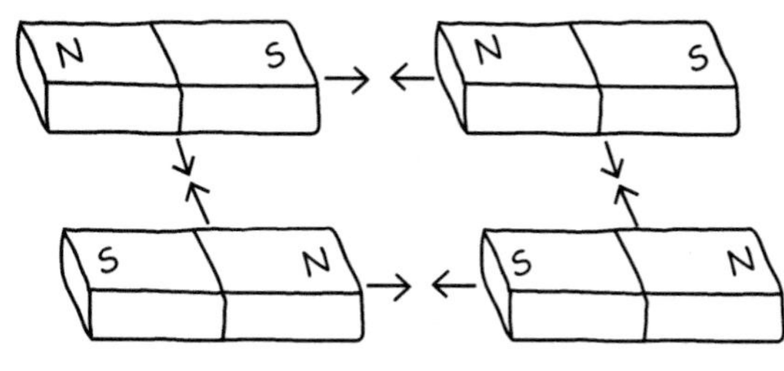

stick together end to end and side to side

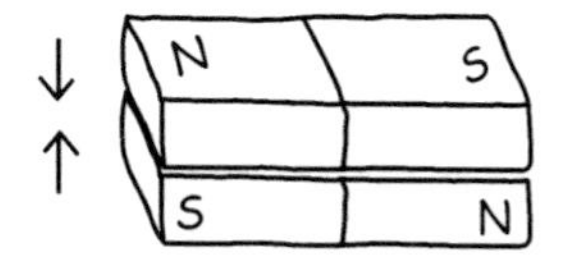

stick together one on top of the other

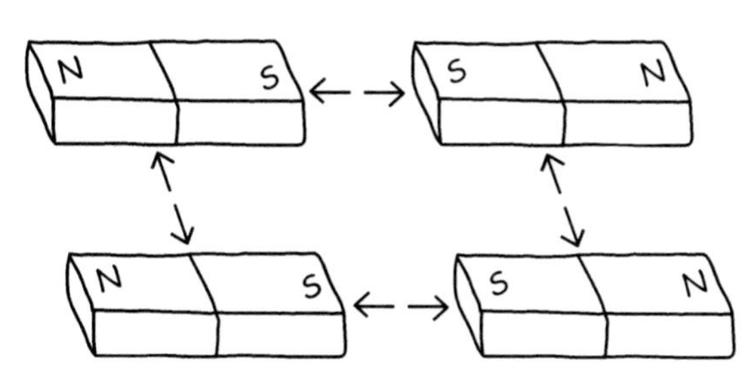

push apart end to end and side to side

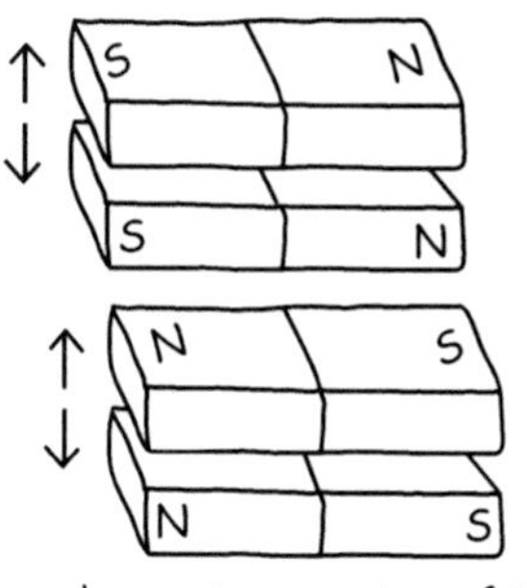

push apart one on top of the other

Examples of student drawing for step 1

Example of car and magnet arrangement for step 4

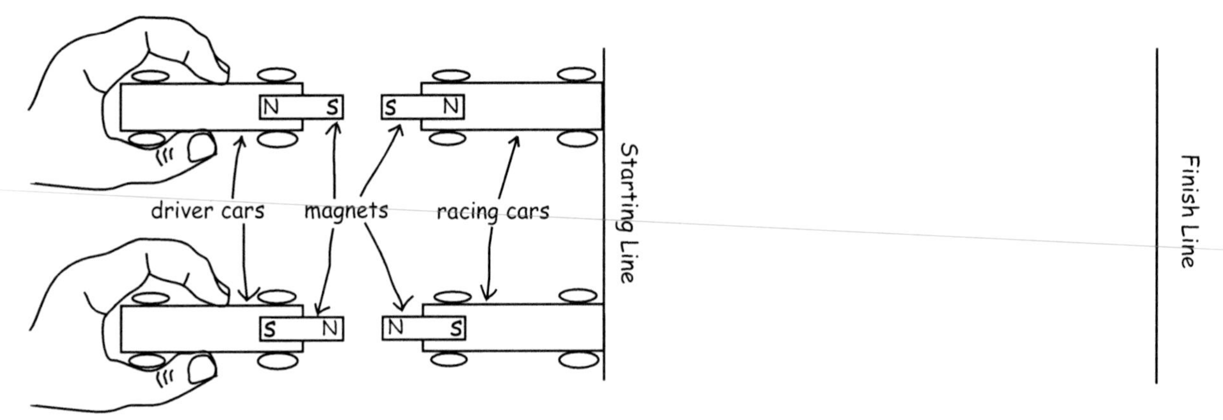

Example of a magnet car game student drawing for step 7

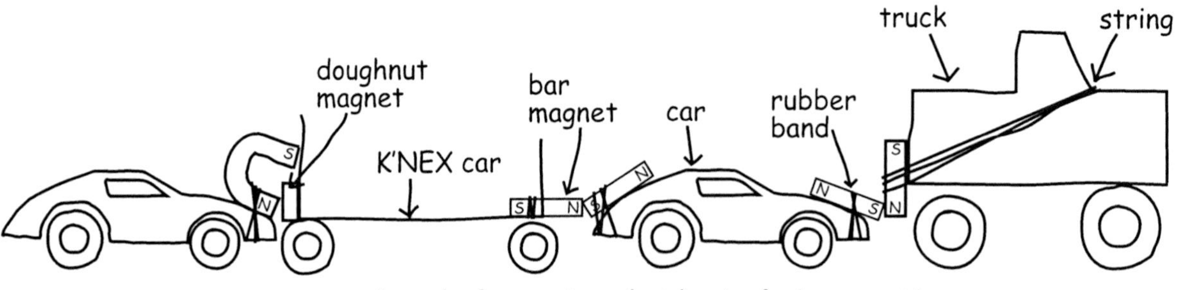

Example of a car train student drawing for Assessment A

Explanation

This section is intended for teachers. Modify the explanation for students as needed.

For young students, a sufficient explanation of magnetism uses the terms "stick together," "pull towards," and "attract" when opposite poles interact and "push away" and "repel" when like poles interact.

Magnetism arises from the motion of electric charges. Because all atoms contain moving electrons, every atom acts like a tiny magnet. However, in most materials, the effects of all of these little magnets just cancel each other out.

Materials are said to be magnetic when the magnetic fields of most of the atoms line up in the same direction. In substances like iron, nickel, and cobalt, the atoms line up in groups creating a magnetic domain. Each domain is made up of millions of aligned atoms. If many domains within a substance are lined up, the magnet is strong. If the domains are not lined up, the magnet is weak. If the atoms in a material cannot form a domain of aligned atoms, there is no net magnetic field and the material is not magnetic.

Magnets exhibit both attractive and repulsive forces. Simple bar magnets are labeled with an "N" for north pointing and "S" for south pointing to designate the opposite magnetic properties. The terms "north pole" and "south pole" come from the observation that a sample of magnetic material suspended from a string will align itself with the earth's magnetic field. One end points toward the earth's geographic north pole, and the other end points toward the earth's geographic south pole. The end that points toward the earth's geographic north is called the north pole of the magnet. The nature of magnetic poles is such that like magnetic poles "push away" from each other and opposite magnetic poles "pull toward" each other.

Cross-Curricular Integration

Language arts:

- Instruct students to use all the characteristics of magnets that they observed in the experiment to write a story based on the following premise: "You woke up this morning and realized that you were magnetic. Tell us about your day and how you got back to normal." Before students begin, you may want to read excerpts from stories in which people find themselves physically changed, such as *Alice in Wonderland* by Lewis Carroll (young adult) and *George Shrinks* by William Joyce (grades 4–7). Have students use a story sequence map as a graphic organizer to help them stay on topic. (See *www.terrificscience.org/physicsez/*.)
- Have the students write a story about where their magnet cars are going.
- Read aloud or suggest that students read the following book:
    - *Magnets and Sparks,* by Wendy Madgwick Raintree (grades 4–7)
      This book introduces basic concepts and uses of electricity and magnetic force and attraction. Simple investigations allow students to explore these concepts.

Art:

- Use a variety of magnets, nuts, bolts, steel balls, paper clips, and other objects to create a magnetic junk sculpture.

References

Althouse, R.; Main, C. *Science Experiences for Young Children, Magnets;* Teacher College: New York, 1975; pp 9–11.

Hewitt, P. *Conceptual Physics*, 9th ed.; Addison Wesley: San Francisco, 2002; pp 458–462.

Explore Magnets

① Play with the two magnets and watch what they do when you put them close to each other. Try arranging them in different positions to see what happens.

On another piece of paper, draw the two magnets in all the ways they stick together. Label the north (N) and south (S) poles on your drawing.

Now draw the two magnets in all the ways they push away from each other. Label the north and south poles on your drawing.

② If you are building your own cars, use K'NEX® pieces to make two cars that can roll.

③ Roll your cars and watch what happens. Observe what happens when the cars are close together.

④ Use rubber bands or tape to attach one magnet to each car so that the two cars can stick together. Now roll the cars around and observe what the cars do when you drive them close to each other.

Compare how the cars with magnets behave when near each other to how the cars without magnets behave in step 3.

....

....

....

⑤ Use rubber bands or tape to attach one magnet to each car so that the two cars can push each other away. Now roll the cars around and notice what the cars do when you drive them close to each other.

Compare how these cars behave when near each other to how the cars behave in step 4.

....

....

....

....

⑥ Play with a variety of differently shaped magnets. Observe how they attract and repel each other as you change the positions of the north and south poles. Try attaching these magnets to cars in different ways. Observe how the cars behave.

✎ Record your observations.

....

....

....

....

....

....

⑦ Invent and play a game that uses magnet cars.

✎ Write the rules of the game.

....

....

....

....

....

✎ On another piece of paper, draw and label a diagram of the game. Show the magnets and label their poles.

✎ Explain how magnetic pushing and/or pulling are part of the game.

....

....

....

....

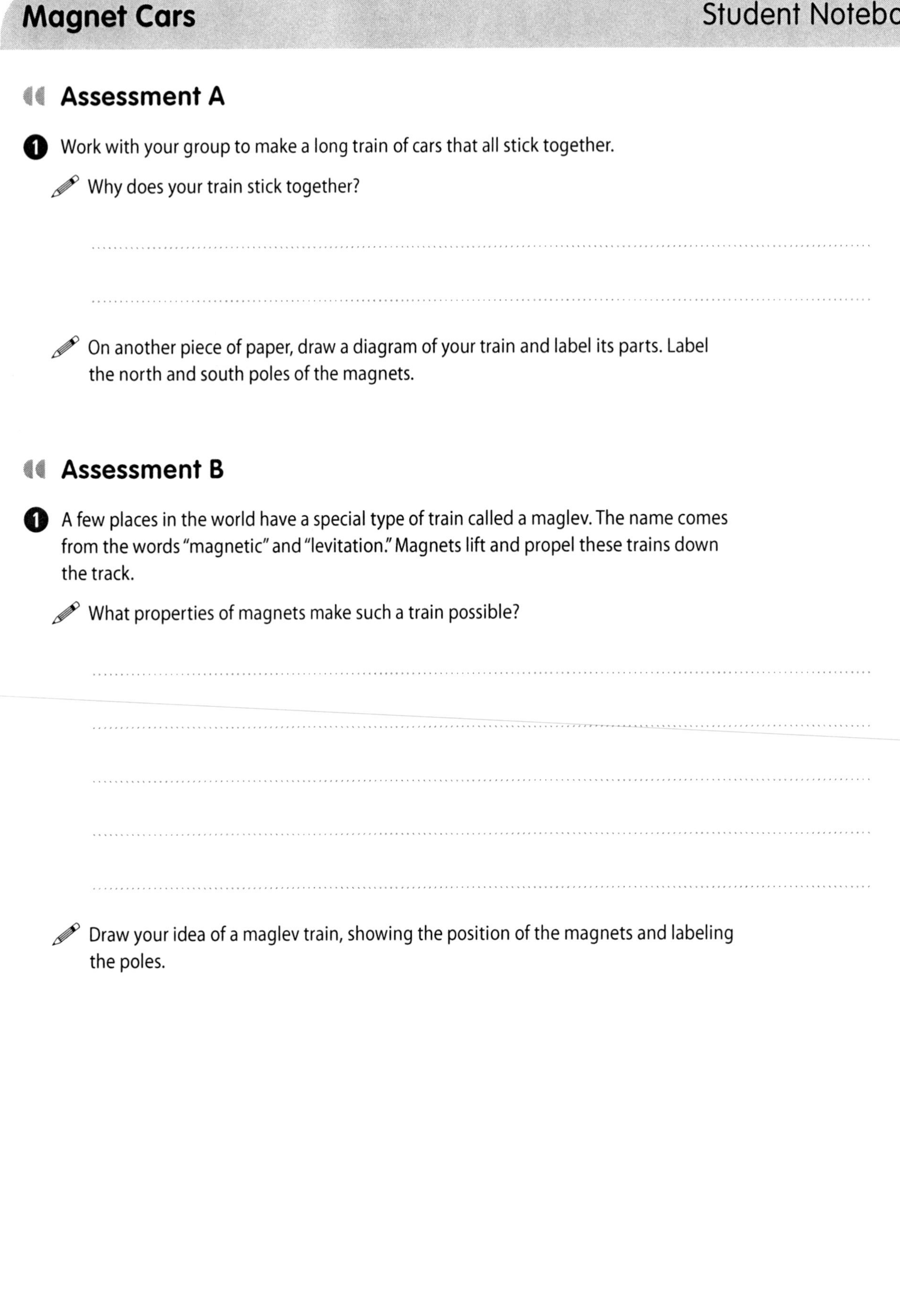

Assessment A

1. Work with your group to make a long train of cars that all stick together.

   Why does your train stick together?

   ....

   ....

   On another piece of paper, draw a diagram of your train and label its parts. Label the north and south poles of the magnets.

Assessment B

1. A few places in the world have a special type of train called a maglev. The name comes from the words "magnetic" and "levitation." Magnets lift and propel these trains down the track.

   What properties of magnets make such a train possible?

   ....

   ....

   ....

   ....

   ....

   Draw your idea of a maglev train, showing the position of the magnets and labeling the poles.

Doc Shock

Use the Operation® game to explore the structure and components of an electric circuit.

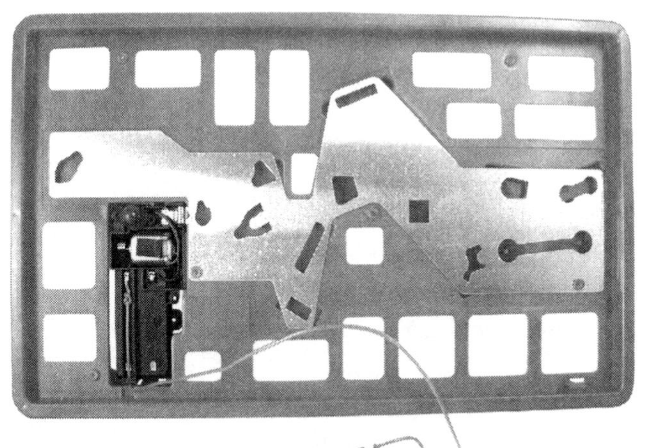

Grade Levels

Science activity appropriate for grades 4–6

Student Background

Students need experience with lighting a bulb with a battery prior to this activity.

Time Required

Setup	5	minutes
Part A	15	minutes
Part B	15	minutes
Part C	15	minutes
Cleanup	5	minutes

Assessment time is not included.

Key Science Topics

- conductors and insulators (nonconductors)
- electricity
- energy transfer
- open and closed circuits
- parts of a circuit
- series and parallel circuits

National Science Education Standards Overview

See *www.terrificscience.org/physicsez/* for details of how these standards relate to the activity.

Science as Inquiry

Abilities Necessary to Do Scientific Inquiry

K–4 *Plan and conduct a simple investigation.*
K–4 *Use data to construct a reasonable explanation.*
K–4 *Communicate investigations and explanations.*

5–8 *Design and conduct a scientific investigation.*
5–8 *Develop descriptions, explanations, predictions, and models using evidence.*
5–8 *Think critically and logically to make the relationships between evidence and explanations.*
5–8 *Recognize and analyze alternative explanations and predictions.*
5–8 *Communicate scientific procedures and explanations.*

Physical Science

K–4 *Light, heat, electricity, and magnetism*

5–8 *Transfer of energy*

Science and Technology

Abilities of Technological Design

K–4 *Evaluate a product or design.*
K–4 *Communicate a problem, design, and solution.*

5–8 *Evaluate completed technological designs or products.*
5–8 *Communicate the process of technological design.*

Materials

For Getting Ready
Per class
- string of nonblinking miniature lights with bulbs that are easily removed
- wire stripper and cutter (for teacher use only)
- utility knife (for teacher use only)
- screwdriver

For the Procedure
Part A, per group
- Operation game by Milton Bradley Company

For supply source suggestions, see www.terrificscience.org/supplies/.

Part B, per group
- assortment of conductors (such as metal paper clips, pennies, and metal binder clips) and insulators (such as pencils, large erasers, and cardboard)
- piece of wire about 10 cm long

Part C, per class
- (optional) Operation Game

Part C, per group
- 2 C or D size batteries
- vinyl electrical tape
- large rubber band
- metal paper clip
- piece of wire from Part B
- 2 pieces of wire about 20 cm long
- 2 miniature lights

For the Assessment
Per group
Materials listed for Part C, plus some groups will need:
- large cardboard box with lid (such as shoe box or shirt box)
- more wire
- aluminum foil strips and pieces
- metal tweezers
- more miniature lights
- masking tape
- scissors
- other items to make a game

Safety and Disposal

No special safety or disposal procedures are required.

Getting Ready

- In Part A, students will need to remove the cardboard surface of the game board to view the game's components. Follow the steps below prior to class so students can more easily investigate the game.
  - Use a utility knife to cut a small square in the cardboard around each plastic peg that holds down the game's cardboard surface. The cardboard can easily be replaced, restoring the game to full usefulness.
  - Use a screwdriver to open the housing for the light and buzzer to reveal the game's components. Replace the cardboard surface of the game without replacing the housing to hide the game's components from the students.

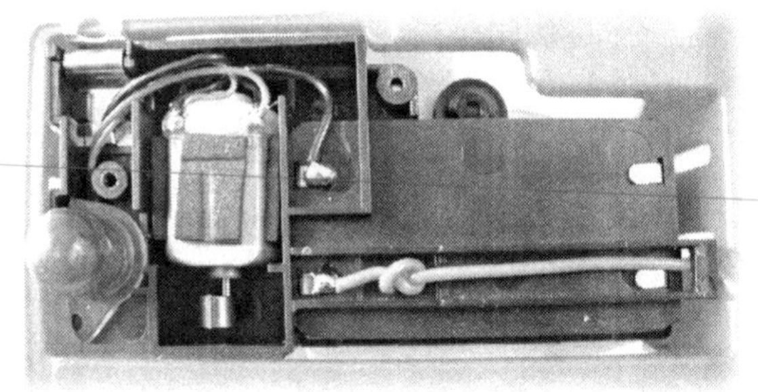

- Use wire cutters to separate a string of miniature lights, leaving about 10-cm lengths of both wires attached to each light. Strip about 1 cm of the coating off the end of each wire.
- After cutting off the miniature lights, the remaining wire can be cut into pieces for Parts B and C. For Part B, cut the wire into 10-cm lengths. For Part C, cut the wire into 20-cm lengths. Strip about 1 cm of the coating off each end of all the wire pieces.

Procedure

This section provides teacher notes corresponding to each step of the student procedure. The procedure without teacher notes is included in the reproducible Student Notebook pages at the end of this activity and at www.terrificscience.org/physicsez/.

	Student Procedure	Teacher Notes
	Activity Introduction	Explain the Operation game and its rules, or allow students who have played it before to explain the game to the class.
	Part A: How the Game Operates	
1	As you play the game, form a hypothesis that explains how the game works. Record your group's ideas.	Allow free exploration for game playing. Some students may realize that when either side of the tweezers touches the metal plate, the circuit path is completed and electric current flows. The current causes the buzzer and light to work.
2	Take the cardboard surface off your game board and further investigate to refine your group's ideas. Make a list of all the internal parts you think are needed to make the buzzer buzz and the bulb light. On another piece of paper, draw a simple diagram to show how the items you listed are attached to each other.	See Sample Answers for student drawing example.
3	All circuits have the following components in common: energy source, energy path, and energy receiver. As a group, answer the following questions: What part of the game plays the role of the energy source? What parts of the game play the role of the energy path? What parts of the game play the role of the energy receiver?	Students should identify the game parts and roles listed below. battery — energy source tweezers, metal pieces of board, wires → energy paths light bulb, buzzer → energy receivers

Student Procedure	Teacher Notes
Class Discussion	Have students share their answers to step 3 in a class discussion. List the various answers on the board.
(4) Look at your group's ideas from steps 1 and 2 about how the game works. Revise this description based on what you now know. Use the terms "energy source," "energy path," and "energy receiver" in your description.	
Class Discussion	Review with your students how the Operation game works. • Point out that an energy source, energy path, and energy receiver are necessary for an electric circuit to work. • If they are all connected and working, the circuit is closed (or completed) and electric current moves around the entire path. If any connections are not made, the circuit is open and electric current can't move around the entire path. An open circuit is like a drawbridge when it is raised—traffic cannot move across. • Illustrate these ideas in the form of a circuit diagram of the game.

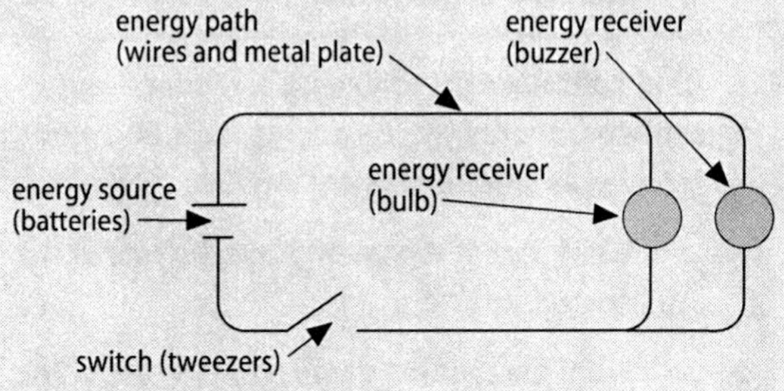

Student Procedure	Teacher Notes
Part B: Further Investigation	
① While the game is disassembled, test other objects to see if they will allow energy transfer. To do this, hold the objects with the tweezers and attempt to close the circuit by touching the objects to the metal plate. ✎ On another piece of paper, keep track of what objects successfully close the circuit (conductors) and what objects don't close the circuit (insulators).	See Sample Answers for a student response example. After students have tested various objects, introduce and explain the terms "conductor" and "insulator." (See Explanation.) Explain that if any game pieces (body parts) touch a cavity's metal edge when playing the game, the light and buzzer don't turn on because the game pieces are insulators. The light and buzzer turn on when the tweezers touch a cavity's metal edge because tweezers are conductors. (The circuit becomes closed.)
② Remove one of the batteries from the game and use a wire to complete the circuit in place of the removed battery. ✎ What happens to the light and buzzer when the circuit is closed? Explain why this happens.	The light gets dimmer and the buzzer's motor runs more slowly because now only half of the energy is available to operate them.
Part C: Parallel and Series Circuits	
① Use all of the materials given to you by your teacher to build an electrical circuit where two lights are connected in parallel as shown. Test to make sure both lights go on when you close the switch. ✎ If you were to remove one bulb from its holder, predict what would happen to the other light if the switch were closed. Explain why you think this would happen.	Explain to students that the diagram showing the circuit is very stylized and does not show exactly how the wires, lights, batteries, and paper clip are connected to each other. Have students use vinyl electrical tape to securely attach the battery and wire connections. Give students ample opportunity to figure out how to use the materials to create the circuit and encourage groups to share solutions. (See Sample Answers.) If students get really stuck, show them an example of a completed circuit.
② Remove one of the lights and close the switch. ✎ What happens to the remaining light and why?	Parallel circuits have branching paths for the current to follow. When two lights are in parallel, unplugging one of the lights and closing the switch will not affect the remaining light. The remaining light will go on since the electric current can move around the other path in the circuit.

Student Procedure	Teacher Notes
③ Now build a circuit where two lights are connected in series as shown. Test to make sure both lights go on when you close the switch. ✎ If you were to remove one bulb from its holder, predict what would happen to the other light if the switch were closed. Explain why you think this would happen.	Have students use vinyl electrical tape to securely attach the battery and wire connections. Give students ample opportunity to figure out how to use the materials to create the circuit and encourage groups to share solutions. (See Sample Answers.) If students get really stuck, show them an example of a completed circuit.
④ Remove one of the lights and close the switch. ✎ What happens to the remaining light and why? ✎ What is a design advantage of parallel circuits?	When two lights are in series, unplugging one of the lights and closing the switch causes the circuit to remain open. Electric current can't move around the entire path, so the remaining light won't turn on. Parallel circuits have an advantage because a break in one branch of the circuit does not interrupt the current flow in the other branches.
Optional	Ask students to come up with a way to test which type of circuit (parallel or series) is used in the Operation game. Students may realize that you can determine if the Operation game's circuit is parallel or in series by disconnecting either the light or the buzzer and observing the effect on the other device. If the remaining device still works, the circuit is in parallel. If the remaining device does not work, the circuit is in series. In reality, the Operation game's design makes it hard to test the circuit without damaging the game. You may want to test the circuit in one game as a class demonstration. Since the light connection is soldered, cut the middle of one of the light's wires to disconnect the light. The buzzer should still work, showing that the circuit is in parallel. After the demonstration, you may be able to splice the light's wire back together to get the light working again.
Assessment	
❶ Design and make a game board using materials given to you by your teacher. Your game must operate based on whether a circuit is open or closed. ✎ On another piece of paper, draw a diagram showing how the parts of your game board are connected together. Explain how your electrical circuit works.	See Sample Answers for a student drawing example. A reasonable game would have students roll the dice to land on a foil circle. After landing, touch the foil circle with the paper clip and earn points if the light lights.

Sample Answers

Where students follow the same procedure using the same materials, these answers are close to answers you can expect. Where students design their own experiment or model, students' results will vary.

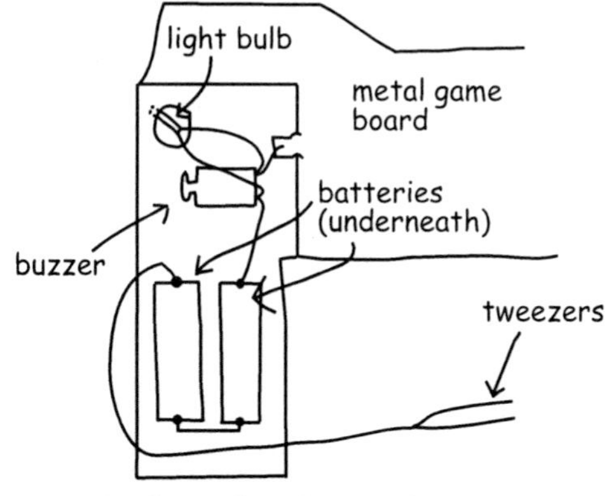

Example of a student drawing for Part A

Object	Conductor?	Insulator?
paper clip	yes	no
pencil	no	yes
index card	no	yes
penny	yes	no

Example of student responses for Part B

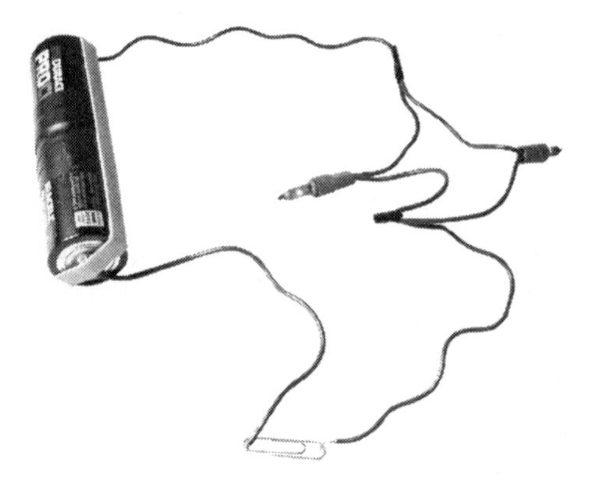

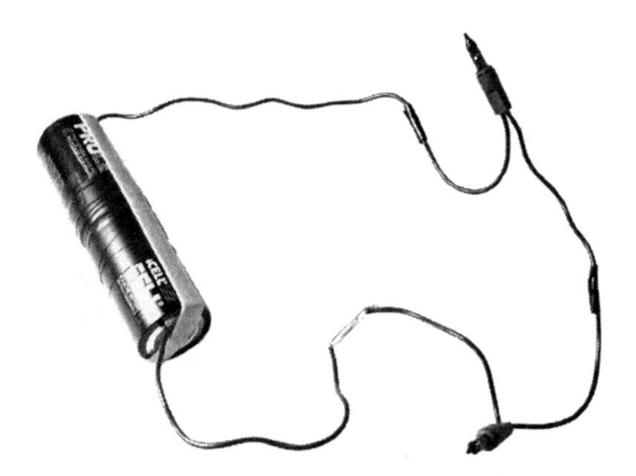

Examples of a parallel circuit (left) and a series circuit (right) for Part C

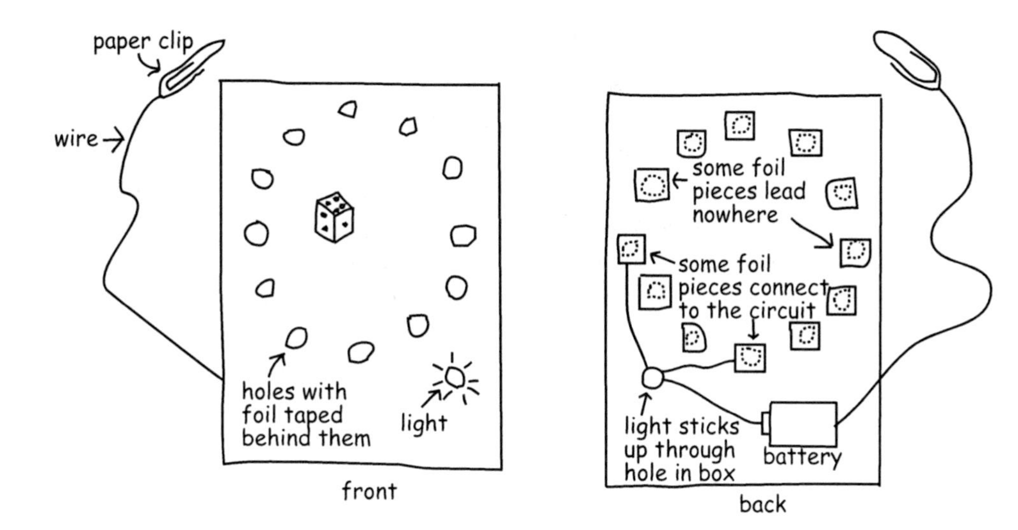

Example of student drawing for Assessment

Explanation

This section is intended for teachers. Modify the explanation for students as needed.

Components of Electric Circuits

Every electric circuit has three essential components: a source of electrical energy, a receiver or user of electric energy, and one or more objects through which the electric current may travel. The most easily recognized energy source is a battery. The source could also be a generator or an AC outlet in the wall through which electric energy comes from a generator that might be far away. The receiver could be an object such as a bell, buzzer, light bulb, toaster, curling iron, or alarm clock. The most common object through which electric current travels easily is a covered copper wire, but it could be a metal plate, water, or a human being. An object that works in this capacity—one through which electric current will travel—is called a conductor. An object that will not allow an electric current to travel through it is called a nonconductor or insulator.

There may be additional components in the circuit, the most common of which is probably the switch. A switch is simply a handy device for opening or closing the circuit without disturbing any of the other connections.

Electric Circuit of the Operation Game

The Operation game uses a simple electric circuit. When the cardboard surface is removed from the game, the circuit's basic components can be seen. Two batteries act as the energy source; the metal plate and the two wires are the path through which the energy flows. The bulb and buzzer are energy receivers and the tweezers act as a switch. When the switch is closed (either side of the tweezers touches the metal plate), the circuit path is closed and the current flows. The figure at the top right shows a diagram of this circuit. A schematic diagram of this circuit would resemble the one shown in Part A of the Teacher Notes.

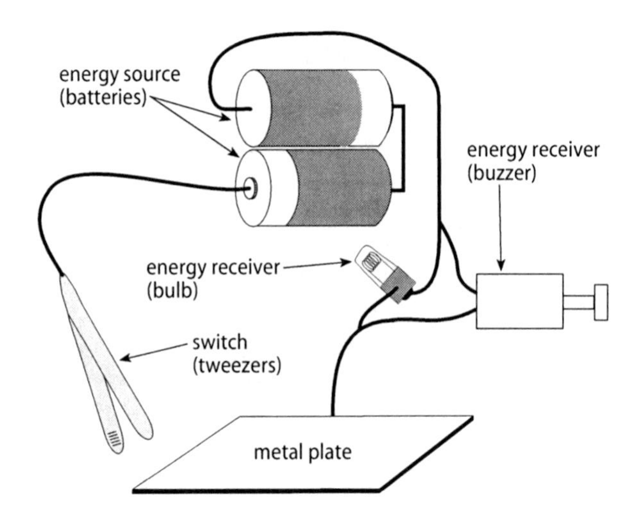

The Operation game uses a simple electric circuit.

Cross-Curricular Integration

Language arts:

- After completing the activity, have the students write a simple descriptive paragraph about the electric circuit. Be sure they include a clear main idea, supporting sentences, and closing summary statement about the circuit. As an alternative, you may wish to have the students write a comparison/contrast paragraph between open and closed circuits or between parallel and series circuits.
- Have students read *Dear Mr. Henshaw,* by Beverly Cleary (grades 4–7). If time doesn't permit reading the entire book, focus on the part of the story where Leigh builds a circuit. A circuit similar to the one Leigh built can be constructed either as a group or class activity. The doorbell described may be hard to find, but a buzzer or small bulb can be substituted. Each group will need a switch, buzzer or light bulb, battery, wire, and lidded container. Have the students demonstrate that they understand how and why switches and electric circuits work. After the group project is completed, students may individually experiment and make their own alarm.

Part A: How the Game Operates

(1) As you play the game, form a hypothesis that explains how the game works.

Record your group's ideas.

(2) Take the cardboard surface off your game board and further investigate to refine your group's ideas.

Make a list of all the internal parts you think are needed to make the buzzer buzz and the bulb light.

On another piece of paper, draw a simple diagram to show how the items you listed are attached to each other.

(3) All circuits have the following components in common: energy source, energy path, and energy receiver. As a group, answer the following questions:

What part of the game plays the role of the energy source?

What parts of the game play the role of the energy path?

What parts of the game play the role of the energy receiver?

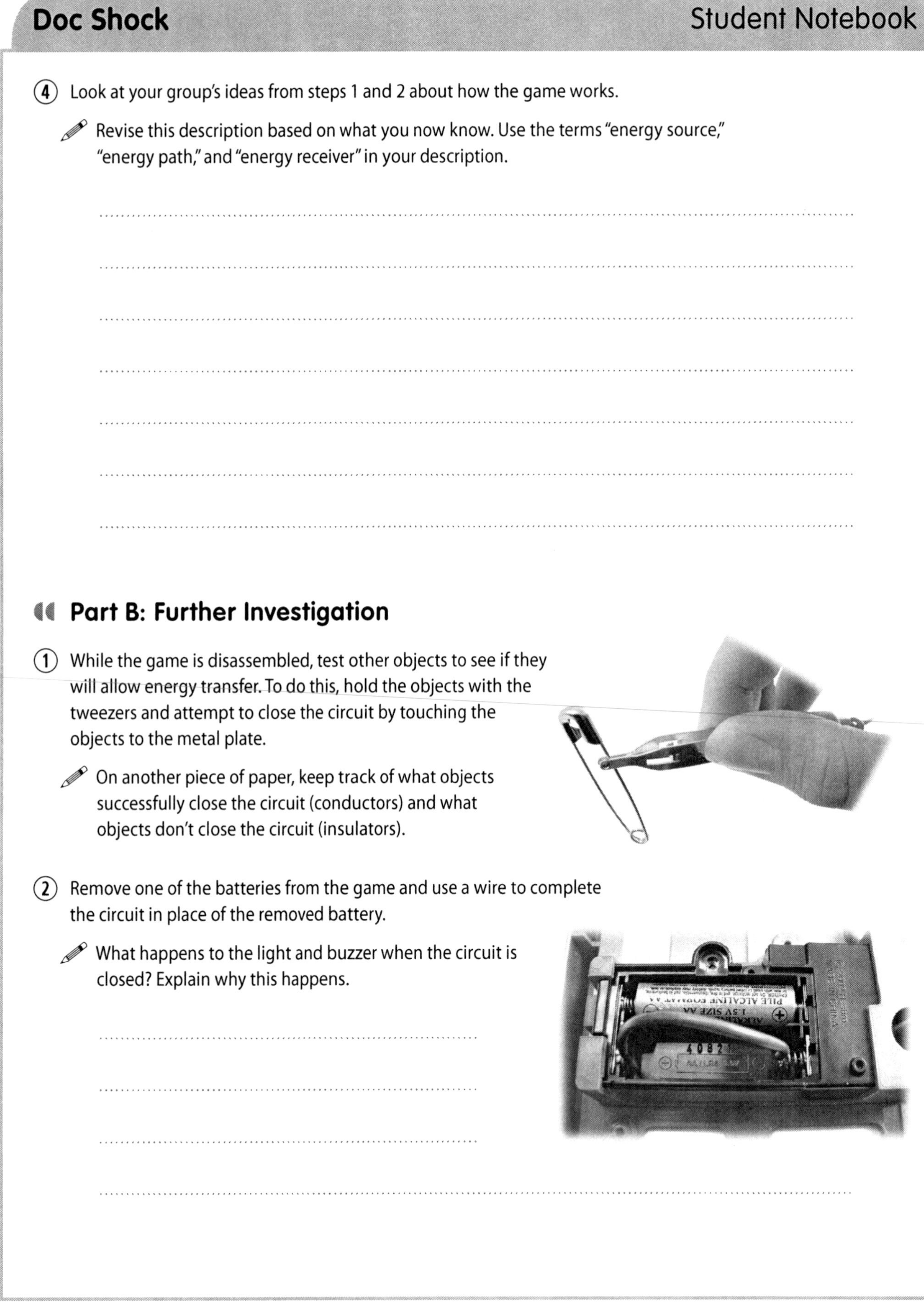

④ Look at your group's ideas from steps 1 and 2 about how the game works.

Revise this description based on what you now know. Use the terms "energy source," "energy path," and "energy receiver" in your description.

Part B: Further Investigation

① While the game is disassembled, test other objects to see if they will allow energy transfer. To do this, hold the objects with the tweezers and attempt to close the circuit by touching the objects to the metal plate.

On another piece of paper, keep track of what objects successfully close the circuit (conductors) and what objects don't close the circuit (insulators).

② Remove one of the batteries from the game and use a wire to complete the circuit in place of the removed battery.

What happens to the light and buzzer when the circuit is closed? Explain why this happens.

Part C: Parallel and Series Circuits

① Use all of the materials given to you by your teacher to build an electrical circuit where two lights are connected in parallel as shown. Test to make sure both lights go on when you close the switch.

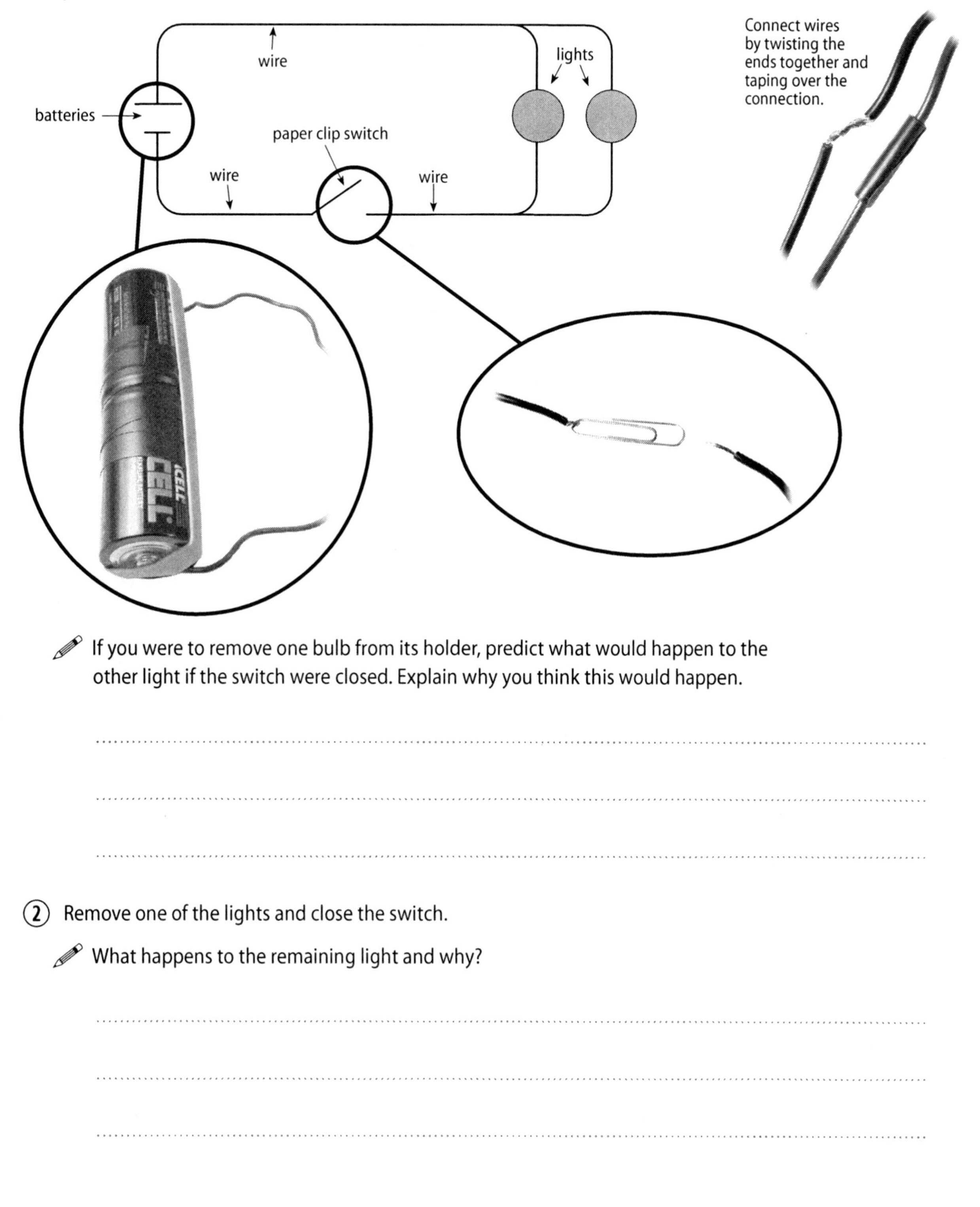

✎ If you were to remove one bulb from its holder, predict what would happen to the other light if the switch were closed. Explain why you think this would happen.

....

....

....

② Remove one of the lights and close the switch.

✎ What happens to the remaining light and why?

....

....

....

③ Now build a circuit where two lights are connected in series as shown. Test to make sure both lights go on when you close the switch.

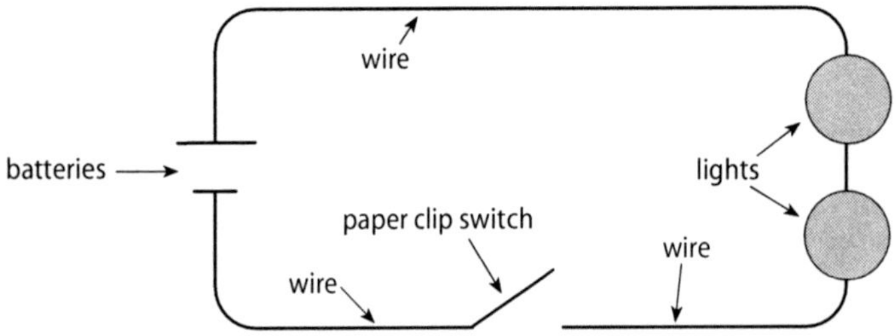

wire
batteries
lights
paper clip switch
wire
wire

If you were to remove one bulb from its holder, predict what would happen to the other light if the switch were closed. Explain why you think this would happen.

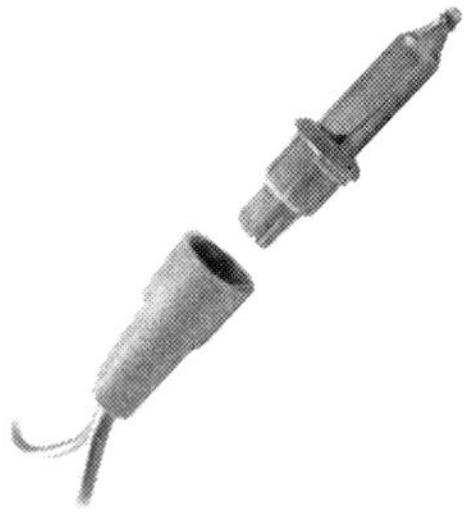

④ Remove one of the lights and close the switch.

What happens to the remaining light and why?

What is a design advantage of parallel circuits?

Assessment

1 Design and make a game board using materials given to you by your teacher. Your game must operate based on whether a circuit is open or closed.

On another piece of paper, draw a diagram showing how the parts of your game board are connected together. Explain how your electrical circuit works.

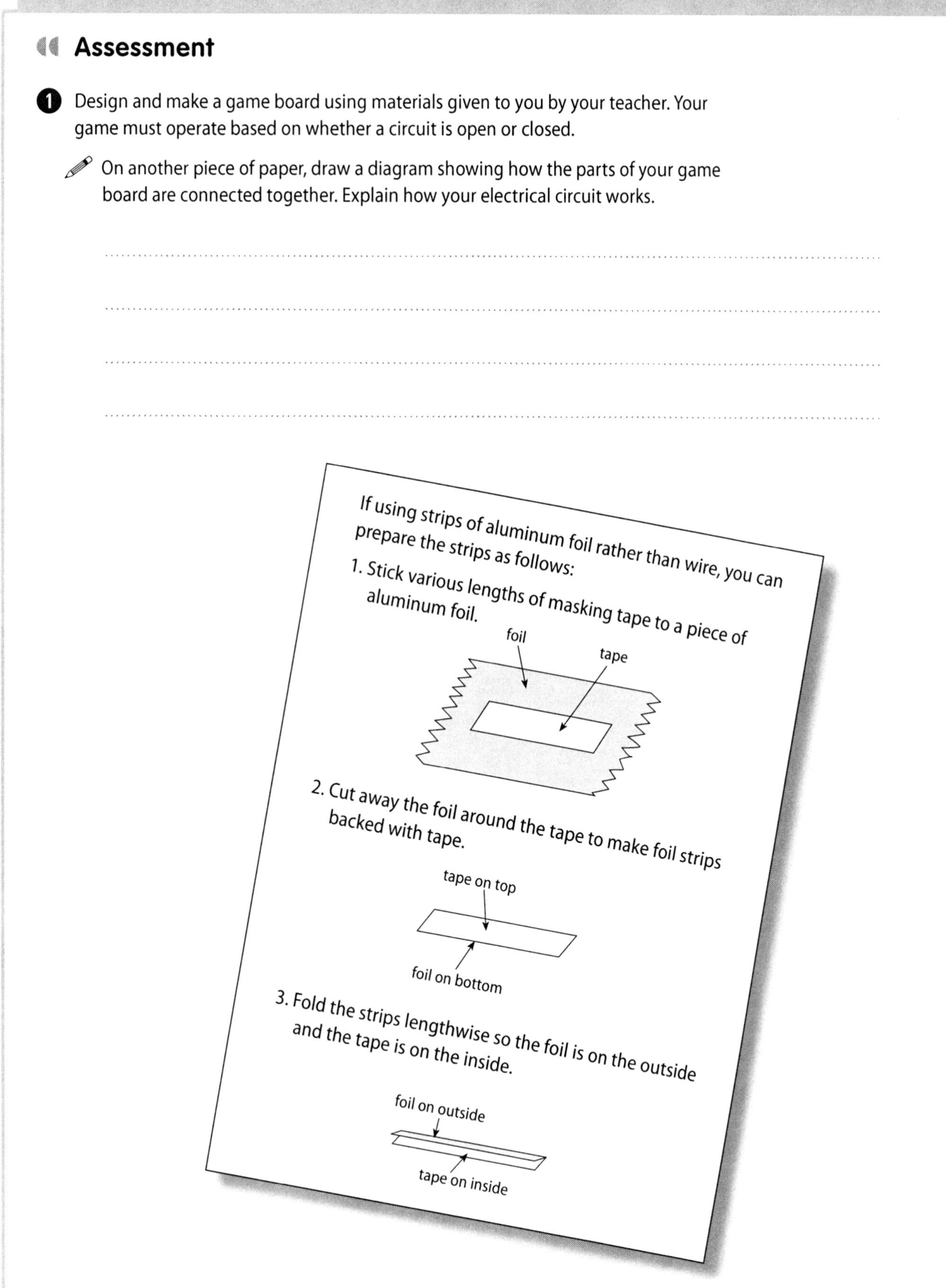

If using strips of aluminum foil rather than wire, you can prepare the strips as follows:

1. Stick various lengths of masking tape to a piece of aluminum foil.

2. Cut away the foil around the tape to make foil strips backed with tape.

3. Fold the strips lengthwise so the foil is on the outside and the tape is on the inside.

From Magnets to Motors

Students make simple electromagnets and electric motors and learn how the two are related. They learn that many toys operate using small electric motors.

Grade Levels

Science activity appropriate for grades 4–9

Student Background

Students should have previous exposure to the concepts of magnetic poles, magnetic fields, and forces of attraction and repulsion. Students should also have experience with simple battery and bulb circuits. These topics are covered in this book in "Magnet Cars" and "Doc Shock," respectively.

Time Required

Setup	20	minutes
Part A	15	minutes
Part B	45	minutes
Cleanup	5	minutes

Assessment time is not included.

Key Science Topics

- electric current
- electric motor
- electromagnet
- magnetic fields and poles
- magnetism

National Science Education Standards Overview

See *www.terrificscience.org/physicsez/* details of how these standards relate to the activity.

Science as Inquiry

Abilities Necessary to Do Scientific Inquiry

K–4 *Use data to construct a reasonable explanation.*
K–4 *Communicate investigations and explanations.*

5–8 *Use appropriate tools and techniques to gather, analyze, and interpret data.*
5–8 *Develop descriptions, explanations, predictions, and models using evidence.*
5–8 *Communicate scientific procedures and explanations.*

Physical Science

K–4 *Light, heat, electricity, and magnetism*

5–8 *Transfer of energy*

9–12 *Interactions of energy and matter*

Science and Technology

Abilities of Technological Design

K–4 *Evaluate a product or design.*

5–8 *Evaluate completed technological designs or products.*

Materials

For Getting Ready

- wire cutters (for teacher use only)

For the Procedure

Activity Introduction, per class

- assorted battery-operated mechanical toys

Try to find at least one toy that you can open up to show students the motor after they have completed Part B. (The Operation® game is a good choice.) Otherwise, keep one toy aside that you have already opened up, even if it had to be broken to do so.

Part A, per class

- 6-volt battery

Part A, per group

- 2-m length of 16- to 22-gauge insulated wire prepared in Getting Ready

Use wire composed of multiple smaller braided or twisted wires. Insulated speaker wire works well. The larger the gauge, the thinner the wire. Wires any thinner than 24-gauge wire will get hot and can possibly burn fingers.

- 3½-inch x ¼-inch steel bolt
- vinyl electrical tape
- small paper clips
- wide rubber band

Rubber bands used to hold fresh vegetables together (such as broccoli and asparagus) work well.

- D size battery (Alkaline is recommended.)
- graph paper

Graph paper masters for copying are available at www.terrificscience.org/physicsez/.

Part B, per class

- several pairs of needle-nose pliers

Part B, per student

- goggles
- 1.5-m length of 22-gauge varnished copper wire
- short length of dowel rod (about 1.5 cm diameter)
- small piece of fine sandpaper
- D size battery
- wide rubber band
- compass
- doughnut-shaped ceramic magnet (about 2.8-cm diameter)
- 2, 15-cm lengths of 14- or 16-gauge unvarnished copper wire
- modeling clay
- (optional) small square of cardboard
- (optional) small DC motor

These small, inexpensive motors are suitable for use in toys. For supply source suggestions, see www.terrificscience.org/supplies/.

Safety and Disposal

Students should wear goggles and take care to work in their own space when bending and winding wire. Keep the wire well away from all eyes. Students should avoid touching the bare wire when the circuit is connected because the wire may get hot. No special disposal procedures are required.

Getting Ready

- For Part A, cut the insulated wire into 2-m lengths so that each group gets one piece. Remove about 2 cm of the insulation from the ends of each piece of wire.
- For Part B, cut the varnished copper wire into approximately 1.5-m lengths so that each student gets one piece. Cut the unvarnished copper wire into 15-cm lengths so that each student gets two.
- If time is limited, shape the varnished and unvarnished wires for each student's mini motor before class. (See "How to Make a Mini Motor" instruction sheet.)

Procedure

This section provides teacher notes corresponding to each step of the student procedure. The procedure without teacher notes is included in the reproducible Student Notebook pages at the end of this activity and at www.terrificscience.org/physicsez/.

	Student Procedure	Teacher Notes
	Activity Introduction	• Give students an opportunity to play with some battery-operated mechanical toys. Ask students how these toys get the energy to move. Use students' ideas to make the point that batteries provide potential energy that eventually becomes kinetic energy. • Ask students if they have any ideas about how this energy transformation happens. Record their ideas and explain that they will be investigating that question.
	Part A: Pick it Up!	
1	**Starting about 10 cm from one end of the wire and about 1 cm from the end of the bolt, tightly wrap the insulated wire around the bolt 20 times. Use a piece of electrical tape to hold the first coils in place as shown. Without the battery connected, touch the bolt to a paper clip.** **What happens?**	Nothing happens. After the bolts have been used as electromagnets, they stay magnetized. If you are conducting this activity in several classes during the day, each class will need to use fresh bolts.
2	**Use a rubber band to hold one end of the wire to the battery's positive terminal (the top part of the battery) and touch the other end of the wire to the negative terminal as shown. This completes the circuit. Bring one end of the bolt near a paper clip and notice what happens. Now, notice what happens when you remove the wire from the negative terminal so that it no longer completes the circuit.** **TIP: Avoid touching the bare wire because it may get hot when the circuit is connected. Keep the circuit connected only when making observations.** **Record your observations.**	When the circuit is closed, students will see that the paper clip is attracted to the bolt. When the circuit is open, students will see that the paper clip is no longer attracted to the bolt. Connecting wires directly to a battery causes the battery to discharge more quickly than when the battery operates a flashlight or other appliance. Therefore, circuits should be closed only when making observations.

	Student Procedure	Teacher Notes
3	**Your group will investigate the relationship between the number of turns in the wire and the strength of the electromagnet. Before the experiment, decide on a standard method for attaching and counting paper clips. Possible methods are shown below.**	To test how many paper clips the bolt can hold for each number of turns, each group must use a standard method for attaching and counting the paper clips. Two examples follow, but other methods are possible. Whatever method each group uses, it must be used consistently in step 4. • Lay a paper clip across the head of the bolt. See how many additional paper clips will cling to the end of it. • Unbend a paper clip to form a hook. Lay this paper clip against the head of the bolt, then hang additional paper clips on it until it falls off.
4	**Using your chosen method, test how many paper clips the bolt can hold with different numbers of turns in the wire. Take data for 20, 30, 40, and 50 turns. (Remember not to touch the wire to the battery's negative terminal while you are adding more turns.)** **TIP: When you come within about 1 cm of the end of the bolt, wind the wire on top of itself.** **Record your results in the data table.**	See Sample Answers for example data.
5	**On graph paper, make a line graph of your data by plotting number of turns on the horizontal axis and number of paper clips on the vertical axis.** **Describe the relationship between the number of turns and the number of paper clips the bolt picks up.**	See Sample Answers for example graph. Students should realize that the number of paper clips the bolt holds increases as the number of turns increases.
	Class Discussion	• Have groups share their graphs. Even though results may vary, point out that everyone should have found that more turns in the electromagnet allow the electromagnet to hold more paper clips. • Also discuss how increasing the current of an electromagnet increases the electromagnet's strength. Demonstrate this by attaching a 6-volt battery to a 40-turn electromagnet. (The higher voltage causes a greater current to flow.) Show that you can pick up more paper clips with the 6-volt battery than with the 1.5-volt D size battery used in step 4.

Student Procedure	Teacher Notes
Class Discussion (continued)	• Explain to students that the magnetic field would be around the coiled wire even without the bolt. (See Explanation.) • Discuss how electromagnets differ from permanent magnets and why electromagnets might be better for some tasks than permanent magnets. (See Explanation.) • You may also want to talk about some of the ways in which electromagnets are used. (See Explanation.)
Part B: From Magnet to Motor	
① **Make a wire coil following the method under Ⓐ of the "How to Make a Mini Motor" instruction sheet.**	
② **As in step 2 of Part A, use a rubber band to hold one coil lead to the battery's positive terminal (the top part of the battery). If necessary, press a finger over the rubber band to hold the coil lead securely to the terminal. (Don't connect the wire to the negative terminal.) Place the coil near the compass.** **What happens?**	Nothing happens.
③ **Attach the other coil lead to the battery's negative terminal by sliding the lead under the rubber band as shown. Place the coil near the compass. Try to move the coil into different positions.** **TIP: Avoid touching the bare wire because it may get hot when the circuit is connected. Keep the circuit connected only when making observations.** **What happens?**	When the coil is placed near the compass, students should see a deflection in the compass needle.
④ **Place a magnet near the compass.** **What happens?** **What is present in the magnet and in the coil during step 3 that causes the compass to behave as it does?**	Students should realize that the magnet and the coil have a magnetic field that causes compass deflection.

	Student Procedure	Teacher Notes
⑤	Attach both coil leads to the battery as in step 3. Position the battery, coil, and magnet as shown. Move the magnet towards the coil and observe. (Don't hold the battery; allow it to move freely.) Flip the magnet over and try again. Observe the coil. ✎ What happens and why?	The magnetic fields of the coil and doughnut magnet affect each other. Depending on how the poles of the doughnut magnet are oriented, the coil will either move toward or away from the magnet. By flipping the doughnut magnet over, students should see both effects. The amount of coil motion may depend on the strength of the battery. If students don't see a reaction from the coil, use a fresh battery or two doughnut magnets stacked together. Make sure students realize that the coil is an electromagnet like the one they made in Part A.
⑥	Straighten out the lead ends of the coil.	You may want to check that the lead ends are symmetrical around each side of the coil and that students have sanded all the varnish off the leads, particularly the varnish close to the coil.
⑦	Follow the method under **B** on the "How to Make a Mini Motor" instruction sheet to prepare the battery of a working electric motor.	Make sure the two wire leads are parallel and on opposite sides of the battery. The loops of the wire leads need to be securely strapped to the battery with a rubber band to provide good contact for the flow of electricity.
⑧	Place the coil lead ends into the open loops of the battery leads to complete the circuit. Adjust the height of the loops until the coil is level. Push the coil gently to start it spinning. Notice how long the coil continues to spin before stopping.	Students will observe that the coil quickly stops spinning. They will compare this to the behavior of the coil in step 10.
⑨	Place the doughnut magnet on the battery under the coil as shown. If the magnet doesn't stay in place, stick it down with a small piece of modeling clay. Adjust the height of the coil if needed. ✎ What happens when you initially put the magnet under the coil? Explain what you think is causing this action.	Students should see the coil vibrate back and forth. Since the coil (electromagnet) is sitting in a magnetic field created by the doughnut (permanent) magnet, forces act on the coil causing it to deflect. If the motor is well balanced to start with, the coil may start to spin in complete circles on its own. If a student's coil has no motion, make sure that: • all varnish is removed from the coil leads, • the coil is sitting level and just above the doughnut magnet, and • the leads are connected securely to the battery. Sometimes the coil won't move, even if everything seems correct.

	Student Procedure	Teacher Notes
(10)	If necessary, push the coil gently to start it spinning. If it does not turn smoothly on the supports, adjust by bending the coil leads. Observe the spinning coil. ✎ Does the coil spin longer than in step 8? ✎ Why do some people have to give the coil a small push to start it in motion? ✎ Do the coil leads remain in constant contact with the loops? ✎ Does the coil receive a steady flow of current? How do you know?	Since the motor is not very strong, it may not start automatically. A small push overcomes the coil's inertia and allows the magnetic forces to create the spinning motion. If the coil doesn't continue to move, try changing the position of the magnet. The coil should spin longer than it did in step 8. The coil will appear to vibrate or jump up and down slightly in the loops, thereby breaking the contact between the coil and the loops. Students should realize that this jumping causes a brief break in the flow of the current.
(11)	After your motor has been spinning awhile, feel the coil and loops. ✎ How can you explain the temperature of the coil and loops?	As current flows through the wire, it experiences a resistance to its flow. Electrical resistance produces heat.
	Class Discussion	Show students the small motor inside one of the battery-operated toys they played with before doing Part A. Explain that the energy transformations in this motor are just like the ones in the mini motor they made. Return to the student ideas about energy transformations in toys. Ask if they can now identify the energy transformations from the battery to the action of the toy. Be sure to discuss the following: • Chemical energy is initially stored in the battery. • When a circuit is completed, the chemical energy is converted to electrical energy. • As electrical energy flows through the coil, a magnetic field is created. • The interaction of the magnetic field of the coil and the magnetic field of the permanent magnet changes the electrical energy into kinetic energy of the coil. • The rotation of the coil turns an axle that moves some part of the toy. • Some of the electrical energy is transformed into thermal energy of the atoms in the wire.

Student Procedure	Teacher Notes
Optional	You may want to open up small DC motors made for toys so that students can see the parts. Although these motors are more complex than the mini motors, you can point out the armature (electromagnet), permanent magnet, and axle. (See Explanation for a labeled figure.)
Assessment A	**These scenarios are presented as a thought problem (rather than a hands-on assessment) because sometimes the mini motors behave unpredictably. For example, coil spin rates can change with no obvious cause, making it difficult to conclude that any changes made to the coil, battery, or permanent magnets were the cause.**
1 **Explain why each of the following changes to the mini motor might create a greater spin rate in the coil.** **Change the mini motor by putting more loops in the coil.** **Change the mini motor by using a higher voltage battery.**	Each of these changes creates a stronger magnetic field, and therefore, a stronger interaction between the coil and the permanent magnet (resulting, in most cases, in an increased spin rate). • **Putting more loops in the coil:** Based on their experience with adding turns of wire to the electromagnet in Part A, students should realize that more loops in the coil would result in a stronger magnetic field and, in some cases, an increase in spin rate. Students may also conclude that increasing the number of loops in the coil will also increase its mass, possibly resulting in a lower spin rate. • **Using a higher voltage battery:** Based on your demonstration in Part A showing that the strength of the electromagnet increases with a higher voltage battery, students should realize that current through the coil would increase. The increased current would result in a stronger magnetic field and, therefore, an increase in spin rate.

Student Procedure	Teacher Notes
Assessment B	**This assessment is presented as a thought problem because, although a motor with two electromagnets could work in theory, the coil electromagnet from Part B is not strong enough to actually make a working motor.**
1 **Plan a design for an electric motor that combines the bolt electromagnet from Part A with the coil electromagnet from Part B. Leave out the permanent (doughnut) magnet.** **Draw and label your design. Explain what interactions will occur.**	See Sample Answers for student drawing example. Look for designs that show the bolt electromagnet from Part A and the coil electromagnet from Part B in close proximity so that their magnetic fields will interact. Students should explain that interaction between the magnetic fields of the two electromagnets will result in the spin of the coil (just as the fields of the coil and the permanent magnet interacted in Part B).

Sample Answers

Where students follow the same procedure using the same materials, these answers are close to answers you can expect. Where students design their own experiment or model, students' results will vary.

How Strong Is the Electromagnet?				
number of turns in the wire	20	30	40	50
number of paper clips	4	11	18	26

Example data for Part A

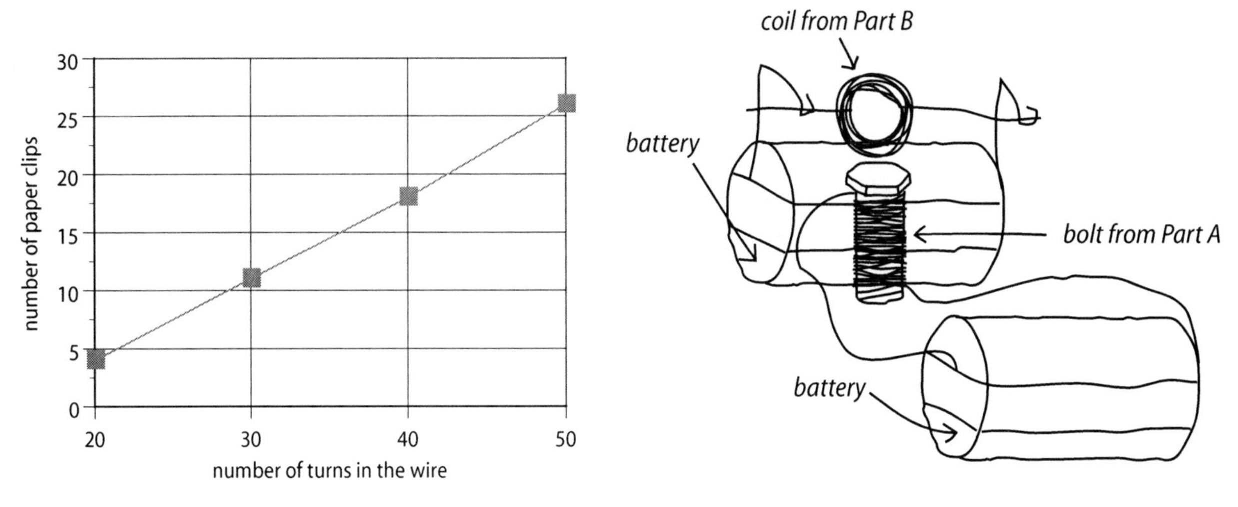

Example graph for Part A

Example of student design for Assessment B

Explanation

This section is intended for teachers. Modify the explanation for students as needed.

Electricity and Magnetic Fields

When electrical current flows through a wire, a magnetic field is created around the wire. Around a straight wire, the field can be visualized as a series of concentric circles or cylinders with the wire down the center. If your right thumb is pointing in the direction of the conventional current, your curled fingers point in the direction of the magnetic field. (See figure below left.) This is sometimes referred to as the "right-hand rule." Conventional current is what is normally labeled on a circuit diagram. This current flows in the opposite direction to the way the electrons are actually moving. Thus, conventional current flows from the positive end of the battery to the negative end.

For a coil of wire (like the one around the bolt), the magnetic field travels straight through the middle. (See figure below right.)

In Part A, the magnetic field would be around the coil even without the bolt. When the iron bolt is inside the coil, areas in the iron called domains align themselves with the field, adding their magnetic fields to the one produced by the current. This greatly increases the strength of the electromagnet. When the current is turned off, most of the domains return to their original random orientation.

The easiest ways to increase the strength of an electromagnet are to increase the current or to increase the number of turns in the coil. The strength of the magnetic field is directly proportional to both. For our electromagnets in Part A, the students cannot vary the current but can only turn it on or off. So to increase the strength of the electromagnet, they must increase the number of coil turns around the bolt. The students' data should show that the number of paper clips held increases as the number of turns in the coil increases.

The demonstration with the 6-volt battery illustrates how an increase in current also increases the electromagnet's strength. You may want to explain that, although voltage and current are not the same thing, increasing the voltage of the battery also increases the current to the electromagnet in this case because you are using the same wire and, therefore, not changing the resistance.

Large electromagnets can produce magnetic fields more than 1,000 times larger than that of a typical classroom magnet. They are particularly useful because they run cleanly and silently, can be started instantly, and can easily be turned on and off. Small electromagnets can be found in doorbells, speakers, and many other appliances. Large ones are used on cranes, in power plants, and in accelerators that scientists use to study nature's smallest particles.

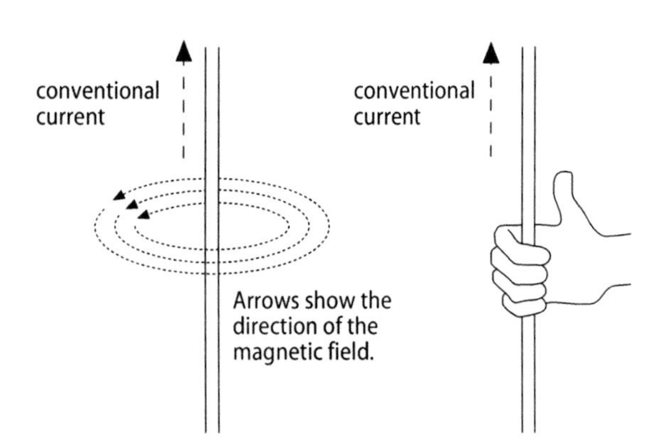

Right-hand rule for current in a straight wire

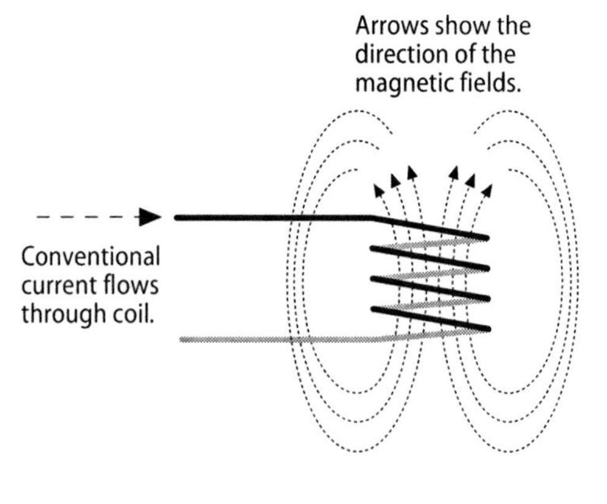

Magnetic field of a coil

Motors

In Part B, students explore an electromagnetic coil that they use as an armature for a simple direct current (DC) motor. Commercial motors may run on either direct or alternating current, may use permanent magnets or electromagnets, and may contain several coils. An electric motor requires three elements to create motion: a source of current (provided in Part B as a direct current by the battery), a magnet (provided in Part B by a doughnut magnet), and one or more loops of wire that are free to turn within the magnetic field of the magnet (provided in Part B by the copper coil). An electric motor works by changing the chemical potential energy of the battery into mechanical energy in the form of a rotating coil.

To understand how an electric motor works, students should be familiar with the concept that unlike poles attract and that like poles repel. In a motor, two magnets are placed so that their opposite poles attract each other. One of the magnets (the armature) is an electromagnet that is placed on an axle so that it is free to rotate. The other magnet (the field magnet) is stationary.

The figures below show an example of how an electric motor works. The magnetic force between the two magnets causes the armature to rotate as its poles align themselves with the opposite poles of the field magnet. However, the torque producing the rotation goes to zero briefly when the armature becomes aligned with the field magnet, then returns but in the opposite direction, so that now it would be slowing the armature down. To keep the armature rotating, some method is needed to interrupt the magnetic field in the armature and let the armature remain free to move. One way to do this is to cause the armature's current to switch off just as the armature's poles become aligned with the poles of the field magnet. (See figure at bottom right.) The armature, which is now nonmagnetic, continues to rotate under its own momentum until it makes a nearly complete turn. At this point, electrical contact is again made with the battery, and the (now magnetized) armature receives another "kick" from the field magnet to go around.

In some motors, the current is not switched off and on to make the armature rotate. Rather, each half-turn of the coil reverses the orientation of the coil leads with respect to the negative and positive terminal leads. This causes the flow of the current through the coil to be reversed with each half-turn. As a result, the force remains in the same direction, pushing the coil forward with every half-turn.

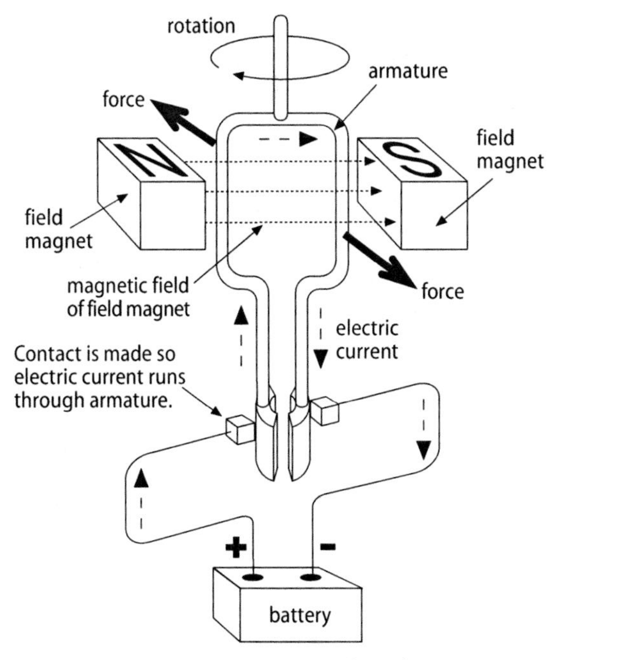

Armature position when electric contact is made with battery

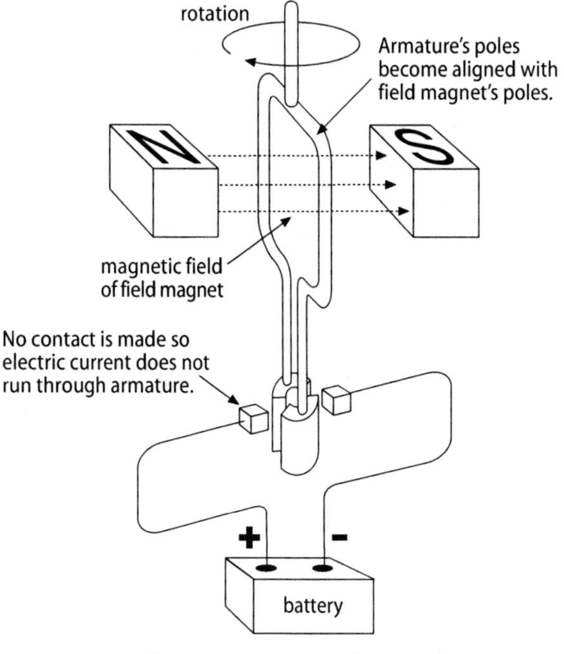

Armature position when no electric contact is made with the battery

How the Mini Motor Works

The mini motor in this activity works because the flow of the current through the coil is periodically interrupted due to the coil's vibration. When the circuit is closed (that is, when the negative and positive leads from the battery come into contact with the leads from the coil), the battery produces a current in the copper wire. A coil carrying a current in a magnetic field has forces on it that cause it to turn. (See figure below.) The force deflects the coil in one direction until the poles align themselves with the opposite poles of the doughnut magnet.

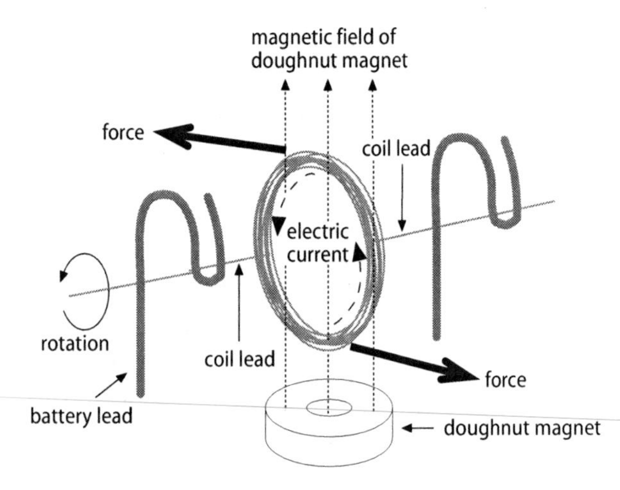

A coil carrying a current in a magnetic field has forces on it that cause it to turn.

Because the coil bounces as it turns, the coil leads periodically move away from the battery leads and break the connection. When the current flow is broken, the force stops pushing the coil, but the coil's inertia keeps the coil turning. Usually the coil's vibration begins the flow of current again and the coil gets another kick.

It is also important to note that the bouncing of the coil that causes the current to turn on and shut off is a random process. On average, the coil is pushed more in one direction than the other, so the coil stays spinning in that direction. However, in some cases, you may see the coil reverse its direction of spin.

In this activity, we removed the varnish from the entire coil lead. At first glance it seems that the motor should not work, because the current can flow through the coil leads without interruption no matter what the orientation of the coil. But, as previously described, the bouncing coil periodically breaks the flow of current. If the coil were perfectly stable and the coil leads remained in contact with the battery leads at all times, our method would not work.

Although the small motors used in the toys are more complex than the mini motor, they depend on the same energy transformations. The potential energy of the battery becomes electrical energy, creating a magnetic field. The interaction of magnetic fields changes the electrical energy to kinetic energy. Students can see the key parts of the motor involved in these steps (the armature and the magnets).

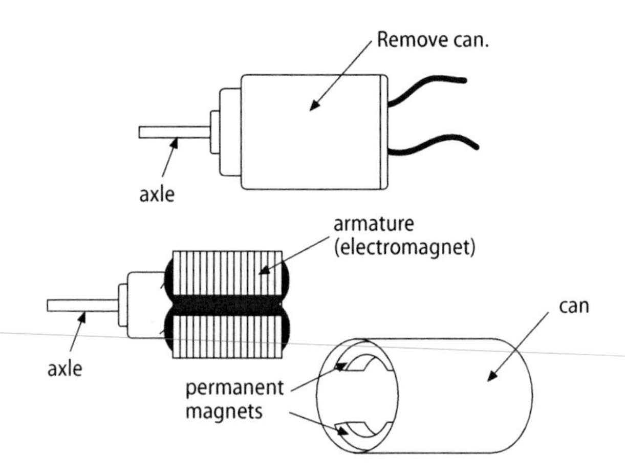

Remove the can from a small motor to reveal the armature and the permanent magnets.

Cross-Curricular Integration

Social studies:

- Have students research and write a report on the history of electric motors or on the impact of the invention of the electric motor on society.

References

HowStuffWorks Website. How Electric Motors Work. http://howstuffworks.com (accessed February 22, 2005).

Macaulay, D. *The Way Things Work;* Houghton Mifflin: Boston, 1988.

Renner, A.G. *How to Make and Use Electric Motors;* Putnam: New York, 1974.

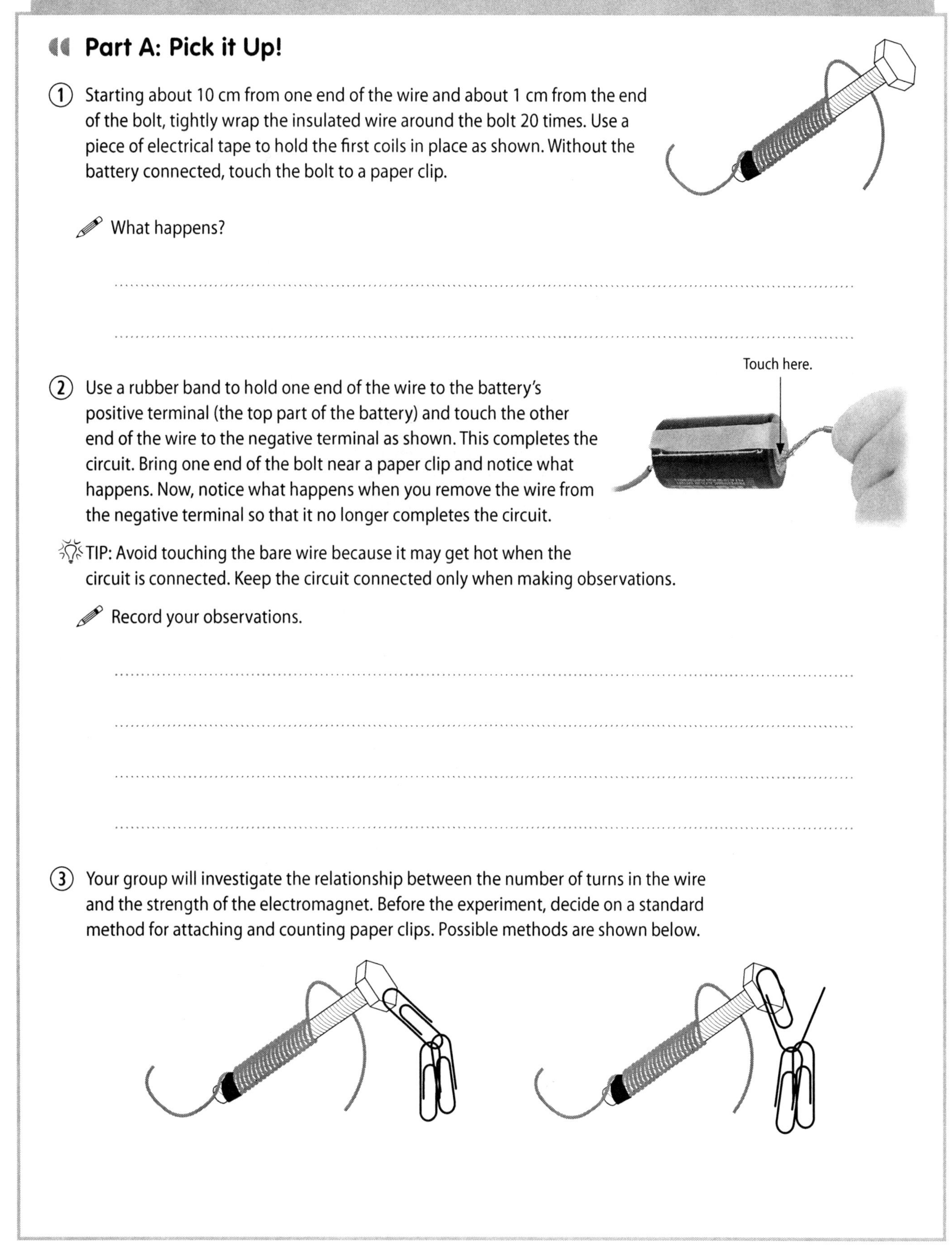

Part A: Pick it Up!

① Starting about 10 cm from one end of the wire and about 1 cm from the end of the bolt, tightly wrap the insulated wire around the bolt 20 times. Use a piece of electrical tape to hold the first coils in place as shown. Without the battery connected, touch the bolt to a paper clip.

What happens?

....

....

② Use a rubber band to hold one end of the wire to the battery's positive terminal (the top part of the battery) and touch the other end of the wire to the negative terminal as shown. This completes the circuit. Bring one end of the bolt near a paper clip and notice what happens. Now, notice what happens when you remove the wire from the negative terminal so that it no longer completes the circuit.

TIP: Avoid touching the bare wire because it may get hot when the circuit is connected. Keep the circuit connected only when making observations.

Record your observations.

....

....

....

....

③ Your group will investigate the relationship between the number of turns in the wire and the strength of the electromagnet. Before the experiment, decide on a standard method for attaching and counting paper clips. Possible methods are shown below.

④ Using your chosen method, test how many paper clips the bolt can hold with different numbers of turns in the wire. Take data for 20, 30, 40, and 50 turns. (Remember not to touch the wire to the battery's negative terminal while you are adding more turns.)

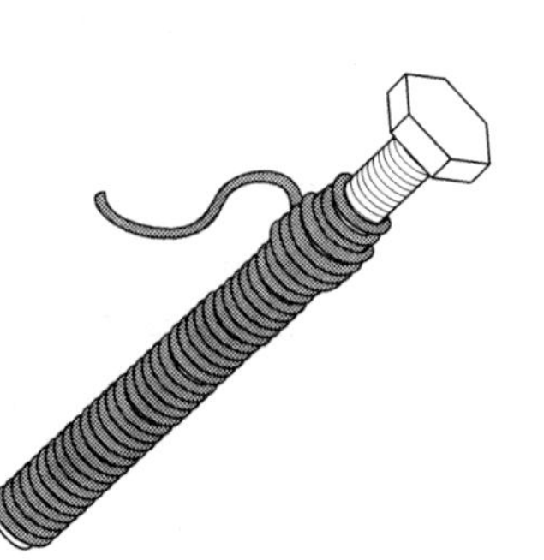

TIP: When you come within about 1 cm of the end of the bolt, wind the wire on top of itself.

Record your results in the data table.

How Strong Is the Electromagnet?				
number of turns in the wire	20	30	40	50
number of paper clips				

⑤ On graph paper, make a line graph of your data by plotting number of turns on the horizontal axis and number of paper clips on the vertical axis.

Describe the relationship between the number of turns and the number of paper clips the bolt picks up.

Part B: From Magnet to Motor

① Make a wire coil following the method under **A** of the "How to Make a Mini Motor" instruction sheet.

② As in step 2 of Part A, use a rubber band to hold one coil lead to the battery's positive terminal (the top part of the battery). If necessary, press a finger over the rubber band to hold the coil lead securely to the terminal. (Don't connect the wire to the negative terminal.) Place the coil near the compass.

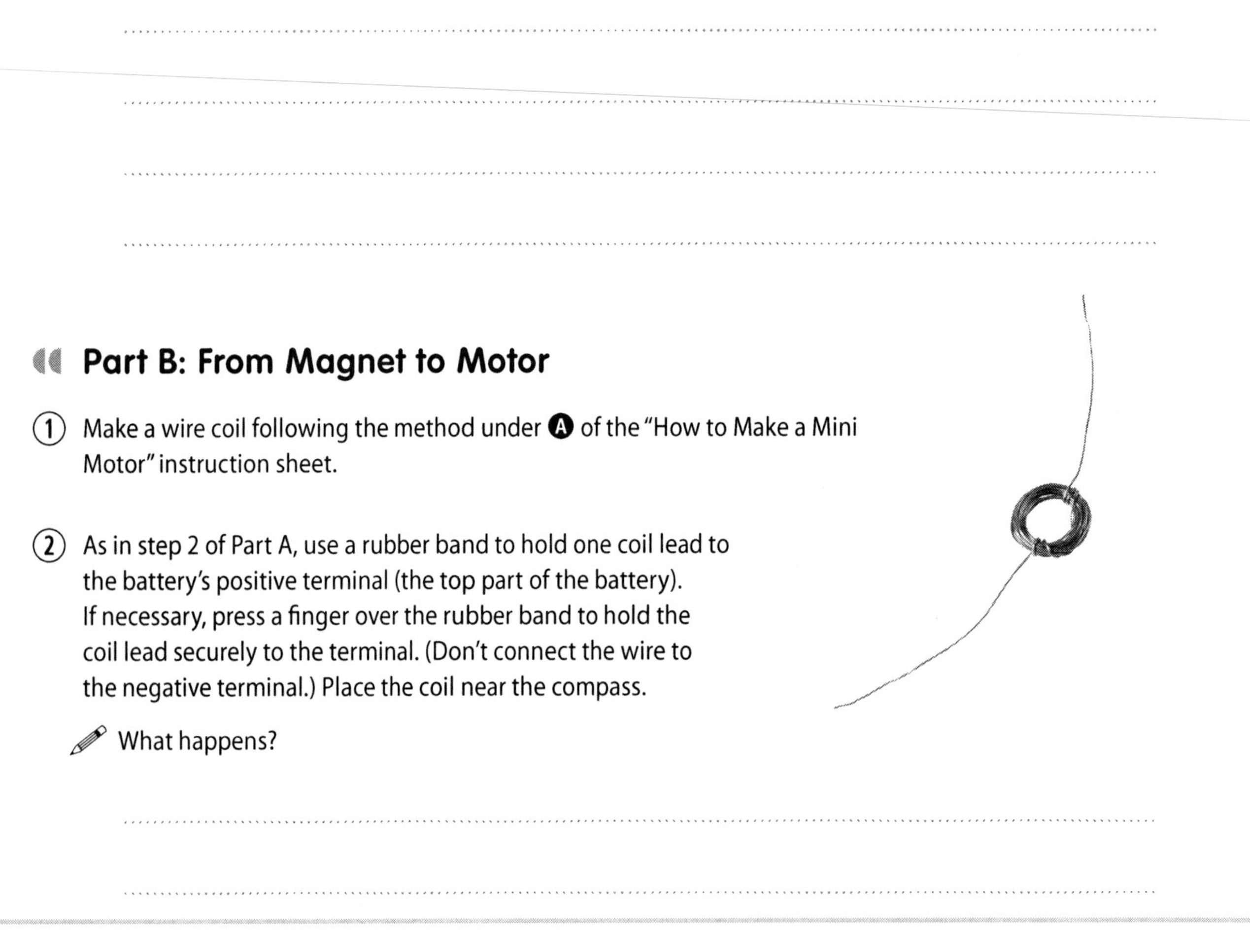

What happens?

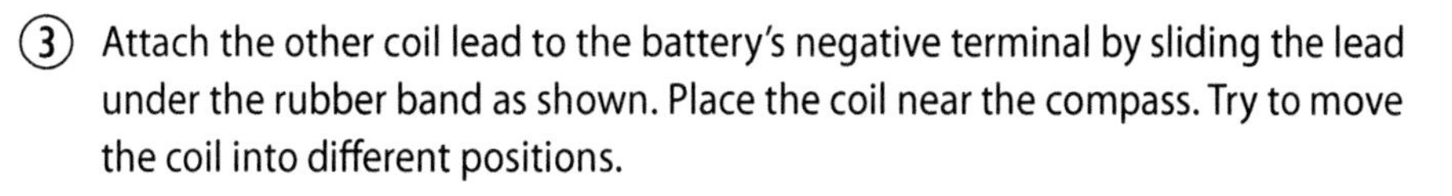

③ Attach the other coil lead to the battery's negative terminal by sliding the lead under the rubber band as shown. Place the coil near the compass. Try to move the coil into different positions.

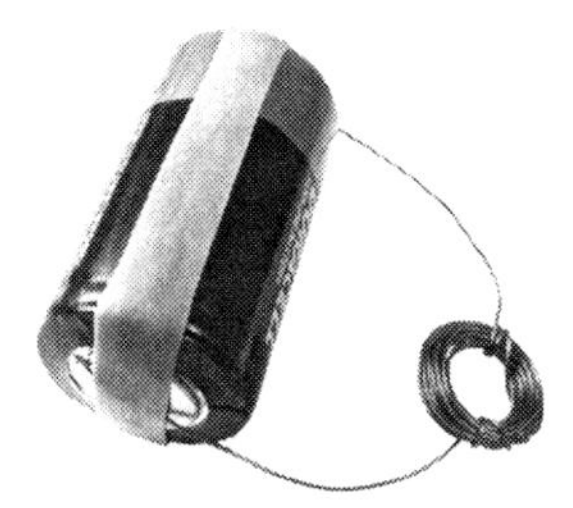

TIP: Avoid touching the bare wire because it may get hot when the circuit is connected. Keep the circuit connected only when making observations.

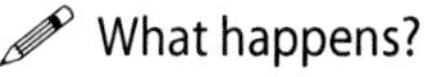

What happens?

....

④ Place a magnet near the compass.

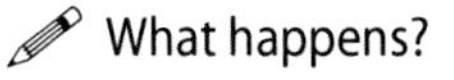

What happens?

....

What is present in the magnet and in the coil during step 3 that causes the compass to behave as it does?

....

....

⑤ Attach both coil leads to the battery as in step 3. Position the battery, coil, and magnet as shown. Move the magnet towards the coil and observe. (Don't hold the battery; allow it to move freely.) Flip the magnet over and try again. Observe the coil.

What happens and why?

....

....

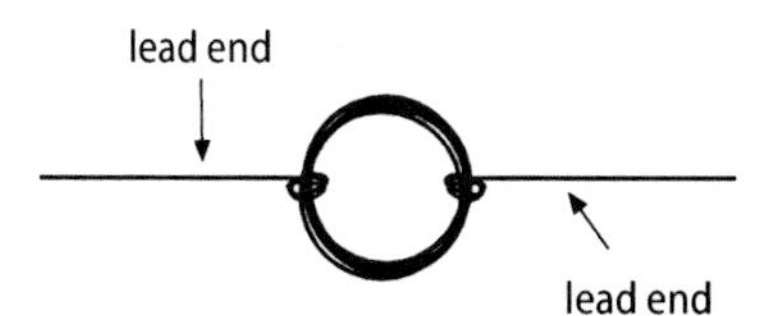

⑥ Straighten out the lead ends of the coil.

⑦ Follow the method under **B** on the "How to Make a Mini Motor" instruction sheet to prepare the battery of a working electric motor.

⑧ Place the coil lead ends into the open loops of the battery leads to complete the circuit. Adjust the height of the loops until the coil is level. Push the coil gently to start it spinning. Notice how long the coil continues to spin before stopping.

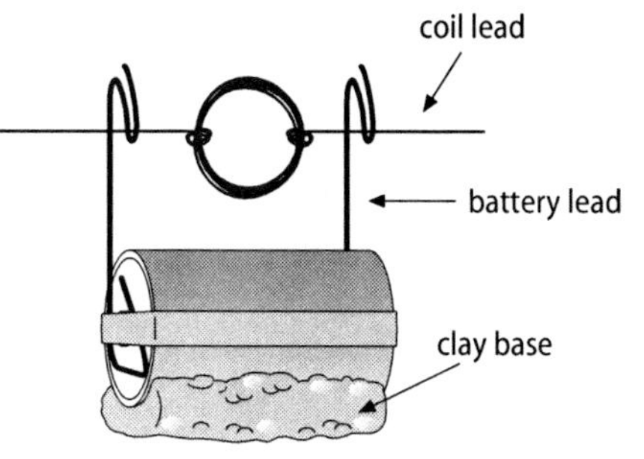

⑨ Place the doughnut magnet on the battery under the coil as shown. If the magnet doesn't stay in place, stick it down with a small piece of modeling clay. Adjust the height of the coil if needed.

What happens when you initially put the magnet under the coil? Explain what you think is causing this action.

⑩ If necessary, push the coil gently to start it spinning. If it does not turn smoothly on the supports, adjust by bending the coil leads. Observe the spinning coil.

Does the coil spin longer than in step 8?

Why do some people have to give the coil a small push to start it in motion?

Do the coil leads remain in constant contact with the loops?

Does the coil receive a steady flow of current? How do you know?

⑪ After your motor has been spinning awhile, feel the coil and loops.

How can you explain the temperature of the coil and loops?

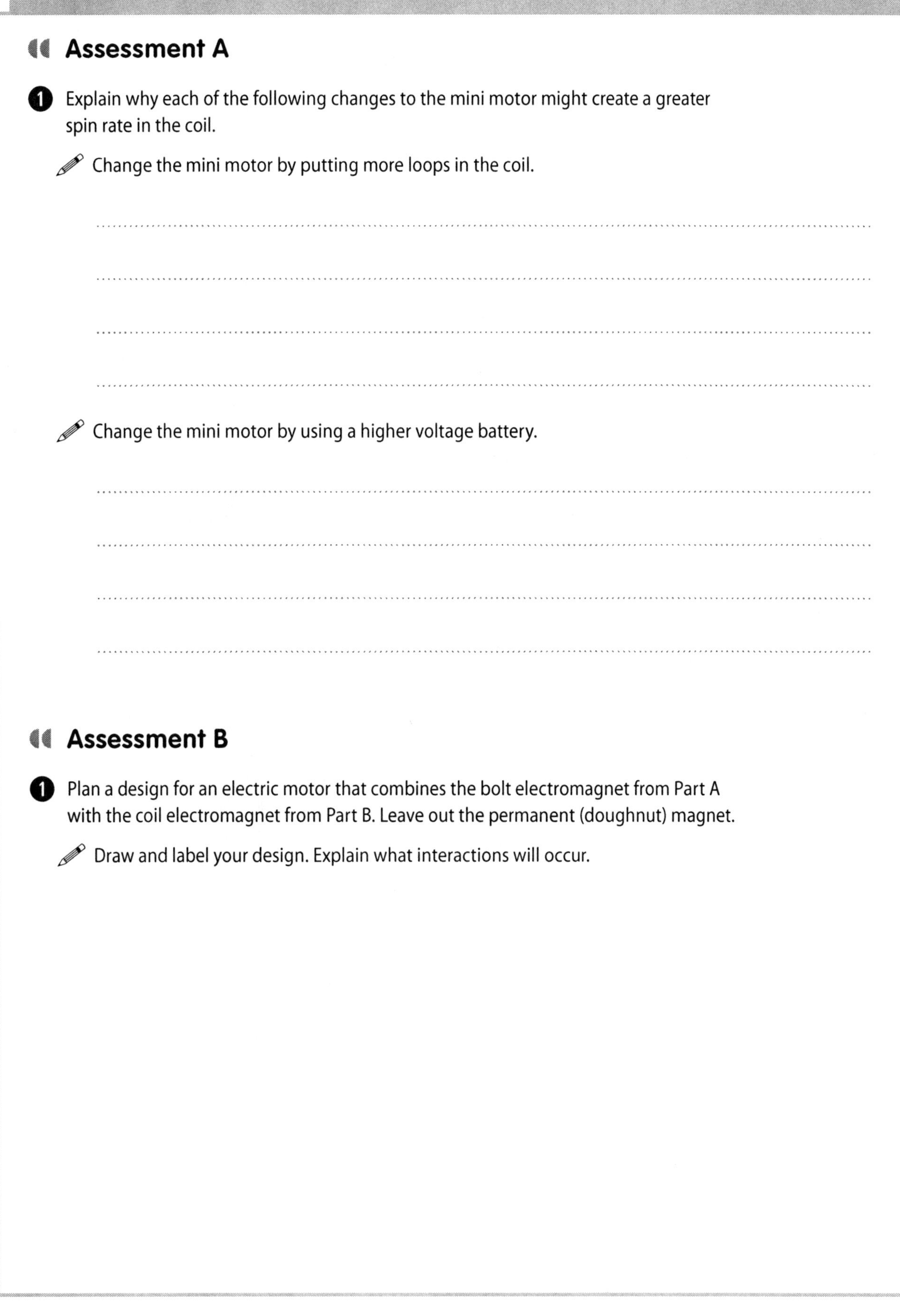

Assessment A

1. Explain why each of the following changes to the mini motor might create a greater spin rate in the coil.

✎ Change the mini motor by putting more loops in the coil.

✎ Change the mini motor by using a higher voltage battery.

Assessment B

1. Plan a design for an electric motor that combines the bolt electromagnet from Part A with the coil electromagnet from Part B. Leave out the permanent (doughnut) magnet.

✎ Draw and label your design. Explain what interactions will occur.

How to Make a Mini Motor

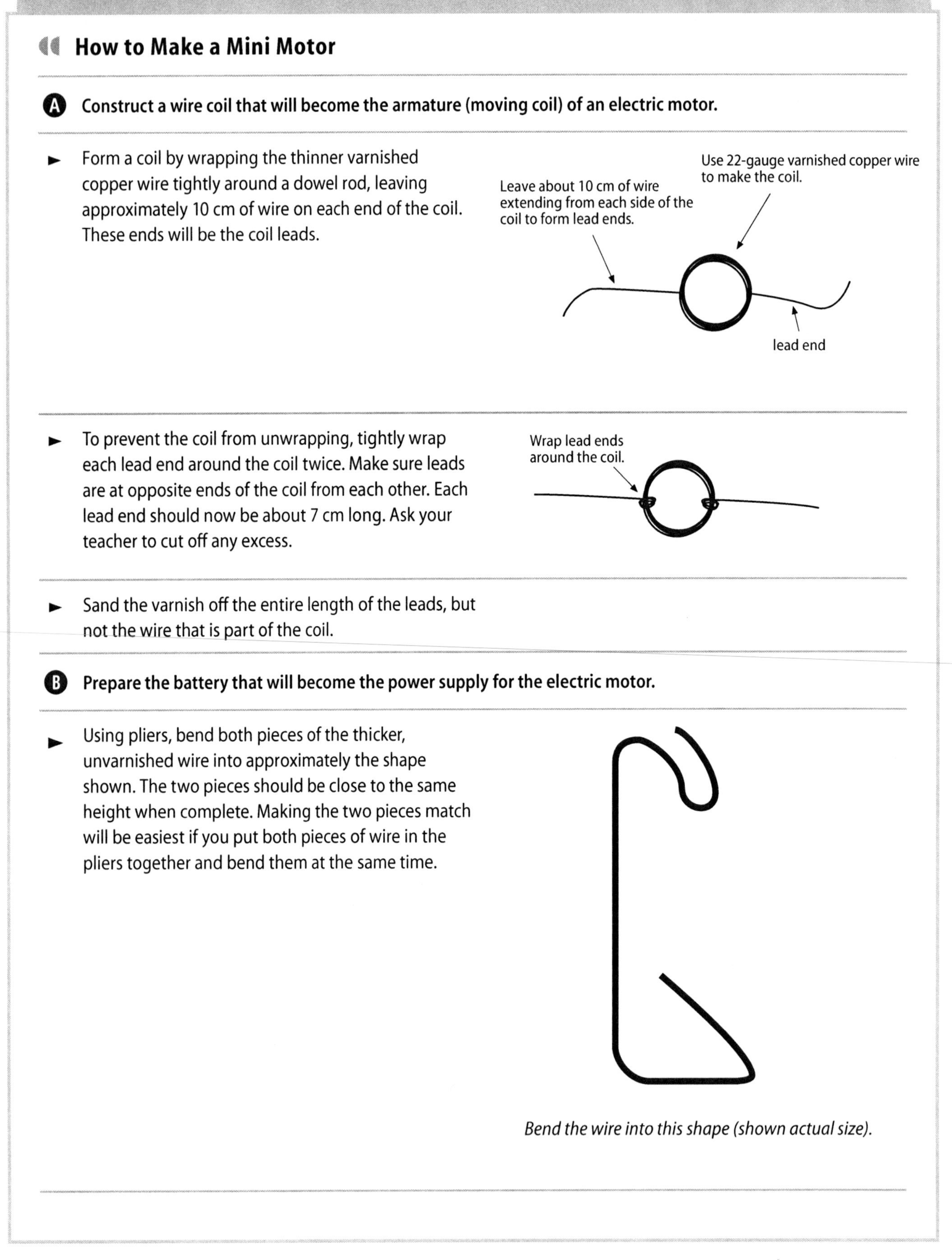

A Construct a wire coil that will become the armature (moving coil) of an electric motor.

- Form a coil by wrapping the thinner varnished copper wire tightly around a dowel rod, leaving approximately 10 cm of wire on each end of the coil. These ends will be the coil leads.

- To prevent the coil from unwrapping, tightly wrap each lead end around the coil twice. Make sure leads are at opposite ends of the coil from each other. Each lead end should now be about 7 cm long. Ask your teacher to cut off any excess.

- Sand the varnish off the entire length of the leads, but not the wire that is part of the coil.

B Prepare the battery that will become the power supply for the electric motor.

- Using pliers, bend both pieces of the thicker, unvarnished wire into approximately the shape shown. The two pieces should be close to the same height when complete. Making the two pieces match will be easiest if you put both pieces of wire in the pliers together and bend them at the same time.

Bend the wire into this shape (shown actual size).

- Use the rubber band to attach the bent wires to the ends of the battery. These posts form a stand to hold the wire coil. Adjust the wires so the loops at the top of the posts are at the same height.

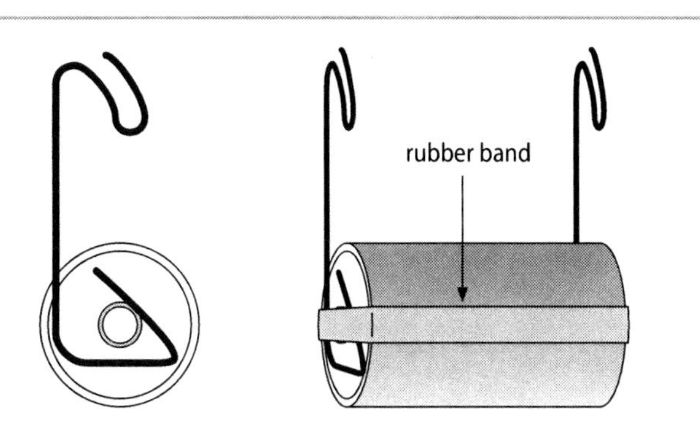

Use a rubber band to attach wires to the ends of the battery.

- Shape a base out of clay. (If desired, press the clay base on a small square of cardboard.) Press the battery into the clay as shown.

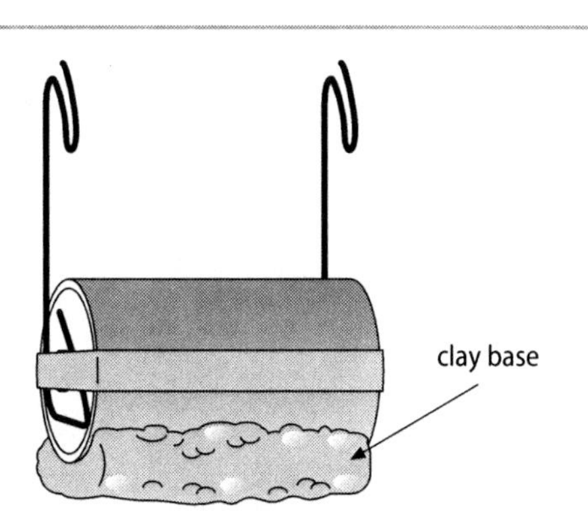

Appendix

Activities Indexed by National Science Education Standards for Grades K–4

	Balance This!	Bounceability	Crash Test	Doc Shock	Exploring Friction	From Magnets to Motors	Gear Up, Gear Down	Magnet Cars	Measuring Mass	Pulley Power Basics	Push and Go	Ramps and Cars	Seesaw Forces	Six-Cent Top	Skyhook	Sound Off	Static Cling	Toy That Returns	Understanding Speed
Science as Inquiry—Abilities Necessary to Do Scientific Inquiry																			
Plan and conduct a simple investigation.		●	●	●	●							●		●		●			●
Employ simple equipment and tools to gather data and extend the senses.		●			●		●		●			●	●	●				●	●
Use data to construct a reasonable explanation.	●	●		●	●	●	●		●				●	●		●	●		●
Communicate investigations and explanations.	●	●	●	●		●	●	●	●	●	●				●	●	●	●	●
Physical Science																			
Properties of objects and materials		●	●						●							●			
Position and motion of objects	●	●	●		●		●	●		●	●	●		●	●	●		●	●
Light, heat, electricity, and magnetism				●		●		●									●		
Science and Technology—Abilities of Technological Design																			
Propose a solution.	●														●				
Implement proposed solutions.	●														●				
Evaluate a product or design.				●		●	●			●	●		●	●	●			●	
Communicate a problem, design, and solution.	●			●	●		●	●						●	●				

Activities Indexed by National Science Education Standards for Grades 5–8

	Balance This!	Bounceability	Car Coaster	Crash Test	Doc Shock	Exploring Friction	From Magnets to Motors	Gear Up, Gear Down	Levers at Work	Magnet Cars	More Pulley Power	Pulley Power Basics	Push and Go	Ramps and Cars	Seesaw Forces	Six-Cent Top	Skyhook	Sound Off	Static Cling	Toy That Returns	Understanding Speed
Science as Inquiry—Abilities Necessary to Do Scientific Inquiry																					
Design and conduct a scientific investigation.		●			●	●								●		●		●			●
Use appropriate tools and techniques to gather, analyze, and interpret data.		●				●	●	●			●			●	●	●				●	●
Develop descriptions, explanations, predictions, and models using evidence.	●	●	●	●	●	●	●		●	●			●	●	●	●	●	●	●	●	●
Think critically and logically to make the relationships between evidence and explanations.	●	●	●	●	●	●		●	●		●						●	●	●	●	●
Recognize and analyze alternative explanations and predictions.					●																
Communicate scientific procedures and explanations.	●	●	●	●	●		●	●	●	●	●	●	●				●	●	●	●	●
Use mathematics in all aspects of scientific inquiry.		●						●	●		●			●		●					●
Physical Science																					
Motions and forces	●	●	●	●		●		●	●	●	●	●	●	●	●	●	●	●		●	●
Transfer of energy		●	●		●	●	●				●	●	●					●		●	
Science and Technology—Abilities of Technological Design																					
Design a solution or product.	●								●	●							●				
Implement a proposed design.	●										●					●	●			●	
Evaluate completed technological designs or products.					●		●	●			●	●	●		●	●	●			●	
Communicate the process of technological design.	●				●	●		●	●	●	●					●	●				

Activities Indexed by National Science Education Standards for Grades 9–12

	Car Coaster	From Magnets to Motors	Levers at Work	More Pulley Power	Push and Go
Science as Inquiry—Abilities Necessary to Do Scientific Inquiry					
Use technology and mathematics to improve investigations and communications.			•	•	
Formulate and revise scientific explanations and models using logic and evidence.	•		•		•
Communicate and defend a scientific argument.	•			•	
Physical Science					
Motions and forces	•		•	•	•
Conservation of energy and the increase in disorder	•				
Interactions of energy and matter		•			
Science and Technology—Abilities of Technological Design					
Implement a proposed solution.				•	
Evaluate the solution and its consequences.				•	
Communicate the problem, process, and solution.			•	•	

LaVergne, TN USA
13 February 2011
216361LV00001B/3/P